Springer Tracts in Advanced Robotics

Volume 71

Hui Xie, Cagdas Onal, Stéphane Régnier,
and Metin Sitti

Atomic Force Microscopy Based Nanorobotics

Modelling, Simulation, Setup Building and Experiments

Springer

Professor Bruno Siciliano, Dipartimento di Informatica e Sistemistica, Università di Napoli Federico II, Via Claudio 21, 80125 Napoli, Italy, E-mail: siciliano@unina.it

Professor Oussama Khatib, Artificial Intelligence Laboratory, Department of Computer Science, Stanford University, Stanford, CA 94305-9010, USA, E-mail: khatib@cs.stanford.edu

Professor Frans Groen, Department of Computer Science, Universiteit van Amsterdam, Kruislaan 403, 1098 SJ Amsterdam, The Netherlands, E-mail: groen@science.uva.nl

Authors

Dr. Hui Xie
Institut des Systèmes Intelligents et de Robotique
Université Pierre et Marie Curie
BC 173, 4 Place Jussieu
75005 Paris
France
E-mail: xie@robot.jussieu.fr

Prof. Stéphane Régnier
Institut des Systèmes Intelligents et de Robotique
Université Pierre et Marie Curie
BC 173, 4 Place Jussieu
75005 Paris
France
E-mail: stephane.regnier@upmc.fr

Dr. Cagdas Onal
Computer Science and Artificial Intelligence Laboratory
Massachusetts Institute of Technology
32 Vassar St. 32-375
02139 Cambridge
USA
E-mail: cagdas@mit.edu

Prof. Metin Sitti
NanoRobotics Lab
Carnegie Mellon University
5000 Forbes Ave
320 Scaife Hall
15213-3890 Pittsburgh
USA
E-mail: sitti@cmu.edu

ISBN 978-3-642-20328-2 e-ISBN 978-3-642-20329-9

DOI 10.1007/978-3-642-20329-9

Springer Tracts in Advanced Robotics ISSN 1610-7438

Library of Congress Control Number: 2011934157

Typeset & Cover Design: Scientific Publishing Services Pvt. Ltd., Chennai, India.

Printed on acid-free paper

5 4 3 2 1 0

springer.com

STAR (Springer Tracts in Advanced Robotics) has been promoted under the auspices of EURON (European Robotics Research Network)

European ROBOTICS Research Network EURON

For our beloved ones

Foreword

Robotics is undergoing a major transformation in scope and dimension. From a largely dominant industrial focus, robotics is rapidly expanding into human environments and vigorously engaged in its new challenges. Interacting with, assisting, serving, and exploring with humans, the emerging robots will increasingly touch people and their lives.

Beyond its impact on physical robots, the body of knowledge robotics has produced is revealing a much wider range of applications reaching across diverse research areas and scientific disciplines, such as: biomechanics, haptics, neurosciences, virtual simulation, animation, surgery, and sensor networks among others. In return, the challenges of the new emerging areas are proving an abundant source of stimulation and insights for the field of robotics. It is indeed at the intersection of disciplines that the most striking advances happen.

The Springer Tracts in Advanced Robotics (STAR) is devoted to bringing to the research community the latest advances in the robotics field on the basis of their significance and quality. Through a wide and timely dissemination of critical research developments in robotics, our objective with this series is to promote more exchanges and collaborations among the researchers in the community and contribute to further advancements in this rapidly growing field.

The monograph written by Hui Xie, Cagdas Onal, Stéphane Régnier and Metin Sitti is a contribution in the area of nanorobotics, which has been receiving a growing deal of attention by the research community in the latest few years. The contents are focused on the use of atomic force miscroscopy as a nanomanipulation system, with emphasis on modelling, simulation, Instrumentation and experiments. Control and teleoperation problems are dealt with and results are presented for a number of significant applications.

STAR is proud to welcome this first volume in the series dedicated to nanorobotics!

Naples, Italy
February 2011

Bruno Siciliano
STAR Editor

Contents

Introduction

Due to the developments in nanotechnology and biotechnology in last two decades, handling nanometer scale entities has become a critical issue. Since the human sensing, precision and size are not sufficient to interact with such nanoscale entities directly, nanorobotics has emerged as a new robotics field to extend our manipulation capabilities to nanometer scale [1, 2]. In such nanorobotic systems, nanoscale grippers or force fields are used to apply external physical forces to manipulate nanoscale entities precisely. By manipulation, it is meant that the entities are pushed or pulled, cut, picked and placed, trapped, indented, bent, twisted, bonded, assembled, etc. by controlling contact or non-contact external forces. The structure of a generic nanorobotic manipulation system using a nanoscale probe tip as a nanomanipulator is displayed in Figure 0.1. The nanomanipulator exerts controlled forces to nanoscale entities such as nanoparticles, carbon nanotubes, nanowires, nanocrytals, and biological entities (DNA, protein, cell, bio-motors, etc.) precisely by applying mechanical, electrical, optical, magnetic, or dielectrophoretic contact or non-contact forces. Far-field (optical microscope or scanning electron microscope (SEM)) and near-field (scanning probe microscopes (SPMs)) imaging devices and integrated nanoscale force sensors are used for teleoperated, semi-autonomous or autonomous control of a high precision, generally piezoelectric, XYZ nanopositioner. For controlling the environmental disturbances and factors, acoustic and vibration isolation systems and humidity and temperature controllers are typically integrated into the system.

Many groups have proposed various nanoscale robotic manipulation systems since the beginning of the 90's. The first revolutionary nanorobotic manipulation demonstration was Eigler et al.'s [3] atomic manipulation work in 1990 in almost absolute zero temperature conditions. This experiment showed the possibility of positioning individual Xenon atoms precisely on a cupper surface using electrical pulses applied by a scanning tunneling microscope (STM) probe tip. Next, using electrostatic forces [4], optical tweezers [5, 6, 7], and dielectrophoretic traps [8], biological nano-entities were manipulated for their mechanical and chemical characterization, sorting, alignment, etc.

H. Xie et al.: Atomic Force Microscopy Based Nanorobotics, STAR 71, pp. 1–11.
springerlink.com

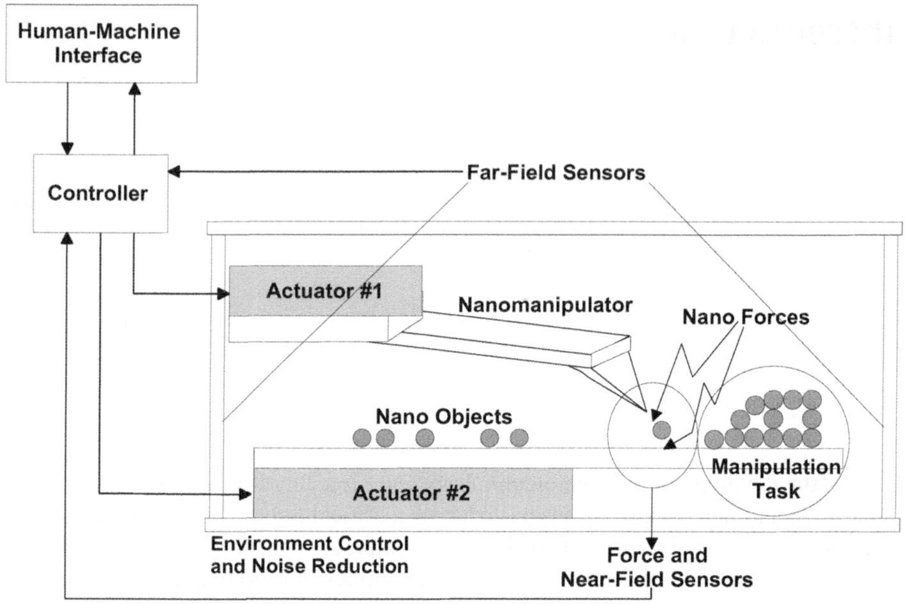

Fig. 0.1 Schematics of a generic nanorobotic manipulation system

Starting in 1995, there has been a significant thrust in atomic force microscope (AFM) based nanorobotic nanomanipulation, and majority of the nanomanipulation studies since then has been using AFM based nanorobotic systems. Such a trend is result of many new advantages that AFM systems have enabled for imaging and manipulation of nanoscale materials after the invention of AFM in 1986 [9] and its commercialization in early 90's. Designed first as an imaging tool to gather three-dimensional (3-D) surface topography with almost atomic resolution, AFM has quickly shown its potential to image other properties of materials, e.g. frictional, electrical, and magnetic properties, at the nanoscale and finally as a manipulator. An AFM basically consists of a microcantilever with a very sharp tip on one end, a high precision 3-D nanopositioning stage, and an optical device to detect the deflections at the end of the cantilever as shown in Figure 0.2.

Advantages of an AFM as a nanorobotic manipulation system can be given as follows:

1. AFM probe can work as a force sensor and a manipulator simultaneously. First, the probe could be used to image the nano-entities using non-contact or tapping mode type of AFM imaging methods without affecting their position and physical and chemical properties. Next, the same probe could be used as a nanomanipulator while the interaction forces on the probe tip during manipulation could be directly measured in real-time, which enables sensory feedback control.
2. Due to its versatility in mechanically interacting with a broad range of materials, an AFM based nanomanipulation task can be conducted in diverse environments such as ambient, vacuum and under liquid conditions for diverse

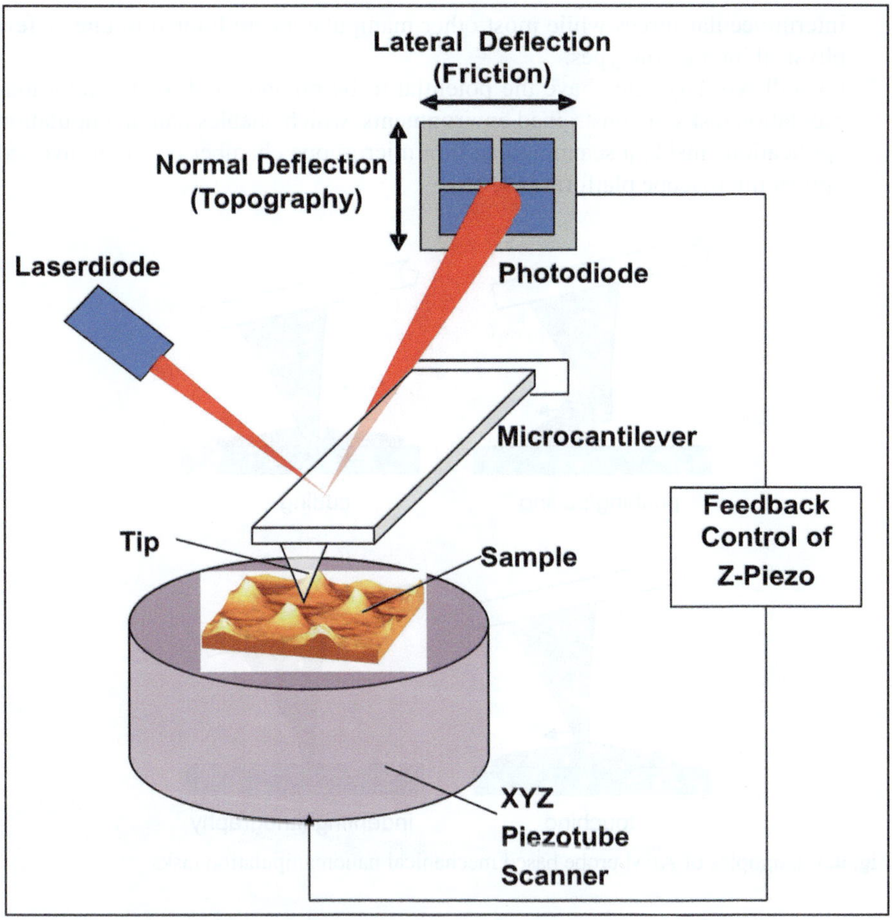

Fig. 0.2 A typical AFM setup with 3-D nanoscale interaction forces on the AFM probe (microcantilever) tip, where these forces vertically bend and laterally twist the probe for topography and friction measurements

nano-entity materials such as non-conductive, conductive, biological, inorganic, and polymeric materials.

3. Since the AFM probes are typically fabricated using optical lithography type of microfabrication processes, it is possible to fabricate and have an array of probes with independent sensing and actuation capabilities for parallel or multi-probe operation [10, 11, 12, 13]. Such probe arrays or multi-probe systems enable multi-functional and high-speed operation. Moreover, multiple probes could be used to conduct simultaneous multi-point measurements or pick-and-place type of complex and 3-D manipulation tasks [54].
4. AFM probes are capable of exerting a broad range of interaction forces during manipulation such as mechanical, electrical, magnetic, optical, and chemical

intermolecular forces while most other manipulators are limited to one or few physical interaction types.

5. Overall AFM systems have the potential to be miniaturized for versatile manipulation tasks in constrained environments, which enables nanomanipulation applications inside a scanning electron microscope chamber or on an inverted optical microscope platform [14, 15].

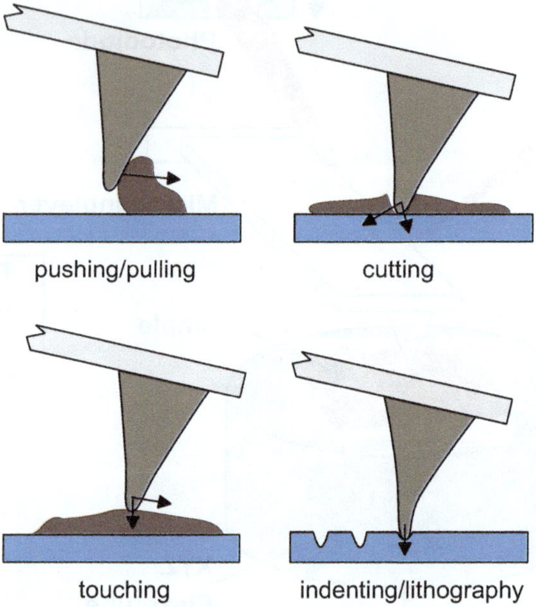

Fig. 0.3 Examples of AFM probe based mechanical nanomanipulation tasks

Most of the AFM based nanomanipulation tasks utilize mechanical contact interactions as depicted in Figure 0.3. An AFM probe can be used as a pushing/pulling [16, 17, 18], cutting [19], indenting [20, 21], touching [22], chemical lithography [23], and pulling type of manipulator. Scientists such as physicists and chemists mostly conducted many of these first mechanical contact nanomanipulation demonstrations. The engineering community started to be active in this area by Hollis et al.'s [24] tele-nanorobotic user interface for atomic topography tactile feedback using an STM system in 1990, human-machine interfaces for AFM based teleoperated nanomanipulation systems [25, 26, 27], two-dimensional (2-D) particle assembly using AFM probe based contact pushing [28, 29], 3-D carbon nanotube manipulation under SEM using multiple AFM probes [30, 31], nanotube attached [32] and customized [33, 34] nano-tweezers for handling nano-entities, and 3-D pick-and-place manipulation of particles and nanowires using two AFM probes [53, 54].

As the simplest mechanical contact nanomanipulation task, an AFM probe with a very sharp tip has been used as an indenter to create nanoscale marks on a given substrate that could be plastically yielded due to high stresses created by the probe tip locally. Using vertical load and speed control, the indentation depth, i.e. the size and depth of the mark, could be precisely controlled. The spatial resolution of such nanoindentation task can be down to tens of nanometers, which make it a promising candidate for high-density data storage systems, exceeding the 1 Tb/in^2 limit potentially [35]. As an example to such data storage application, an AFM probe is used to indent soft surfaces with holes as small as 37 nm (0.48 Tb/in^2 density) for the development of ultrahigh density data storage devices at IBM Zurich [10]. In this indentation process, the local heat generated by the probe tip is also utilized to facilitate and increase the control of the nano-mark size [36]. Moreover, such local heat control method is used to erase the created marks on the polymer substrate to reset the data. With the areal densities of conventional magnetic recording technologies eventually reaching well-known physical limits, the ultimate confinement of interaction that the local probe provides renders probe-based recording a natural candidate for extending the storage density roadmap. A potentially commercial scanning-probe based data storage system is codenamed Millipede at the IBM Zurich. The Millipede exploits parallel operation of very large two-dimensional AFM cantilever arrays (an array of 32 $\times$ 32 probes currently) with integrated tips and write/read/erase functionality. Data storage speed with a single probe is very limited for commercial data storage rate requirements, resulting in the development of parallel processing with an array of probes. Information is stored as sequences of indentations or no indentations formed by the tips on a nanometer-thick polymer film using thermo-mechanical nanomanipulation. The polymer film is integrated in a micro-scanner with XYZ motion capabilities and a lateral scanning range about 100 m. Several challenges of this nanorobotic system include polymer recording media endurance, long term probe tip and media wear, tip uniformity in probe array chip, shock resistance, and form factor integration of channel electronics.

Next, mechanical pulling or pushing has been one of the most popular nanomanipulation task realized by an AFM based nano-robotic system since 1995. The most pushed nano-entity has been nanoparticles [17, 18, 28, 29] since the particle-tip (sphere-sphere) and the particle-substrate (sphere-plane) surface and contact forces can be analytically modeled using continuum nanoscale physics and mechanics models for improved understanding of the nanomanipulation interactions, and nanoparticles also enable new optical or electronic nanodevice prototyping and nanosoldering type of applications. In such particle nanomanipulation demonstrations, first, particles are scattered randomly on an atomically smooth and planar substrate using capillary force effects (evaporation of the suspension liquid of the particles) and semi-absorbed on the substrate using different surface forces. For example, for gold nanoparticles, a mica substrate with a poly-L-lysine coating has been typically used to semi-absorb the particles using electrostatic forces [28]. Such semi-absorption makes the non-contact or tapping mode imaging of the nano-entities stable without affecting their position before the manipulation task. Next, the probe tip is aligned with the center of the particle and the tip pushes, pulls, or rolls the particle after turning the imaging servo control off. The particle motion behavior depends on the interaction forces on the tip-particle and the particle-substrate interface. At the nanoscale, it is observed that the particles tend to slide more than roll due to increased moment of roll resistance resulting from the surface forces

[37]. During the pushing operation, most of the studies have never used any sensory feedback; they pushed blindly and then looked by imaging to see where the particle went after pushing. Such a blind push-and-look operation triggered three main issues: reliability, speed and precision of the nanomanipulation system. To solve such issues, recent studies [38] used force feedback during the pushing operation to detect the particle contact loss and local scanning and detection of the particle center. Moreover, full automation of the pushing process of multiple particles has been used by several studies [38, 39] to minimize these issues. To enable such fully automated manipulation systems, these studies proposed drift compensation methods to position the AFM probe tip laterally with very high precision where the piezoelectric nanopositioner of AFM systems drift thermally [40]. These nanoparticle pushing based manipulation demonstrations have been used for different applications in nanoscale photonics, plasmonics, and soldering. As one of the most successful and cited demonstrations, Maier et al. used precisely located silver nanoparticles to create an optical photonic waveguide [41]. In this waveguide, electromagnetic energy was guided below the optical diffraction limit along chains of closely spaced metal nanoparticles that converted the optical mode into non-radiating surface plasmons. Due to the precise positioning of gold nanoparticles using AFM based nanomanipulation, direct experimental evidence for energy transport along plasmon waveguides was possible for the first time.

Finally, different nano-entities such as carbon nanotubes (CNTs) and nanowires have been pushed on surfaces to understand their nanomechanical and nanotribological properties [16] or prototype new nanoelectronic devices such as single-electron transistors [42]. In these manipulations, nanotubes or nanowires can roll, slide or spin. Moreover, AFM probe tips have been used to conduct nanolithography on surfaces by removing material from the substrate. Several groups have used the material removal from a substrate by scratching it mechanically using an AFM probe tip as an alternative nanolithography method [43]. On the other hand, multiple probes were used to manipulate CNTs under a SEM to characterize their nanomechanical properties [44] or assemble particles or nanowires in 3-D using pick-and-place methods [55].

As electrical interaction based AFM nanomanipulation methods, anodic oxidation based nanolithography is the most widely applied technique to write nanoscale features on surfaces [45]. The AFM-tip induced oxidation process is based on negatively biasing the tip with respect to the substrate under ambient conditions, which can be either a semiconductor or a metal. The substrate locally oxidizes upon moving the tip in contact mode across the surface. The oxidant for the chemical reaction is provided by OH- ions in the water droplet that is formed between the tip and the sample. Thus, the lateral resolution of the AFM oxidation process depends strongly on the humidity in the air since the oxidation process utilizes the presence of a water-bridge between the tip and the substrate under ambient conditions [46]. Next, as a new method, conductive AFM tips have been recently used to induce local electric field based field emission in a chemical vapor deposition (CVD) chamber to pattern surfaces to create nanowires and nanodots to prototype new nanosensors, single electron transistors, and other nanodevices [47, 48].

As an example chemical interaction based AFM nanomanipulation method, AFM probe tips coated with chemicals or biological samples such as proteins are used to write/deposit them precisely on a substrate while moving in contact with the substrate. Mirkin et al. [23] deposited chemical ink on to an AFM probe tip for writing molecular thickness nanoscale lines on gold surfaces. This technique was named as 'Dip-Pen Lithography' and commercialized by NanoInk [49].

In overall, there are many nanomanipulation method demonstrations and possibilities using AFM based nanorobotic systems. Current status and future directions of such nanorobotic systems can be summarized as shown in Table 0.1. To enable these future targets, many basic research challenges exist. Such challenges could be summarized as follows:

- Although there are many attempts for continuum modeling [50, 51, 52], real-time realistic nanophysical simulators for nanorobotic system design and control are still not available. There is a wide range of nanoscale materials, geometries and environments for which there are no realistic continuum nanophysics simulator tools yet.
- Current nanomanipulation tasks are generally controlled in one- or two-dimension. There are only few studies on three-dimensional (3-D) nano-scale manipulation and assembly [54], and more 3-D nanomanipulation methods are indispensible for heterogeneous integration of complex nanodevices.
- Real-time *in situ* far-field visualization during nanomanipulation is challenging in ambient and liquid conditions. Therefore, utilizing real-time force sensing capability of AFM systems could be very beneficial for feedback control.
- Intuitive human-machine user interfaces with real-time visual and haptic displays are required for teleoperated nanomanipulation applications in time varying and complex environments such as biological systems. Here, developing Augmented Reality systems with integrated real-time nanophysical modeling, scaling nanoscale forces to the macroscale, robust and stable teleoperation controller design, and compensating the time-scale differences between nano and macro worlds are issues that remain unsolved.
- Miniaturizing the current large scale AFM systems down to few centimeters scale is crucial for future portable, robust, modular, massively parallel, and cost-effective nanorobotic manipulation systems.
- Serial nanomanipulation systems could only enable low volume prototyping applications. However, high volume and high-speed nanomanipulation systems are indispensable for future nanotechnology products. Therefore, autonomous and massively parallel AFM systems are required.

The first chapter will introduce general topics related to the AFM-based nanorobotic system, as well as challenges and opportunities of the state-of-art research of nanomanipulation. The second chapter will be dedicated to the instrumentation issues for the design or the use of an AFM based nanorobotic system. The fourth chapter gives a review of existing models aiming at modeling micro or nanomanipulation tasks. In the fifth chapter, particularities of AFM based teleoperations are reviewed, and classical control schemes are anlysed and compared in terms

Table 0.1 Current status and future directions of nanorobotic manipulation systems

Key Features	Current State-of-the-art	Future Directions in 10-15 years
Nano Physics Modeling	- Heuristic Models or No Modeling - Simplified Continuum Models	- Advanced Continuum Models - Molecular Dynamics Modeling
Manipulation capability	- 1D or 2D - Homogenous Nano-Building Blocks	- 3D - Heteregenous Integration
Control	- Direct Teleoperation - Semi-Autonomous	- Autonomous and Smart - Swarm of Nanorobots
User Interface	- Visual and Haptic Displays	- Augmented Reality Systems
Architecture	- Serial	- Parallel
Robot Size	- Tens of Centimeters	- Few Centimeters
Outcomes	- Prototyping Nano-Devices - Scientific Tool	- Nanomaterial Characterization Tools - 3D Nano-Manufacturing Systems - Miniature Micro/Nano Factory
Cost	\$50,000-\$200,000	\$5,000-\$30,000

of transparency, stability, and ease of manipulation as well as limits of human force sensing. This chapter will focus on how AFM could be used to automatically manipulate micro and nano-objects. The last chapter deals with potential applications of atomic force microscopy based robotic system.

References

1. Sitti, M.: Microscale and nanoscale robotics systems - Characteristics, state of the art and grand challenges. IEEE Robotics and Automation Magazine 14(1), 53–60 (2007) 1
2. Requicha, A.: Nanorobots, NEMS and Nanoassembly. Proc. of the IEEE 91(11), 1922–1933 (2003) 1
3. Eigler, D.M., Schweitzer, E.K.: Positioning single atoms with a scanning electron microscope. Nature, 524–526 (April 1990) 1
4. Washizu, M., Kurosawa, O.: Electrostatic manipulation of DNA in microfabricated structures. IEEE Trans. on Industry Applications 26(6), 1165–1172 (1990) 1
5. Cui, Y., Bustamente, C.: Pulling a single chromatin fiber reveals the forces that maintain its higher-order structure. Proc. Nat. Acad. Sci. 97(1), 127–132 (2000) 1
6. Liphardt, J., Onoa, B., Smith, S., Tinoco, I., Bustamente, C.: Reversible unfolding of single RNA molecules by mechanical force. Science 292, 727–730 (2001) 1
7. Sitti, M.: Survey of nanomanipulation systems. In: Proc. of the IEEE Nanotechnology Conference, Maui, USA, pp. 75–80 (Novenber 2001) 1
8. Fuhr, G., Gimsa, J., Hagedorn, R., Muller, T., Wagner, B., Benecke, W.: Linear motion of dielectric particles and living cells in microfabricated structures induced by traveling electric fields. In: Proc. of the IEEE Micro Electro Mechanical Systems, pp. 259–264 (January 30- February 2, 1991) 1
9. Binnig, G., Quate, C.F., Gerber, C.: Atomic force microscope. Phys. Rev. Lett. 56(9), 930–933 (1986) 2
10. Eleftheriou, E., Eleftheriou, E., Antonakopoulos, T., Binnig, G.K., Cherubini, G., Despont, M., Dholakia, A., Durig, U., Lantz, M.A., Pozidis, H., Rothuizen, H.E., Vettiger, P.: Millipede - a MEMS-based scanning probe data-storage system. IEEE Trans. on Magnetics 39(2), 938–945 (2003) 3, 5

11. Bullen, D., Liu, C.: Electrostatically actuated dip pen nanolithography probe arrays. Sensors and Actuators A: Physical 125(2), 504–511 (2006) 3
12. Lutwyche, M., Andreoli, C., Binnig, G., Brugger, J., Drechsler, U., Haberle, W., Rohrer, H., Rothuizen, H., Vettiger, P., Yaralioglu, G., Quate, C.: 5x5 2D AFM cantilever arrays a first step towards a Terabit storage device. Sensors and Actuators A: Physical 73, 89–94 (1999) 3
13. Wang, X., Liu, C.: Multifunctional Probe Array for Nano Patterning and Imaging. Nano Letters 5(10), 1867–1872 (2005) 3
14. Brufau, J., Puig-Vidal, M., Lopez-Sanchez, J., Samitier, J., Snis, N., Simu, U., Johansson, S., Driesen, W., Breguet, J.-M., Gao, J., Velten, T., Seyfried, J., Estana, R., Woern, H.: MICRON: Small Autonomous Robot for Cell Manipulation Applications. In: Proc. IEEE International Conference on Robotics and Automation, Barcelona, Spain, pp. 844–849 (2005) 4
15. Fatikow, S., Wich, T., Hulsen, H., Sievers, T., Jahnisch, M.: Microrobot System for Automatic Nanohandling Inside a Scanning Electron Microscope. IEEE/ASME Trans. on Mechatronics 12(3), 244–252 (2007) 4
16. Falvo, M., Taylor II, R.M., Helser, A., Chi, V., Brooks, F.P., Washburn, S., Superfine, R.: Nanometer-scale rolling and sliding of carbon nanotubes. Nature 397, 236–238 (1999) 4, 6
17. Schafer, D., Reifenberger, R., Patil, A., Andres, R.: Fabrication of two-dimensional arrays of nanometric-size clusters with the Atomic Force Microscopy. Appl. Physics Letters 66, 1012–1014 (1995) 4, 5
18. Junno, T., Deppert, K., Montelius, L., Samuelson, L.: Controlled manipulation of nanoparticles with an Atomic Force Microscopy. Appl. Physics Letters 66, 3627–3629 (1995) 4, 5
19. Thalhammer, S., Stark, R.W., Müller, S., Wienberg, J., Heckl, W.M.: The Atomic Force Microscope as a New Microdissecting Tool for the Generation of Genetic Probes. J. Struct. Biol. 119, 232–237 (1997) 4
20. Drechsler, U., Durig, U., Gotsmann, B., Haberle, W., Lantz, M.A., Rothuizen, H.E., Stutz, R., Binnig, G.K.: The millipede-nanotechnology entering data storage. IEEE Trans. on Nanotechnology 1, 39–55 (2002) 4
21. Sitti, M.: Teleoperated and automatic control of nanomanipulation systems using atomic force microscope probes. In: Proc. of the IEEE Conf. on Decision and Control, Hawaii, pp. 2118–2123 (December 2003) 4
22. Sitti, M., Hashimoto, H.: Teleoperated touch feedback of surfaces at the nanoscale: Modeling and experiments. IEEE/ASME Trans. on Mechatronics 8(2), 287–298 (2003) 4
23. Piner, R.D., Zhu, J., Xu, F., Hong, S., Mirkin, C.A.: "Dip-Pen" Nanolithography. Science 283(5402), 661–663 (1999) 4, 7
24. Hollis, R.L., Salcudean, S., Abraham, D.W.: Toward a telenanorobotic manipulation system with atomic scale force feedback and motion resolution. In: Proc. of the IEEE Int. Conf. on MEMS, pp. 115–119 (1990) 4
25. Falvo, M., Superfine, R., Washburn, S., Finch, M., Taylor, R.M., Brooks, F.P.: The nanoManipulator: A teleoperator for manipulating materials at the nanometer scale. In: Proc. of the Int. Symp. on the Science and Technology of Atomically Engineered Materials, Richmond, USA, pp. 579–586 (November 1995) 4
26. Sitti, M., Hashimoto, H.: Tele-nanorobotics using atomic force microscope as a robot and sensor. Advanced Robotics 13(4), 417–436 (1999) 4
27. Li, G., Xi, N., Yu, M., Fung, W.K.: Development of Augmented Reality System for AFM-Based Nanomanipulation. IEEE/ASME Transactions on Mechatronics 9(2), 358–365 (2004) 4

28. Ramachandran, T.R., Baur, C., Bugacov, A., Madhukar, A., Koel, B.E., Requicha, A., Gazen, C.: Direct and controlled manipulation of nanometer-sized particles using the noncontact atomic force microscope. Nanotechnology 9(3), 237–245 (1998) 4, 5
29. Sitti, M., Hashimoto, H.: Controlled pushing of nanoparticles: Modeling and experiments. IEEE/ASME Trans. on Mechatronics 5, 199–211 (2000) 4, 5
30. Yu, M., Kowalewski, T., Ruoff, R.: Investigation of the radial deformability of individual carbon nanotubes under controlled indentation force. Physical Review Letters 86, 87–90 (2000) 4
31. Dong, L., Arai, F., Fukuda, T.: 3D nanorobotic manipulation of nano-order objects inside SEM. In: Proceedings of the International Symposium on Micromechatronics and Human Science, pp. 151–156 (October 2000) 4
32. Kim, P., Lieber, C.M.: Nanotube nanotweezers. Science 286, 2148–2150 (1999) 4
33. Boggild, P., Hansen, T.M., Molhave, K., Hyldgard, A., Jensen, M.O., Richter, J., Montelius, L., Grey, F.: Customizable nanotweezers for manipulation of free-standing nanostructures. In: Proc. of the IEEE Nanotechnology Conference, October 28-30, pp. 87–92 (2001) 4
34. Hashiguchi, G., Fujita, H.: Micromachined nanoprobe and its application. Proceedings of the IEEE 2, 922–925 (2002) 4
35. Hosaka, S.: SPM based recording toward ultrahigh density recording with trillion bits/inch2. IEEE Transactions on Magnetics 37(2), 855–859 (2001) 5
36. Yang, F., Wornyo, E., Gall, K., King, W.P.: Nanoscale indent formation in shape memory polymers using a heated probe tip. Nanotechnology 18, 285–302 (2007) 5
37. Sumer, B., Sitti, M.: Rolling and Spinning Friction Characterization of Fine Particles using Lateral Force Microscopy based Contact Pushing. Journal of Adhesion Science and Technology 22, 481–506 (2008) 6
38. Onal, C., Ozcan, O., Sitti, M.: Automated 2-D Nanoparticle Manipulation using Atomic Force Microscopy. IEEE Trans. on Nanotechnology (2010), doi:10.1109/TNANO.2010.2047510 6
39. Mokaberi, B., Requicha, A.A.G.: Drift compensation for automatic nanomanipulation with scanning probe microscopes. IEEE Transactions on Automation Science and Engineering 3, 199–207 (2006) 6
40. Krohs, F., Onal, C., Sitti, M., Fatikow, S.: Towards Automated Nanoassembly With the Atomic Force Microscope: A Versatile Drift Compensation Procedure. ASME Journal of Dynamic Systems, Measurement, and Control 131(6) (November 2009) 6
41. Maier, S.A., Kik, P.G., Atwater, H.A., Meltzer, S., Harel, E., Koel, B.E., Requicha, A.A.G.: Local detection of electromagnetic energy transport below the diffraction limit in metal nanoparticle plasmon waveguides. Nature Materials 2, 229–232 (2003) 6
42. Roschier, L., Penttil, J., Martin, M., Hakonen, P., Paalanen, M., Tapper, U., Kauppinen, E.I., Journet, C., Bernier, P.: Single-electron transistor made of multiwalled carbon nanotube using scanning probe manipulation. Appl. Phys. Lett. 75, 728 (1999) 6
43. Liu, Z., Li, Z., Wei, G., Song, Y., Wang, L., Sun, L.: Manipulation, dissection, and lithography using modified tapping mode atomic force microscope. Microscopy Research and Technique 69(12), 998–1004 (2006) 6
44. Yu, M., Dyer, M.J., Skidmore, G.D., Rohrs, H.W., Lu, X., Ausman, K.D., Von Her, J.R., Ruoff, R.S.: Three-dimensional manipulation of carbon nanotubes under a scanning electron microscope. Nanotechnology 10, 244 (1999) 6
45. Dagata, J.A., Schneir, J., Harary, H.H., Evans, C.J., Postek, M.T., Bennett, J.: Modification of hydrogen passivated silicon by a scanning tunneling microscope operating in air. Applied Physics Letters 56 (1990) 6

46. Garcia, R., Calleja, M., Perez-Murano, F.: Local oxidation of silicon surfaces by dynamic force microscopy: Nanofabrication and water bridge formation. Applied Physics Letters 72(18) (1998) 6
47. Liu, J.F., Miller, G.P.: Field-assisted nanopatterning of metals, metal oxides and metal salts. Nanotechnology 20, 055303 (2009) 6
48. Malshe, A.P., Rajurkar, K.P., Virwanic, K.R., Taylord, C.R., Bourelle, D.L., Levy, G., Sundaramg, M.M., McGeough, J.A., Kalyanasundarama, V., Samant, A.N.: Tip-based nanomanufacturing by electrical, chemical, mechanical and thermal processes. CIRP Annals - Manufacturing Technology 59(2), 628–651 (2010) 6
49. `http://www.nanoink.net/` 7
50. Fearing, R.S.: Survey of sticking effects for micro parts handling. In: Proc. of the IEEE/RSJ International Conference on Intelligent Robots and Systems, August 5-9, pp. 212–217 (1995) 7
51. Arai, F., Fukuda, T.: Adhesion-type micro end effector for micromanipulation. In: Proc. of the IEEE Robotics and Automation Coference, April 20-25, pp. 1472–1477 (1997) 7
52. Sitti, M., Hashimoto, H.: Tele-Nanorobotics Using Atomic Force Microscope. In: Proc. of the IEEE/RSJ Int. Conference on Intelligent Robots and Systems, Victoria, Canada, pp. 1739–1746 (October 1998) 7
53. Xie, H., Régnier, S.: Journal of Micromechanics and Microengineering 19(7), 075009(p. 9) (2009) 4
54. Xie, H., Haliyo, D.S., Régnier, S.: A versatile atomic force microscope for three-dimensional nanomanipulation and nanoassembly. Nanotechnology 20, 215301 (2009) 3, 4, 7
55. Xie, H., Régnier, S.: IEEE/ASME Transactions on Mechatronics 16(2), 266–276 (2011) 6

Chapter 1
Descriptions and Challenges of AFM Based Nanorobotic Systems

Nanorobotics means literally the study of robots that are nanoscale in typical size, i.e. nanorobots, which have yet to be realized. Generally, nanorobots are large robots capable of manipulation nanoscale objects with nanometer resolution, e.g. a AFM-based nanorobotic manipulation system and a scanning electron microscope (SEM) equipped with a nanomanipulator. When studying nanorobotics, we first have to understand physics that underlies interactions at the nanoscale. At microscale, some basic micromanipulation problems attributed to the scale affects have been identified. We have seen how the surface effects, instead of volume effects, dominate the physical phenomena at this scale. Most of these scaling laws are still available at the nanoscale. However, the scale affects become more severe at the nanoscale due to the additional three orders of magnitude in size reduction, and it becomes much more difficult to predict and control because of more scale effects and uncertainties introduced when the nanomanipulation performed in the nanoworld.

The atomic force microscope (AFM) has gave nanotechnology a significant boost by providing it with a powerful tool for understanding physical and chemical phenomena from the nanoscale to atomic scale, as well as for performing engineering operations on nanoscale objects, molecules and atoms. The first successful manipulation of atoms was accomplished using STM about two decades ago [1]. After that, various nanomanipulation and nanoassembly schemes and systems related to the AFM have been developed. AFM has gone beyond imaging and has been by far the most widely applied for the manipulation or characterization of nanomaterials and biology samples. In this chapter, descriptions and challenges of the AFM-based nanomanipulation will be described in detail.

H. Xie et al.: Atomic Force Microscopy Based Nanorobotics, STAR 71, pp. 13–29.
springerlink.com

1.1 Descriptions

1.1.1 The AFM as a Nanorobot–Beyond Imaging

Typically, when manipulating an nanoobject, e.g. moving an nanoobject across a surface or even picking-up an nanoobject from one surface and then releasing it on another surface, we may need effective tools to complete two aspects of processes of such nanomanipulation tasks:

- We need an imaging tool, with nanoscale resolution, for simultaneous positioning of nanoobjects to be manipulated and the nano-environment surround the nanoobjects.
- We need effective tools with very tiny size comparable with the nanoobjects and sufficient feedback is available to monitor the manipulation process.

For the nanoscale imaging, three types of microscopies are generally considered, including transmission electron microscope (TEM), scanning electron microscope (SEM) and AFM. Fig. 1.1 shows a comparison of the imaging "length scale" of four types of microscopes. In the respect of the imaging resolution, the AFM can be compared to TEM, as low as 0.05 nm that is much higher than the SEM. However, the AFM has a greater measuring dimensions than the TEM in the horizontal axis. Compared with the electron microscopes, the AFM balances the resolution well with the measuring dimensions. Moreover, the AFM can also be used for making measurements in the vertical axis to a surface. In that way, the AFM has a advantage to magnify the nanoworld in three dimensions.

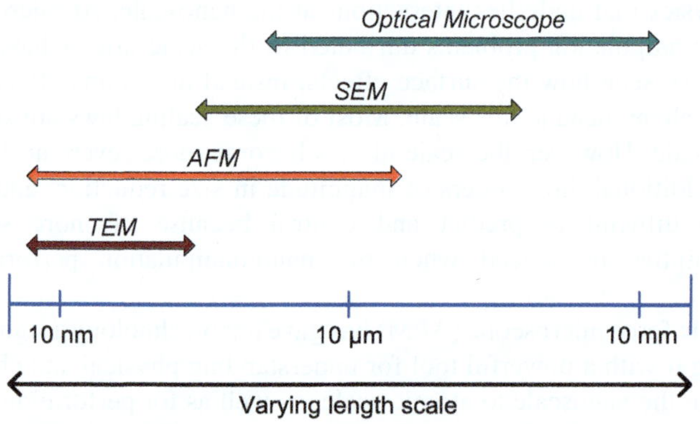

Fig. 1.1 Comparison of the imaging "length scale" of four often-used imaging instruments.

Nanostructures have been manipulated,assembled and characterized by nanorobots equipped with manipulators in SEMs and TEMs [2, 3, 4, 5, 6, 7, 8, 9], in which nanoscale grasping can be performed due to the vacuum environment and the visual feedback. A liquid medium enables noncontact grasping where the adhesion forces are greatly reduced, e.g., optical [10] and magnetic tweezers [11]. Table 1.1

shows a comparison of the some of the major factors of these three microscopes. Unlike the SEM and TEM, the AFM has much more flexibility since it can be applied to different nanoobjects in various environments (air, liquid and vacuum). Especially the application in liquid that is quite practical for life science. Although the imaging time required for the TEM/SEM is typically less than the AFM, the amount of time required to get meaningful images might be similar. This is due to substantial time often required to prepare a sample with the SEM/TEM. In contrast, much less time is required for sample preparation with the AFM.

Table 1.1 Comparisons between the AFM, SEM and TEM

Descriptions	TEM/SEM	AFM
Samples	conductive	not required
Environment	Vacuum	Vacuum/Air/Liquid
Horizontal Resolution	0.2 nm (TEM) 5 nm (SEM)	0.2 nm
Vertical Resolution	n/a	0.05 nm
Field of View	100 nm (TEM) 1 mm (SEM)	100 μm
Time for image	0.1∼1 minute	1∼5 minutes
Depth of Field	Good	Poor

The AFM has been originally developed for the nanoscale image measurement. The image is generated by controlling the motion of an sharp probe as it is scanning across a surface. With the AFM, it is possible to directly measure more than physical dimensions of features on the surface having a few nanometer-sized dimensions including single atoms and molecules, since there are direct physical interactions between the probe and the surface. This gives scientists and engineers an ability to directly "feel" and "touch" the nanoworld, and then use the probe as an tool to complete physical characterizations of the nanoobjects, modification physical characters of the surface and manipulation of atoms or molecules.

As one of significant aspects of AFM applications, it is can be used to directly move nanoobjects across the surface by pushing, pulling, even pick-and-place with the probe. With such methods, it is possible to create 2D nanopatterns and 3D nanostructure or nanodevices. To summarize, nanomanipulation with the AFM benefits from followings:

- By AFM image scanning, nanoobjects, as well as the probe that simultaneously acts as manipulation tool, can be accurately located. This function makes it possible to perform nanomanipulation without visual feedback (normally SEM or TEM vision) and additional end-effectors. Moreover, AFM-based nanomanipulation can be performed in various environments, e.g. vacuum, air, and liquid.
- The AFM tip is very tiny (typically with an apex radius of 10 nm or less) compared with the normal size of the nanoobject to be manipulated, which leads to smaller adhesive forces that favor the release operation. In addition, with chemical modifications of the probe, it is possible to achieve pick-and-place nanomanipulation with a single probe.

- With interactions measurement occurring in the AFM-based manipulation, such as force, amplitude, frequency or phase shift of the probe, manipulation process can be well monitored with automatic strategies or user interfaces.

1.1.2 AFM-Based Nanomanipulation

As seen in Fig. 1.2, nanomanipulation can be generally classified into three types:

- Two-dimensional lateral pushing or pulling, including contact and noncantact modes;
- Three-dimensional pick-and-place nanomanipulation with a single nanotip with the help of external force field;
- Three-dimensional pick-and-place nanomanipulation with a multi-tip nanotweezer.

AFM pushing or pulling nanoobjects on a surface are typical lateral nanomanipulation methods, as shown in Fig. 1.2(a). A amount of research work have been carried out with these methods for making two-dimensional patterns [12, 13, 14, 15, 16], characterizing nanofriction by sliding or rolling nanoparticles, bending nanotubes or nanowires [17, 18, 19, 20, 21], exploring nanophysical phenomena [22, 23, 24, 25], and material testing by bending or breaking [26, 27, 28, 29]. The pick-and-place is significant for 3D nanomanipualtion since it is an basic process to assemble nano blocks into three-dimensional structures or devices. Recently, pick-and-place nanomanipulation tasks shown in Fig. 1.2(b) and Fig. 1.2(c) have been demonstrated by an coaxial atomic force microscope [30], electro-enhanced capillary forces [31] and an 3D manipulation force microscope [32], respectively. Main difficulties in achieving 3D nanomanipulation are fabricating sharp end-effectors with enough grasping force, as well as the capabilities of force sensing while controlling interactions between the nanoobject and the tool or the substrate.

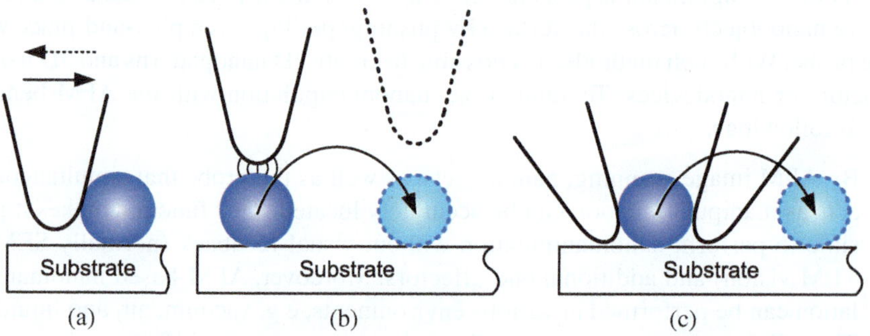

Fig. 1.2 Fundamental nanomanipulation strategies with the AFM. (a) Lateral contact nanomanipulation with pushing/pulling strategies. (b) Pick-and-place with a single AFM probe. (c) Pick-and-place with a dual-probe nanotweezer.

Figure 1.3 shows the general overall layout of the AFM-based nanomanipulation system. An AFM with an AFM cantilever is used as the nanomanipulator. A *X-Y-Z* piezo-scanner is used as the positioner in the AFM for image scan and driving the manipulation. The three axes of the piezo stage are controlled by its dedicated controller or directly by the main control PC. With certain control strategies, nanomanaipulation, e.g. pushing or pulling expressed in the diagram, can be controlled through lateral or normal signals from the AFM cantilever, including force F, amplitude A, phase φ. In addition, a user interface with a haptic device can be integrated into the AFM to facilitate nanomanipulation with interactive feedback, haptic devices and virtual reality interfaces were introduced into AFM-based nanomanipulation systems [33, 34, 35, 36, 37], thereby enabling an operator to directly interact with the real nanoworld to control the manipulation process.

To practically increase the efficiency for mass output, automation is also another significant issue of AFM-based nanomanipulation and nanoassembly from the very start. However, automation on the AFM based nanorobot is not a trivial extension from the macroworld to the nanoworld. This is because of lack of sufficient feedback/information to accurately predicate uncertainties, such as thermal drift, scanner hysteresis and creep, etc, introduced during nanomanipulation. Methods to estimate and compensate such uncertainties and control strategies related to the conventional AFM have been introduced for the sake of automation [38, 39, 40, 41]. However, achieving high efficiency and mass output of AFM based nanomanipulation and nanoassembly is still a long-term objective.

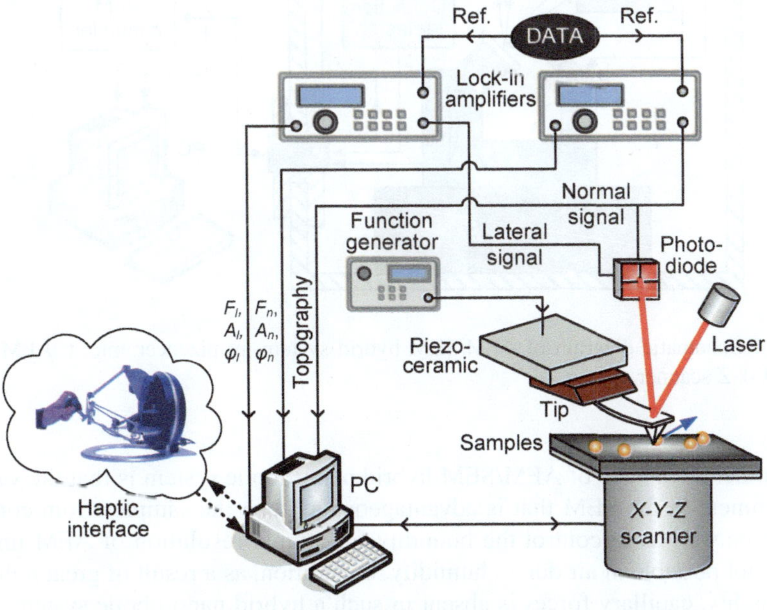

Fig. 1.3 Schematic diagram of pushing/pulling nanomanipualtion with a conventional AFM.

1.1.3 AFM/SEM Hybrid Nanorobotic Systems

It is well known that the conventional SEM can resolve features to several nanometers and its high depth-of-field enables three-dimensional imaging of samples. Moreover, its big chamber make it possible to integrate a AFM inside which gives the advantage of combining two microprobes, as show in Fig. 1.4. A hybrid AFM/SEM enable users to image samples conventionally by the SEM as well as investigate local features more accurately by the AFM with image scanning [42, 43]. With this hybrid nanorobotic system, nanomanipulation can also be completed with vision/force hybrid control strategies, thereby enabling users to analyze a certain area of interest on a sample and then locate the AFM tip to this typical area under the monitoring of vision feedback. In the next step, the local area is zoomed by the AFM and viewed nanoobjects are then manipulated by the AFM tip under the monitoring of force feedback from the AFM.

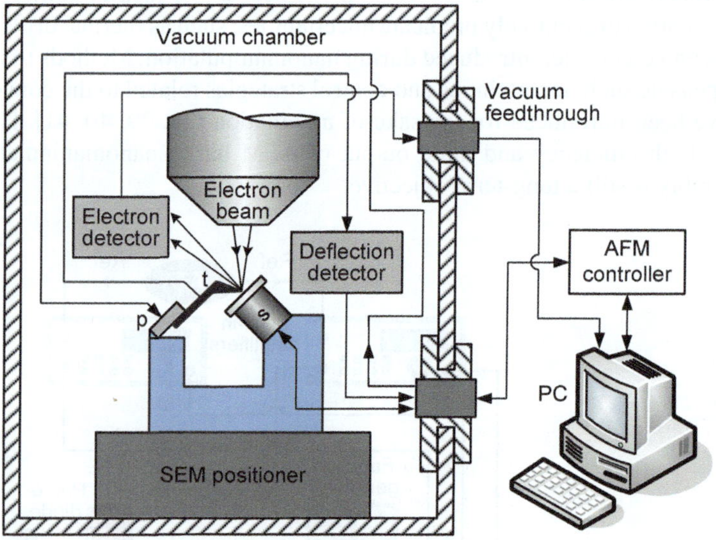

Fig. 1.4 Schematic diagram of AFM/SEM hybrid system. p: pizeoceramic; t: AFM tip; s: AFM *X-Y-Z* scanner.

Another advantage of AFM/SEM hybrid nanorobotic system is that the vacuum environment in the SEM that is advantageous to keep the samples from contamination, as well as to control the humidity for atomic resolution of AFM imaging that is not possible in air due to humidity. In addition, as a result of great reduction of humidity, capillary forces is absent in such a hybrid nanorobotic system that is favorable to nanoobject manipulation, especially release operation.

In addition, due to the subangstrom imaging resolution of the TEM, physical properties are practically characterized inside a TEM since sample deflections during manipulation can be accurately detected. A compact nanrobotic manipulator with several degree-of-freedom can be equipped inside the narrow ultra-high vacuum specimen chamber of the TEM. The manipulator is generally mounted on the TEM holder that is opposite to the specimen. One or several AFM probes are mount on the end of to serve as end-effectors. With this hybrid system, physical properties of nanosamples, such as carbon nanotube, nanobelt and nanowires, can be characterized by bending, buckling and stretching manipulation [7, 8].

1.2 Challenges and Opportunities

AFM-based manipulation has two inherent limitations. One of the limitations is that AFM-based nanomanipulation has usually been restricted to building two-dimensional nanopatterns or in-plane nanomaterial characterization through pushing or pulling manipulation on a single surface. In addition, the manipulation involves an insufficient scan-manipulation-scan process, which makes mass production impossible. So we would say nanofabrication with the AFM is still in infancy. To build three-dimensional nanostructures or nanodevices and then realize mass production with AFM-based nanomanipulation and nanoassembly systems, research on enhancing manipulation capabilities and improving efficiency has to be addressed.

1.2.1 Two-Dimensional Applications

After the first achievement of atom manipulation using a scanning tunneling microscope (STM) in 1990 [1], various nanomanipulation and nanoassembly schemes and systems related to the AFM have been developed. Up to now the conventional AFM has been adopted widely by researchers and engineers for the manipulation or characterization of nanomaterials from 0-dimension to three-dimension and biology samples. However, almost all the applications are restricted on the general pushing or pulling operations within a single surface. Rarely is three-dimensional nanomanipulation completed by a conventional AFM. As shown in Fig. 1.5(a), lateral AFM pushing of a nanoparticle will be blocked by a side wall of a bank that with a comparable height h with the nanoparticle radius R. Three-dimensional nanomanipulation, seen in Fig. 1.5(b), is similarly difficult to be obtained with a way of single-probe grabbing .

On the other hand, as discussed in section 1, nanostructures have been manipulated, assembled and characterized within the three-dimensional space with the help of nanomanipulators or nanogrippers integrated into the SEM and the TEM. Both the SEM and the TEM provide a vacuum environment where the van der Waals force is the main force to be overcome during the manipulation. 3D nanomanipulation

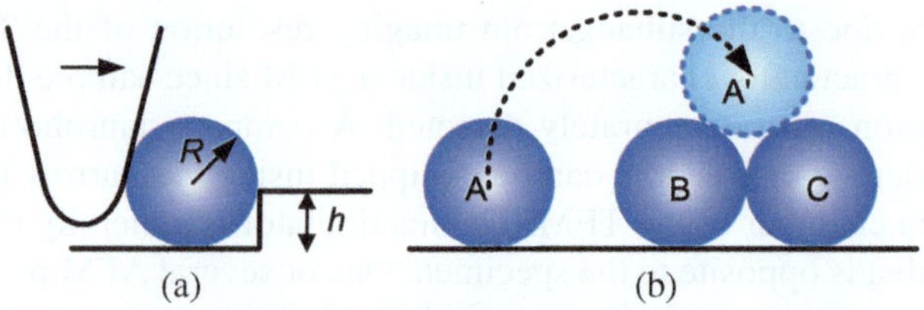

Fig. 1.5 Nanomanipulation limited on the plane with the AFM. (a) Lateral pushing is blocked by a bank. (c) Three-dimensional assembly is difficult to be carried out by the conventional AFM.

could be also achieved with optical tweezers in liquid, where the adhesion forces are greatly reduced.

Pick-and-place nanomanipulation is no doubt a promising technique in 3D nanostructure fabrication since it is an indispensable step in the bottom-up building process. It can be used to overcome limitations of bottom-up and top-down methods of nanomanufacturing and further combine advantages of these two methods to fabricate complex 3D nanostructures or nanodevices. To achieve three-dimensional nanomanipulation and nanoassembly, the main difficulties are fabricating practical end-effectors to produce enough grasping force within a very tiny contact/noncontact interface during nanomanipulation, as well as capabilities of force sensing while controlling interactions between the nanoobject and the tool or the substrate.

Approaching on concept of a two-tip nanogripper was proposed for 3D nanomanipulation as a future work [44]. Until recently, a prototype of three-dimensional manipulation force microscope have been developed (3DMFM) [32]. The 3DMFM consists of two individually actuated cantilevers with protruding tips that are facing each other, constructing a dual-probe nanotweezer for the pick-and-place nanomanipulation, as seen in a schematic diagram Fig. 1.6. Manipulation capabilities of this system were demonstrated by grabbing and manipulating silicon nanowires to build 3D nanowire crosses. More recently, a single-probe tweezer with coaxial electrodes was introduced for three-dimensional particle grabbing. Experiments validate that the coaxial tweezer can perform three dimensional assembly by picking up a specified silica microsphere deposited on substrate. However, this tweezer might not be suitable for nanoscale pick-and-place because of its relatively large truncated tip that makes the release operation more difficult.

Fabricating an effective nanomanipulation tool is a crucial issue for achieving three-dimensional manipulation. Such nanomanipulation tool should provide controllable interactive forces that are favorable to nanoscale grabbing and smooth releasing, as well as protecting the tool and nanoobjects from damage. Possible approaches can enable AFM probes based 3D nanomanipulation systems:

- Designing and fabricating AFM tips that can modify their physical or chemical surface properties to actively or passively change their surface forces, or that can

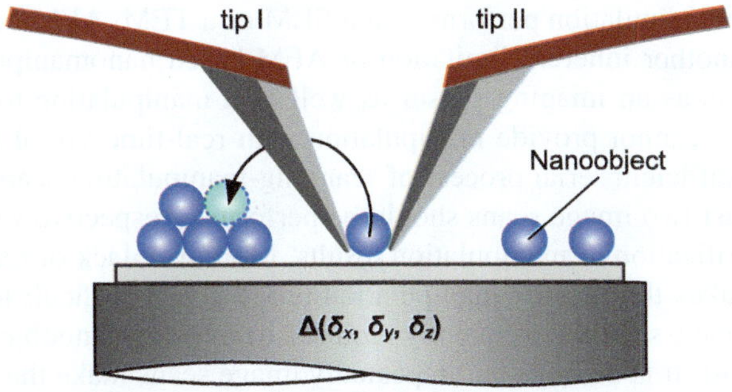

Fig. 1.6 Pick-and-place nanomanipulation using a nanotweezer formed by two AFM cantilevers with protruding tips.

apply controllable interaction forces on nanoobjects. By these means, adhesion on the contact interface can be increased or reduced in a controlled manner for successful grabbing and releasing operations.

- Successful grabbing and releasing operations can be achieved by modifying physical or chemical of surfaces of samples or substrates to obtain controllable adhesion forces on the contact interface.
- Back to the conventional mechanical grabbing, multi-tip nanotweezers might be another practical method to realize pick-and-place nanomanipulation. For instance, an ideal nanotweezer can be formed with AFM cantilevers since sharp tips of the AFM cantilevers and also manipulating forces can be monitored if deflections of the cantilevers are measured as a convectional AFM by optical levers. In addition, pick-and-place nanomanipulation becomes more controllable with the multi-tip nanotweezer if interactive forces among the nanotweezer, samples and substrate can be controlled by surface modification,

1.2.2 Manipulation Efficiency

Although the AFM has been proved to be a powerful manipulating tool for understanding nanoscale physical and chemical phenomena as well as for performing fabrication of nanostructures or nanodevices, it is well known that AFM-based nanomanipulation is very insufficient and definitely influence its practical applications. Three aspects might contribute to the low efficiency of AFM based nanomanipulation and nanoassembly:

- Despite the AFM has been demonstrated to be a powerful tool in exploring the nanoworld, the approach itself is severely limited in imaging speed. A typical AFM image of a samples with 256 by 256 pixels takes at least tens of seconds, and often minutes, to scan. It is clear that mapping large areas of a surface can be particularly laborious. Therefore, low-speed image scanning involved make a AFM-based nanomanipulation task insufficient.

- Unlike nanomanipulation performed in a SEM or a TEM, AFM-based nanomanipulation another inherent limitation of AFM-based nanomanipulation is that the AFM acts as an imaging sensor as well as a manipulating tool simultaneously, and so cannot provide manipulation with real-time visual feedback, but rather an insufficient serial process of scanning-manipulating-scanning. As seen in 1.7, at least two image scans should be performed respectively for task planning and verification of manipulation results. Moreover, lack of real-time visual feedback makes the "blind" manipulation process very difficult to control and then local image scan are frequently required to relocate nanoobjects when manipulation lost. It is obvious that repeatedly image scans make the manipulation process fairly lengthy.

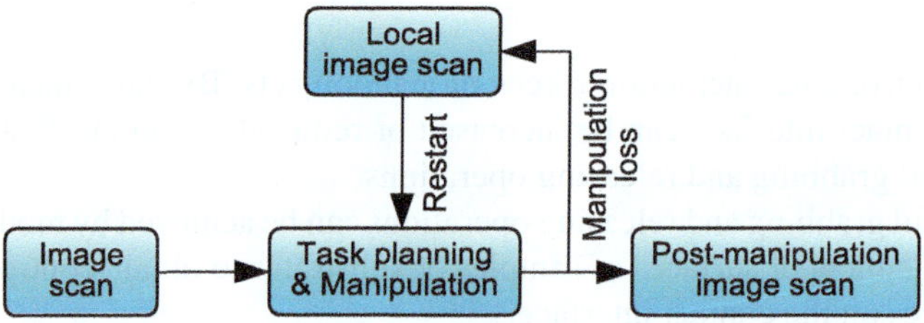

Fig. 1.7 A serial procedure of conventional AFM based nanomanipulation.

In order to conquer limitations above-mentioned on visual feedback and facilitate nanomanipulation, haptic devices and virtual reality interfaces have been introduced into AFM-based nanomanipulation systems 13, 14, thereby enabling an operator to directly interact with the real nano-world. The virtual reality generally provide a user with a static nanoenvironment. In this case, the user is still "blind" since the changes in the nanoenvironment is unavailable. Although behaviors of the AFM tip and nanoobjects being manipulated can be theoretically predicted by interactive models and real-time force feedback from the manipulation, many uncertainties, e.g. thermal drift, scanner hysteresis and creep, sticking phenomena between the tip and nanoobjects as well as some others unpredicted behaviors of nanoobjects, can not be accurately predicted and modeled. Thus, haptic devices and virtual reality interfaces can not provide perfect solutions on lack of visual feedback.

Back to the image rate, high-speed AFM might be a promising way to improve the efficiency of the AFM nanomanipulaion. High-speed AFM was originally developed to study and track dynamic behaviors of biology samples, e.g. protein molecules, cirrus. New developments of the high-speed AFM can provides users with a real video rate reaching to tens of frames that is sufficient for visual feedback [45]. However, the time-consuming scanning-manipulating-scanning operation is still required, making mass production impossible. On the other hand, although high-speed AFMs have succeeded in raising the scanning efficiency 15-18, the excellent imaging potential of high-speed AFM will be greatly reduced if it is used for nanomanipulation with such a serial procedure.

The central problem in speeding up AFM-based nanomanipulation is in fact to develop sufficient harmony between image scan and manipulation processes. Like manipulation under optical microscopes or electron microscopes, the classical parallel imaging/manipulation method might offer promise for high efficiency of AFM-based nanomanipulation. New protocols and novel system designs might provide means to improve the manipulation efficiency of AFM-based nanomanipulation.

1.2.2.1 Simultaneous Manipulation Protocol

A simultaneous manipulation protocol is a promise method to get high-efficiency nanomanipulation on the platform of the conventional AFMs [46]. By increasing scanning speed on the fast scan line when the tip approach a nanoparticle so that it can be pushed on the slow scan line and the motion of the nanoparticle can be monitored by a trace on the image scan. With this method, imaging and manipulation process are performed by a single AFM tip at the same time, thus task steps is great shrinked by one-third from a common serial procedure.

However, the simultaneous manipulation with a single tip has two shortcomings. Due to the very small energy dissipation on the oscillating tip, one of the shortcomings is that this method is limited to manipulate small particles with weak adhesion from a substrate. It might be difficult to fulfill manipulation of large nanoparticle or nanowire/tubes. Another shortcoming is that nanoparticle pushing on the fast scan line could not be deliberately performed or the nanoparticle motion is difficult to control. Therefore, such a simultaneous nanomanipulation with a single AFM tip can not push a nanoparticle to any points on the surface and thus can not arbitrarily produce two-dimensional nanoparticle nanopatterns.

1.2.2.2 Two-Tip Parallel Manipulation AFM

As we discussed, performing imaging and manipulation in parallel is the key to speed up AFM based nanomanipulation. A prototype of parallel imaging/manipulation force microscopy has been developed on the principle of distributing the imaging and manipulation tasks separately on two individually actuated cantilevers with protruding tips. One cantilever acts as an imaging sensor by scanning nanoobjects and tip of the other cantilever that is used as a manipulating tool. During manipulation, the task is on-line planed by the image data form the imaging tip, and the manipulating tip is controlled to perform manipulation at opportune moments. This parallel manipulation scheme can save much time, as compared with the serial imaging/manipulation operation.

Imaging speed is also crucial issue in parallel nanomanipulation scheme. One disadvantage of this scheme with the normal-speed image scan is that environment-based motion planning is unavailable during the manipulation, even the parallel imaging/manipulation can be perfectly performed as a manipulation objective is defined before the operation. For the high-speed image rate, the manipulation process is monitored by the high-speed visual feedback as the manipulation performed in the

SEM. The manipulation is undoubtedly more stable and efficient than the manipulation monitored by the normal-speed image scan. Thus it can be seen that, high-speed image scan is one of crucial to improve efficiency of AFM-based nanomanipulation.

1.2.2.3 Tip Array for Parallel Manipulation

The manipulation tool possessing more than two tips to an tip array might be potentially used for mass production. It can be imaged that if a tip array, consisting of hundreds of tips on a single chip, is used for nanomaipulation, hundreds of targets can be manipulated correspondingly at one time. The tip array greatly increase the manipulation efficiency by hundreds of times as compared with single point manipulation. The AFM tip array was firstly introduced by IBM for the purpose of batch fabrication of high-density data storage [47]. Apart from data storage in polymers or other media, such a tip array is also promising for other areas of nanoscience and nanotechonolgy, such as lithography, large-scale image scan and nanomanipulation.

Concerning applications of nanomanipualtion, uniform tip height is the key to fabricate such a tip array. Height errors among tips should smaller than feature size of nanoobjects being manipulated, or nanoobjects under the shorter tips will be missing from the manipulation. In addition, an disadvantage, or a distinct characteristic of the tip array nanomanipulation is that the manipulation results, e.g. a 2D nanoparticle pattern, are always formed as an symmetrical array to the tip array. This is because this tool are originally developed for high-density data storage applications.

1.2.3 Automation

The commonly used extensive user interaction results in low throughout and severely limits the construction of complex nanostructures with reasonable time and labor. Automation, thus, is quite necessary for increasing of the throughput in AFM-based nanomanipulation. The main goal of automation is to modify the conventional AFM-based manipulation system and manipulation schemes, aiming to automated manipulating nanoobjects with a wide range from atom to tens of nanometers. Corresponding research of this project mainly includes highly-precise nanoobject positioning, modelling and control of tip-nanoobject interactions, and real-time tracking in manipulation.

1.2.3.1 Highly-Precise Positioning of Nanoobjects

Even though the distribution of nanoobjects is known by AFM image scanning before manipulation, automation remains hard to reach for manipulating nanoobjects with sizes on the order of 10 nm. This is primarily because of positioning errors due to spatial uncertainties that generated from scanner creep and hysteresis as well as thermal drift between the tip and the nanoobjects. The errors can reach tens of nanometer in the ambient environment, which is comparable to sizes of the nanoobjects. In actual use, scanner creep and hysteresis could be compensated by precise models or feedback control, but not for the tip drift, which is associated with the time and environmental temperature, and the drift velocity might not be a constant. Therefore, to realize automation on the AFM based manipulation, the spatial uncertainties should be accurately characterized and compensated.

1.2.3.2 Modeling and Control of Tip-Nanoobject Interactions

The interactions between the tip and nanoobjects are fundamental information that for the control of the manipulation, either in the contact or non-contact mode. The interaction information includes the bending force and torsional force, amplitude, frequency and phase shifts, which can be used to control the whole manipulation process, including the control of the contact point and interaction forces between the tip and the nanoobject. Especially the control of the contact point, which is a most significant cause for successful manipulation associated with the sticking problem during the manipulation. In addition, using the model analyses and the feedback data, the manipulation can be effectively controlled by the feedback loop as well as the control interaction forces between the tip and the sample for minimum tip and sample damage.

1.2.3.3 Real-Time Tracking Control in Manipulation

During the manipulation, either pushing or pulling, real-time tracking of the operated nanoobjects is a key issue for successive automated AFM-based nanomanipulation, mainly because the nanoobjects often rotate on the plane, resulting in the missing problem in the operation. Real-time tracking using the feedback information of tip-sample interaction should be employed to relocate the real-time positions of the nanoobjects, and restart the manipulation when the tip loses the nanoobject. As described in the second section, the feedback includes the interaction force and frequency/amplitude/phase shifts. Thus, certain tracking modes or strategies will be built for robust control of the manipulation process.

A force controlled strategy has been widely used for pushing/pulling nanomanipulation. During manipulation, tip-nanoobject contact induced normal forces is observed to monitor the motion of the nanoobject [14]. Successful pushing/pulling operations can be observed if the normal force keep constant or changes periodicity. With constant normal force feedback, the motion of the nanoobject can be considered as static sliding, and the rolling motion occurs if the normal force is with periodic intervals. Contact loss is detected if the normal force reduced sharply to zero and there no periodic intervals observed, e.g. due to in-plane rotation/spinning of the nanoobject as the result of a positioning error on the contact point or non-uniform interactions between the nanoobject-substrate contact. In this case, manipulation is stopped and the image scan is used to re-located the lost nanoobject by searching vicinity of the point where the nanoobject has been lost.

A dynamic pushing method has been proposed to track the motion of the nanoobject in real time and simultaneously achieve high-efficiency and stable pushing/pulling operations, utilizing the advantages of the contact and noncontact pushing control methods [12, 48]. With this method, a constant distance between the tip and substrate is kept during the pushing/pulling manipulation. The tip-substrate separation is more or less equal to the height of the nanoobject, which should provides enough maximum effective deflection to overcome strong static friction at the beginning of the push. while the tip slides up the nanoparticle. Once contact with the nanoobject is established, the amplitude of the oscillating tip reduces to zero and the contact pushing is started. When the tip-nanoobject contact may be lost, the tip returns to oscillating with the same amplitude before contact; this, can be used to detect the contact loss automatically to restart the manipulation.

1.3 Conclusion

This chapter is dedicated to general topics related to the AFM-based nanorobotic system, as well as challenges and opportunities of the state-of-art research of nanomanipulation. Two main shortcomings, including manipulation capability limited in two-dimensional applications and low efficiency of the AFM-based nanomanipulation have been discussed. To achieve three-dimensional nanomanipulation, methods for making interactive forces controllable among a tool, nanoobjects and the nanoworld, and multi-probe nanotweezers have been recommended. To improve nanomanipulation efficiency, the simultaneous manipulation protocol, the dual-probe parallel manipulation scheme and the tip array for parallel manipulation have been proposed. Concerning automation on the AFM-based nanomanpulation, methods to compensate spatial uncertainties, accurate modeling and control of tip-nanoobject interactions and real-time tracking control have been introduced. However, enhancing manipulation capabilities and increasing throughput of AFM-based nanomanipulation are considered to be long-term goals.

References

1. Eigler, D.M., Schweizer, E.K.: Positioning single atoms with a scanning tunneling microscope. Nature 344, 524–526 (1990) 13, 19
2. Fukuda, T., Arai, F., Dong, L.X.: Assembly of nanodevices with carbon nanotubes through nanorobotic manipulations. Proc. IEEE 91, 1803–1818 (2003) 14
3. Dong, L.X., Arai, F., Fukuda, T.: Electron-beam-induced deposition with carbon nanotube emitters. Appl. Phys. Lett. 81, 1919–1921 (2002) 14
4. Dong, L.X., Arai, F., Fukuda, T.: Destructive constructions of nanostructures with carbon nanotubes through nanorobotic manipulation. IEEE/ASME Trans. Mechatronics 9, 350–357 (2004) 14

5. Fahlbuscha, S., Mazerolleb, S., Breguetb, J.-M., Steineckerc, A., Agnusd, J., Pérezd, R., Michlera, J.: Nanomanipulation in a scanning electron microscope. J. Materials Processing Technology 167(2-3), 371–382 (2005) 14
6. Molhave, K., Wich, T., Kortschack, A., Boggild, P.: Pick-and-place nanomanipulation using microfabricated grippers. Nanotechnology 17, 2434–2441 (2006) 14
7. Nakajima, M., Arai, F., Fukuda, T.: In situ measurement of Young's modulus of carbon nanotubes inside a TEM through a hybrid nanorobotic manipulation system. IEEE/ASME Trans. Nanotechnology 5, 243–248 (2006) 14, 19
8. Dong, L.X., Tao, X.Y., Zhang, L., Nelson, B.J., Zhang, X.B.: Nanorobotic spot welding: controlled metal deposition with attogram precision from copper-filled carbon nanotubes. Nano Lett. 7, 58–63 (2007) 14, 19
9. Andersen, K.N., Petersen, D.H., Carlson, K., Molhave, K., Sardan, O., Horsewell, A., Eichhorn, V., Fatikow, S., Boggild, P.: Multimodal Electrothermal Silicon Microgrippers for Nanotube Manipulation. IEEE/ASME Trans. Nanotechnology 8, 76–85 (2009) 14
10. Leach, J., Sinclair, G., Jordan, P., Courtial, J., Padgett, M.J., Cooper, J., Laczik, Z.J.: 3D manipulation of particles into crystal structures using holographic optical tweezers. Opt. Express 12, 220–226 (2004) 14
11. de Vries Anthony, H.B., Krenn, B.E., van Driel, R., Kanger, J.S.: Micro Magnetic Tweezers for Nanomanipulation Inside Live Cells. Biophysical Journal 88, 2137–2144 (2005) 14
12. Martin, M., Roschier, L., Hakonen, P., Parts, Ü., Paalanen, M., Schleicher, B., Kauppinen, E.I.: Manipulation of Ag nanoparticles utilizing noncontact atomic force microscopy. Appl. Phys. Lett. 73, 1505–1507 (1998) 16, 26
13. Sitti, M., Hashimoto, H.: Controlled pushing of nanoparticles: modeling and experiments. IEEE/ASME Trans. Mechatron. 5(2), 199–211 (2000) 16
14. Resch, R., Lewis, D., Meltzer, S., Montoya, N., Koel, B.E., Madhukar, A., Requicha, A.A.G., Will, P.: Manipulation of gold nanoparticles in liquid environments using scanning force microscopy. Ultramicroscopy 82, 135–139 (2000) 16, 25
15. Albrecht, P.M., Lyding, J.W.: Lateral manipulation of single-walled carbon nanotubes on H-passivated Si(100) surfaces with an ultra-highvacuum scanning tunneling microscope. Small 3, 146–152 (2007) 16
16. Tranvouez, E., Orieux, A., Boer-Duchemin, E., Devillers, C.H., Huc, V., Comtet, G., Dujardin, G.: Manipulation of cadmium selenide nanorods with an atomic force microscope. Nanotechnology 20, 165304 (2009) 16
17. Sitti, M.: Atomic force microscope probe based controlled pushing for nanotribological characterization. IEEE/ASME Trans. Mechatron. 9, 343–349 (2004) 16
18. Palacio, M., Bhushan, B.: A nanoscale friction investigation during the manipulation of nanoparticles in controlled environments. Nanotechnology 19, 315710 (2008) 16
19. Dietzel, D., Monninghoff, T., Jansen, L., Fuchs, H., Ritter, C., Schwarz, U.D., Schirmeisen, A.: Interfacial friction obtained by lateral manipulation of nanoparticles using atomic force microscopy techniques. J. Appl. Phys. 102, 084306 (2007) 16
20. Conache, G., Gray, S.M., Ribayrol, A., Fröberg, L.E., Samuelson, L., Pettersson, H., Montelius, L.: Friction measurements of inAs nanowires on silicon nitride by AFM manipulation. Small 5, 203–207 (2009) 16
21. Dietzel, D., Monninghoff, T., Jansen, L., Fuchs, H., Ritter, C., Schwarz, U.D., Schirmeisen, A.: Interfacial friction obtained by lateral manipulation of nanoparticles using atomic force microscopy techniques. J. Appl. Phys. 102, 084306 (2007) 16
22. Xu, B., Tao, N.J.: Measurement of single-molecule resistance by repeated formation of molecular junctions. Science 301, 1221–1223 (2003) 16

23. Grill, L., Rieder, K., Moresco, F., Stojkovic, S., Gourdon, A., Joachim, C.: Exploring the interatomic forces between tip and single molecules during STM manipulation. Nano Lett. 6, 2685–2689 (2006) 16
24. Mougin, K., Gnecco, E., Rao, A., Cuberes, M.T., Jayaraman, S., McFarland, E.W., Haidara, H., Meyer, E.: Manipulation of gold nanoparticles: influence of surface chemistry, temperature, and environment (vacuum versus ambient atmosphere). Langmuir. 24, 1577–1581 (2008) 16
25. Whittaker, J.D., Minot, E.D., Tanenbaum, D.M., McEuen, P.L., Davis, R.C.: Measurement of the adhesion force between carbon Nanotubes and a silicon dioxide substrate. Nano Lett. 6, 953–957 (2006) 16
26. Li, X., Gao, H., Murphy, C.J., Caswell, K.K.: Nanoindentation of silver sanowires. Nano Lett. 3, 1495–1498 (2003) 16
27. Bordag, M., Ribayrol, A., Conache, G., Fröberg, L.E., Gray, S., Samuelson, L., Montelius, L., Pettersson, H.: Shear Stress measurements on InAs nanowires by AFM manipulation. Small 3, 1398–1401 (2007) 16
28. Wu, B., Heidelberg, A., Boland, J.J.: Mechanical properties of ultrahigh-strength gold nanowires. Nature Materials 4, 525–529 (2005) 16
29. Ni, H., Li, X.D.: Youngs modulus of ZnO nanobelts measured using atomic force microscopy and nanoindentation techniques. Nanotechnology 17, 3591–3597 (2006) 16
30. Brown, K.A., Aguilar, J.A., Westervelt, R.M.: Coaxial atomic force microscope tweezers. Appl. Phys. Lett. 96, 123109 (2010) 16
31. Toset, J., Gomila, G.: Three-dimensional manipulation of gold nanoparticles with electro-enhanced capillary forces. Appl. Phys. Lett. 96, 043117 (2010) 16
32. Xie, H., Haliyo, D.S., Régnier, S.: A versatile atomic force microscope for three-dimensional nanomanipulation and nanoassembly. Nanotechnology 20, 215301 (2009) 16, 20
33. Sitti, M., Hashimoto, H.: Teleoperated touch feedback from the surfaces at the nanoscale: Modeling and experiments. IEEE/ASME Trans. Mechatron. 8, 287–298 (2003) 17
34. Li, G.Y., Xi, N., Yu, M.M., Fung, W.K.: Development of augmented reality system for afm-based nanomanipulation. IEEE/ASME Trans. Mechatron. 9, 358–365 (2004) 17
35. Li, G.Y., Xi, N., Chen, H.P., Pomeroy, C., Prokos, M.: 'Videolized' atomic force microscopy for interactive nanomanipulation and nanoassembly. IEEE Trans. Nanotechnol. 4, 605–615 (2005) 17
36. Vogl, W., Ma, B.K.-L., Sitti, M.: Augmented reality user interface for an atomic force microscope based nanorobotic system. IEEE Trans. Nanotechnol. 5, 397–406 (2006) 17
37. Kim, S.G., Sitti, M.: Task-based and stable telenanomanipulation in a nanoscale virtual environment. IEEE Trans. Autom. Sci. Eng. 3(3), 240–247 (2006) 17
38. Mokaberi, B., Requicha, A.A.G.: Drift compensation for automatic nanomanipulation with scanning probe microscopes. IEEE Trans. Autom. Sci. Eng. 3, 199–207 (2006) 17
39. Mokaberi, B., Requicha, A.A.G.: Compensation of scanner creep and hysteresis for AFM nanomanipulation. IEEE Trans. Autom. Sci. Eng. 5, 197–206 (2008) 17
40. Xie, H., Haliyo, D.S., Régnier, S.: Characterizing piezoscanner hysteresis and creep using optical levers and a reference nanopositioning stage. Rev. Sci. Instrum. 80, 046102 (2009) 17
41. Krohs, F., Onal, C., Sitti, M., Fatikow, S.: Towards Automated Nanoassembly With the Atomic Force Microscope: A Versatile Drift Compensation Procedure. J. Dyn. Sys. Meas. Control 131, 061106 (2009) 17
42. Ermakov, A.V., Gatfunkel, E.L.: A novel AFM/STM/SEM system. Rev. Sci. Instrum. 65, 2853–2854 (1994) 18

43. Thomas, Ch., Heiderhoff, R., Balk, L.J.: Acoustic near-field conditions in an ESEM/AFM hybrid system. J. Physics: Conference Series 61, 1180–1185 (2007) 18
44. Sitti, M.: Teleoperated 2-D micro/nanomanipulation using an atomic force microscope. PhD Thesis, University of Tokyo, Japan (1999), `http://www.cs.cmu.edu/msitti/pub.html` 20
45. Ando, T., Uchihashi, T., Fukuma, T.: High-speed atomic force microscopy for nano-visualization of dynamic biomolecular processes. Progress in Surface Science 83, 337–437 (2008) 22
46. Kim, S., Ratchford, D.C., Li, X.: Atomic force microscope nanomanipulation with simultaneous visual guidance. ACS Nano 3, 2989–2994 (2009) 23
47. Ettiger, P., Despont, M., Drechsler, U., Durig, U., Haberle, W., Lutwyche, M.I., Rothuizen, H.E., Stutz, R., Widmer, R., Binnig, G.K.: The 'Millipede'–More than thousand tips for future AFM storage. IBM J. Research and Development 44, 323–340 (2000) 24
48. Xie, H., Régnier, S.: High-efficiency automated nanomanipulation with parallel imaging/manipulation force microscopy. IEEE Trans. Nanotechnol. doi:10.1109/TNANO.2010.2041359 26

Chapter 2
Instrumentation Issues of an AFM Based Nanorobotic System

While the atomic force microscope (AFM) was mainly developed to image the topography of a sample, it has been discovered as a powerful tool also for nanomanipulation applications within the last decade. A variety of different manipulation types exists, ranging from dip-pen and mechanical lithography to assembly of nanoobjects like carbon nanotubes (CNTs), deoxyribonucleic acid (DNA) strains, or nanospheres. The latter, the assembly of nanoobjects, is a very promising technique for prototyping nanoelectronical devices that are composed of DNA-based nanowires, CNTs, etc. But, pushing nanoobjects in the order of a few nanometers nowadays remains a very challenging, labor-intensive task that requires frequent human intervention. To increase throughput of AFM-based nanomanipulation, automation can be considered as a long-term goal. However, automation is impeded by a large nulber of uncertainties existing in every AFM system.

This chapter is dedicated to the instrumentation issues for the design or the use of an AFM based nanorobotic system. The first part deals with the calibration of the normal and the lateral stifnesses of the cantilever. Calibration needs to be done accurately because the sensing force greetly depends on the stifness. The second part will introduce the problem of the cross-talk compensation for the measurement of the interaction forces. The last part will describe the problem of the thermal drift of the piezo scanning stage.

2.1 Force Calibration Issues in AFM (Normal Force and Lateral Force Calibration)

Techniques for the reliable and precise calibration of atomic force microscopes (AFM) have been significant issues since AFM was developed more than two decades ago [1]. There are two categories of AFM calibration: normal force calibration and lateral force calibration. The quantitative determination of absolute values of normal and lateral force conversion factors generally involves two steps: the calibration of photodiode responses and the measurement of cantilever spring constants.

H. Xie et al.: Atomic Force Microscopy Based Nanorobotics, STAR 71, pp. 31–86.
springerlink.com

The spring constants can be calculated from the geometric and physical properties of the cantilevers [2, 3, 4] or modeled by finite-element analysis [5, 6, 7]. However, these methods are approximate as ideal models of cantilevers are used (e.g., ideal geometry, coatings not factored in, etc.). Moreover, minor errors in dimension measurements can induce substantial stiffness errors, especially the thickness measurement. Thus, there has been a tendency to determine the cantilever spring constants experimentally.

2.1.1 Normal Force Calibration

To measure the normal spring constant, the most commonly adopted method was developed by Cleveland who measured frequency shifts due to the known mass loaded on the free end of the cantilevers [8], although it is thought to be time-consuming. Ruan et al. used a stainless steel spring sheet with known stiffness to measure the spring constants of the cantilevers [9]. Recently, Sader et al. developed a method to calculate the cantilever's normal spring constant from resonant frequencies induced by thermal fluctuations [10]. More recently still, several methods have been reported for the calibration of the normal constant [11, 12, 13, 14, 15]. A newly reported "piezosensor" uses a piezoresistive cantilever as an active force sensor to calibrate the normal spring constants of the AFM cantilevers [16]. A normal force applied to the cantilever's tip can be easily calculated by multiplying the cantilever's vertical deflection to its normal spring constant. Thus, the normal conversion factor can be easily determined by experimental results.

In this part, we show the basics of the model dedicated to evaluate the normal stiffness of a cantilever. The Euler-Bernoulli theory is the most popular method for calculation of spring constants of microcantilevers based on experimental results.

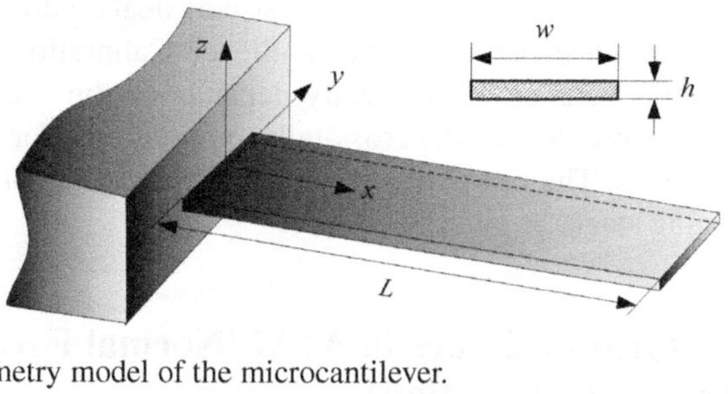

Fig. 2.1 The geometry model of the microcantilever.

As shown in Fig. 2.1, the cantilever is built in at one end, free at the other end as-sumed to deform in the linear elastic range. L, w, and h are the length, width and the thickness of a rectangular cantilever, respectively. The coordinates are defined as follows: the origin is located on the centre of the cross section of the built-in end, the x-axis is along its length, and the z-axis and y-axis are along its thickness and the y-axis, respectively. The motion of flexural vibrations is a function of x:

$$EI_y(x)\frac{\partial^4 v}{\partial x^4} + \eta \rho S\frac{\partial v}{\partial t} + \rho S\frac{\partial^2 v}{\partial t^2} = 0 \tag{2.1}$$

where E is Young's modulus of the silicon cantilever, $I_y = \frac{wh^3}{12}$ is its moment of inertia about the z-axis, $S = w \times h$ is its cross-section area, η is a damping constant and deflection function $v(z,t) = V(z)e^{i\omega t}$, ρ is the density of cantilever. Taking no account on the effect of the damping, the solutions of the equation is:

$$(K_nL)^3\left[\cos(K_nL)\cosh(K_nL) + 1\right] = 0 \tag{2.2}$$

where K_n is the nth wave number of the cantilever ($K_1L \approx 1.8751$) for the first flexural mode), The resonant frequencies are further obtained using the dispersion relation for nth flexural frequency in the cantilever:

$$\omega_n = (K_nL)^2\sqrt{\frac{EI_y}{\rho SL^4}} \tag{2.3}$$

where ω_n is the nth flexural resonant frequency. If ω_n is known, so the thickness h of cantilever can be calculated by:

$$h = \frac{\omega_n}{K_n^2}\sqrt{\frac{12\rho}{E}} \tag{2.4}$$

Thus the stiffness k of the cantilever can be also calculated by:

$$E = \frac{Ewh^3}{4L^3} \tag{2.5}$$

The equation of the motion of the torsional modes is de-fined as :

$$GJ\frac{\partial^2\theta}{\partial x^2} = \rho I_p\frac{\partial^2\theta(x,t)}{\partial t^2} + c\frac{\partial\theta(x,t)}{\partial t} \tag{2.6}$$

where $\theta(x,t)$ is the rotation angle of the cantilever, G is the shear modulus, ρ is the density of the cantilever, c is the coefficient of viscous damping, $I_p = (wt^3 + w^3t)/12$ is the polar area moment of inertia and J is the torsional constant. For a rectangular cantilever, J can be obtained by :

$$J \approx \frac{1}{3}wh^3\left[1 - 0.63\frac{h}{w} + 0.052\left(\frac{h}{w}\right)^5\right] \tag{2.7}$$

The nth torsional frequency is obtained by

$$\omega_n = \frac{(2n-1)\pi}{2L}\sqrt{\frac{GJ}{\rho I_p}} \qquad n = 1,2,.... \tag{2.8}$$

2.1.2 Lateral Force Calibration

The conversion of lateral force and photodiode signal is more challenging than the normal calibration. Normally, two kinds of methods are used to calibrate the lateral conversion factor: a two-step method and a direct method. The former involves the calibration of the lateral stiffness of the cantilever and the measurement of the lateral photodiode response. This method is not straightforward and is limited in application. Unlike the calibration of the normal constant, the lateral sensitivity of the photodiode is more difficult to determine because the lateral contact stiffness between the AFM tip and the sample surface is proportional to the contact radius [17] and often comparable to the lateral stiffness of the cantilever and the tip [18], which significantly reduces the calibration result of the lateral sensitivity of the photodiode19, 20. In order to overcome this limitation, several methods have been put forward for lateral sensitivity measurement [20, 21, 22]. A test probe with an attached colloidal sphere was successfully used to determine lateral photodiode sensitivity by loading the colloidal sphere laterally against a vertical sidewall [20]. However, this kind of method is also limited in application because of difficulties in characterization of the lateral stiffness of the V-shaped cantilevers [23].

In contrast with the two-step method, a one-step direct method, named wedge method, developed by Ogletree et al. is the most commonly accepted method in current use [24]. This method gets round the difficulties in the separate measurement of the lateral stiffness of the cantilever and lateral sensitivity of the photodiode. An improved wedge method developed by Varenberg et al. utilizes a commercially available calibration grating with a well-defined slope instead of the friction loops generated from different slopes [25], which enables calibration of all types of probes, including probes integrated with sharp or colloidal tips. A newly reported method based on direct force balances on surfaces with known slopes considers detector cross-talking and off-centered tip problems and reduces tip wear during the calibration [26]. Direct force loading methods, including small glass fibers [27, 28, 29], a specially fabricated microelectromechanical device (MEMS) [30] and magnetic forces [31, 32] have been also used to directly calibrate AFM lateral force measurement. Compared with the wedge method, the accuracy of calibration results with these methods is greatly affected by friction forces [27, 28, 29], and the system setup is more technically complex for the experiments [30, 32].

This part presents a new method to calibrate the lateral force measurement of the atomic force microscope using a commercially available, accurately calibrated piezoresistive force sensor. It consists of a piezoresistive cantilever and accompanying electronics, providing a standard force applied on the AFM tip for lateral force calibration. Before use, the spring constant of the piezoresistive cantilever and sensitivity of the accompanying electronics were accurately calibrated. This method may be used to directly calibrate the factor between the lateral force and the photodiode signal for cantilevers with a wide range of spring constants, regardless of their size, shape, material or coating effects. Three rectangular cantilevers with normal spring constants from 0.092 to 1.24N/m (lateral stiffness from 10.34 to 101.06) were calibrated. Moreover, we compared the calibration results with the theoretically calculated results based on the beam mechanics, which would yield the best results when the cantilever has a simple geometry and uniform physical properties, and the photodiode has an ideal symmetry of the normal and lateral output.

2.1.2.1 Calibration of the Piezoresistive Force Sensor

A piezoresistive cantilever fabricated by the standard silicon bulk micromachining technology with low scatter and drift of sensitivity can be used as a portable micro-force calibration standard [33]. The piezoresistive cantilever (Nascatec GmbH, Germany) and its accompanying electronics used in our work are commercially available. Microscopy images of the cantilever are shown in Fig. 2.2.

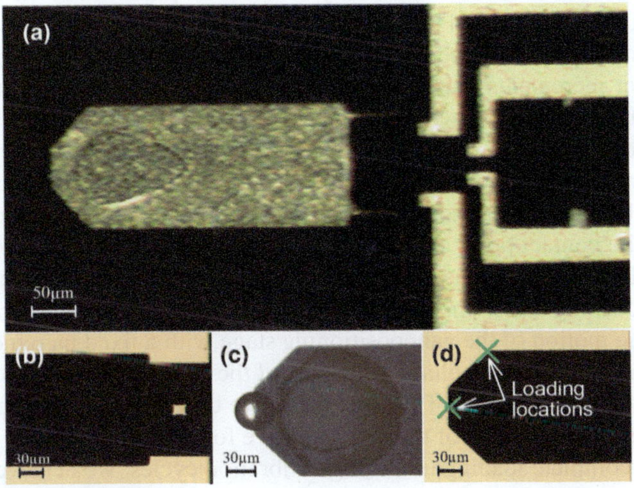

Fig. 2.2 Optical microscopy images of the piezoresistive cantilever used in the calibration of the lateral force measurements. (a) Top image of the piezoresistive cantilever. (b) The shape of the clamping end of the piezoresistive cantilever with a step shape and a hole in its center. (c) An image obtained after a glass microsphere was placed on the tip of the piezoresistive cantilever. (d) A magnified image of the tip in which two loading locations for the lateral calibration are marked. These two locations are close to the end of the side edge and the center point of the top edge on the back of piezoresistive cantilever respectively.

Dimensions of the piezoresistive cantilever were measured as 525.8 μm in length and with an average width of 152.7 μm using microscopic image processing (under an optical microscope Olympus BX50WI with a 50× objective and Sony XC-711P CCD, providing a resolution of 0.22 μm/pixel). The top view Fig. 2.2(b) shows that the clamping end of the piezoresistive cantilever has a step shape with a difference of 12.5 μm on the width and a hole with a length of 15 μm on square (maybe for stress enhancement), so it is not convenient to directly calculate its normal spring constant. Therefore, in our experiment, the piezoresistive cantilever stiffness k_p was calibrated using Cleveland's mass loading method [8]. We used six glass microspheres with diameters from 25.6μm to 64.4μm measured under the optical microscope, and used a glass density of $2.4\mathrm{g/cm}^3$. As shown in Fig. 2.2(c), the glass microspheres were released on the free end of the piezoresistive cantilever and their centers were measured for stiffness compensation due to position errors [15, 16].

Experiments showed that the adhesion force between the glass microspheres and the backside of the piezoresistive cantilever was strong enough to hold the microbeads during the first mode of vibration with very low amplitude (under a humidity of 50% − 60%). The first natural resonant frequency of the piezoresistive cantilever is 37.463kHz. The stiffness of the piezoresistive cantilever was calibrated at $k_p = 18.209 \pm 0.471\mathrm{N/m}$.

The next step is the force calibration of the piezoresistive sensor. The piezoresistive cantilever was mounted horizontally on a 3 DOF platform, so the force applied on the cantilever was normal to its longitudinal axis. A glass substrate was attached on a Z nanopositioning stage with a resolution of 1.8nm, which was used for the displacement increments during the calibration. On the surface of the glass substrate, a glass microsphere with a diameter of 50μm was glued near the substrate edge used for the point of contact with the piezoresistive cantilever during the force loading. First, the nanopositioning stage was adjusted by 100nm increments until the contact between the piezoresistive cantilever tip and the glass microsphere was achieved. The contact point on the horizontal plane was controlled by microscopy vision, while the Wheatstone bridge voltage output was used to detect the contact on the approaching direction. After the contact had been setup, a program was used to control the motion of the nanopositioning stage with a fixed increment (20nm in our experiment) while the voltage output V_p of the electronics was recorded. It was found that the displacement of the piezoresistive cantilever tip was approximately 5.7μm across the full range of the piezoresistive force sensor output.

After 20 complete loading/unloading calibration cycles, we achieved a piezoresistive force sensor sensitivity $S_p = 10.361 \pm 0.267\mu\mathrm{N/V}$. The sensitivity $S_p = \frac{dF}{dV_p}$ is defined as the gradient of the applied force F versus the voltage output V_p plot. During the force calibration of the piezoresistive force sensor, four contact points were used to test whether a position change on the width affects its sensitivity.

As shown in Fig. 2.3, the first three contact points were located on the hemline of the trapezoidal head on the free end, on which there were two points: one on each of the left and right bottom corners, and another in the center of the line. The fourth contact point was located on the very tip of the free end. Plots of the voltage output versus displacement (applied force) have the same gradient except for the fourth contact point, which has a lower gradient because of a lower stiffness on the tip. This experimental result demonstrated that the Wheatstone resistance bridge is not sensitive to the torsion loading applied on the piezoresistive cantilever's longitudinal axis. Thus, the points on the side edges as well as the points on the tip edge of the piezoresistive cantilever can be used as loading locations for the AFM cantilever calibration. This will be discussed below.

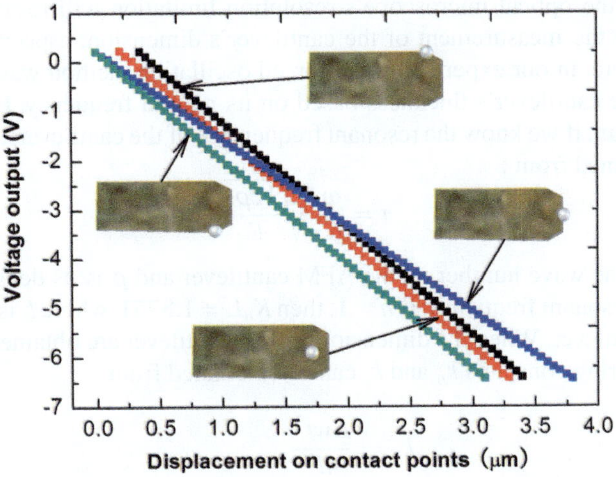

Fig. 2.3 Examples of the sensitivity calibration on four different contact points (see four corresponding scaled images) on the cantilever. In which three curves with the same gradient present the contact points on the hemline of the trapezoid head on the free end of the cantilever. The fourth curve with a lower gradient denotes the calibration result when the contact point is on the tip of cantilever.

2.1.2.2 Testing AFM Cantilevers

Three types of AFM cantilevers with rectangular cross sections and normal force constants from 0.092 to 1.24 N/m (shown in table 2.1) were used: LFMR (NANO World), ContAL and Multi75AL (Budget Sensors). Although dimensions of the cantilevers were provided by manufactures, the optical microscope (with 50 and 100 lenses) was used to determine the cantilever's length, width and tip height.

Table 2.1 Descriptions of the cantilevers based on the beam mechanics and the experimental results. The cantilever's length L, the width w and the tip height h were measured using the optical microscope. The first flexure resonant frequencies f_0 were used to determine the thickness of the AFM cantilevers from Eq. 2.9. The normal spring constant k_n and lateral spring constant k_l were calculated from Eq. 2.20. a_0 calculated based on the beam mechanics from Eq. 2.20, a measured from the proposed top loading and side loading methods are listed in the last three columns respectively.

Tip No.	$L(\mu m)$	$h(\mu m)$	$w(\mu m)$	$t(\mu m)$	f_0(kHz)	k_n(N/m)	k_l(N/m)	$a_0(\mu N/V)$	$a(\mu N/V)$	$a(\mu N/V)$
1	228	15.5	47.8	0.83	21.84	0.092	10.34	1.61 ± 0.53	2.29 ± 0.26	2.32 ± 0.26
2	449	17.4	53.4	2.01	13.64	0.19	62.27	11.24 ± 3.71	12.04 ± 1.35	12.31 ± 1.38
3	229	17.6	31.4	2.28	59.49	1.24	101.06	18.79 ± 6.20	25.84 ± 2.89	26.2 ± 2.93

However, the optical microscope's resolution limitation will result in a significant error in the measurement of the cantilever's dimension, especially its thickness. Therefore, in our experiment, the forced oscillation method was employed to determine the cantilever's thickness based on its natural frequency. For the Euler-Bernoulli beam, if we know the resonant frequencies of the cantilevers, the thickness t can be obtained from :

$$t = \frac{\omega_n}{K_n^2}\sqrt{\frac{12\rho}{E}} \tag{2.9}$$

where K_n is the wave number on the AFM cantilever and ρ is its density, ω_n is the nth flexural resonant frequency. If $n = 1$, then $K_nL = 1.8751$, where L is the length of the AFM cantilever. When the dimensions of the cantilever are obtained, the normal and lateral spring constants k_n and k_l can be calculated from:

$$k_n = \frac{Ewt^3}{4l^3} \tag{2.10}$$

$$k_t = \frac{Gwt^3}{3L(h+t/2)^2} \tag{2.11}$$

where w, t and h are the width, thickness and tip height of the AFM cantilever respectively.

2.1.2.3 Experimental Methods

Once the piezoresistive force sensor had been calibrated, it was used as a force standard to determine the lateral force conversion factor a of the AFM cantilevers. The experiments described below were performed on a combined AFM/optical microscope system. Although the experiment procedure in this work may be not available on all commercial AFMs, this method could be widely used after some adjustment.

For its actual use, the piezoresistive cantilever was fixed into a metal harness with four contact clips. The cantilever and the harness were integrated onto a thin circuit board, which was attached to the AFM stage using a fixture. The electronics, including amplifier, signal filter and power supply, was separated in another unit. Considering the limitations of the manipulation space in the AFM, two ways were recommended for mounting the piezoresistive cantilever. The first involves mounting the piezoresistive cantilever vertically on the AFM stage along its longitudinal axis (see Fig. 2.4(a)), termed top loading, which reduces the mounting area at the cost of height space. In this way, the AFM cantilever tip contacts with the top edge of the piezoresistive cantilever during the calibration (see the loading location in Fig.2.2 (d)). Nevertheless, if the piezoresistive cantilever is too high (8mm in our experiments), the piezoresistive cantilever can be mounted in the second way: horizontally on the AFM stage (see Fig. 2.4(b)), termed side loading. Note that in this case, shoulders of the piezoresistive cantilever substrate might be in the way of the reflected laser beam, so an angle θ to the vertical plane that is through the longitudinal axis of the AFM cantilever was deliberately mounted ($\theta = 15^\circ$ in our

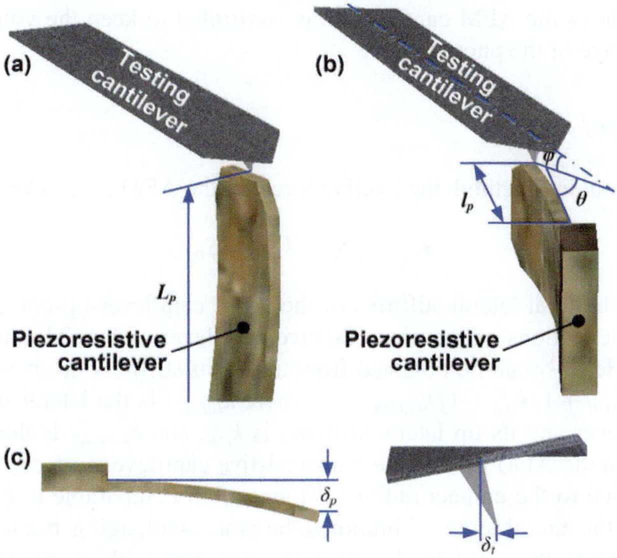

Fig. 2.4 Diagrams of the experimental configurations for the calibration of the AFM cantilevers with a piezoresistive force sensor. Two methods, termed top loading (a) and side loading (b), in which L_p and l_p are distances from the contact points to the clamping end of the piezoresistive cantilever, ϕ is the mounting angle of AFM cantilever on the vertical plane that is through its longitudinal axis, and θ is the mounting angle of the piezoresistive cantilever on the horizontal plane. (c) The deflections of the piezoresistive cantilever and AFM cantilever tip are δ_p and $delta_t$ respectively.

experiments). The loading location in the second method was very close to end of the side edge (see Fig. 2.2(d)).

Lateral force calibration was started once the whole setup was ready. At first, we had to find the loading locations. For the top loading, after the AFM cantilever was brought into contact with the top surface of the piezoresistive cantilever, the contact mode was used to scan the top side edge to identify its center point. Then the AFM cantilever was moved 2μm away from the scanned side edge. In order to ensure the AFM tip was reliably in contact with the top side edge, the AFM cantilever was moved down with a displacement $\Delta h = 0.5 - 0.8$μm before being moved back to the contact location. For the side loading, the procedure was largely the same; the only difference being the position of the loading location. As shown in Fig. 2.2(d), the side loading was positioned at the end of the top side edge, in fact a bottom corner of the trapezoid head on the free end. After the loading locations had been determined, the AFM tip was moved laterally to the contact location with a step of 10nm. Under the programmed control, voltage V_p and V_l, outputs of the piezoresistive force sensor and the photodiode began to be recorded at a frequency of 5Hz when V_l reached the defined preload value 0.01V. During the measurement, the deflection of the AFM cantilever was controlled to keep the voltage output in the linear range of the photodiode.

2.1.2.4 Data Analysis

For the top loading method, the loading force on the AFM tip can be presented as:

$$F_t = k_p \delta_p = k_t \delta_t = S_p V_p \tag{2.12}$$

where k_t is the total lateral stiffness of the AFM cantilever-tip-contact system, δ_p and δ_t are deflections of the piezoresistive cantilever and AFM cantilever tip respectively. Here, k_t can be obtained from a sum of stiffness inverses of each part: $k_t = (1/k_{lateral} + 1/k_{tip} + 1/k_{contact})^{-1}$, where $k_{lateral}$ is the lateral stiffness of the AFM cantilever and its tip lateral stiffness is k_{tip}, and $k_{contact}$ is the contact stiffness between the AFM tip and the piezoresistive cantilever surface. $k_{contact}$, which is proportional to the contact radius [17] and often comparable to $k_{lateral}$ and k_{tip} [18], causes the lateral force calibration to be more challenging, because in this case dt is not equal to the real lateral deflection associated with the photodiode output. Fortunately, in our method, the lateral conversion factor is directly provided by the ratio of the applied force on the AFM tip and the photodiode's voltage output, regardless of any knowledge of the cantilevers and the laser measuring system. For the top loading method, the force $F_t = S_p V_p$ is applied on the AFM tip, so the conversion factor a can be simply obtained from:

$$\alpha = \frac{F_t}{V_l} = S_p \frac{V_p}{V_l} \tag{2.13}$$

For the side loading method, in order to reduce the effects of friction force on the contact point, the loading direction is perpendicular to the piezoresistive cantilever. In this case, the lateral force conversion factor a under the side loading is determined from:

$$\alpha = \frac{F_s}{V_l} = S_p \frac{V_p L_p \cos\theta}{V_l l_p} \tag{2.14}$$

where L_p and l_p are distances from the contact points to the clamping end of the piezoresistive cantilever for top loading and side loading, respectively (shown in Fig. 2.4).

However, considering that the lateral force is loaded on the side of the tip, not on the tip head of the AFM cantilever, a simple linear transformation of the factor a is given by:

$$\alpha\prime = \alpha \left(1 + \frac{\Delta h}{h + t/2 - \Delta h}\right) \tag{2.15}$$

where $\Delta h = 0.5 - 0.8\mu\text{m}$ is the distance from the contact point to the tip head of the AFM cantilever, t is the cantilever thickness. For the cantilevers used in our experiments $h = 15.5 - 17.6\mu\text{m}$ and $t = 0.83 - 2.28\mu\text{m}$, so an error of the factor a generated from Δ_h is $2.7\% - 5\%$, which is just within an acceptable range.

We quantitatively compared our method with the theoretical method based on the beam mechanics. For the convenient comparison of the normal and lateral conversion factors, the photodiode sensitivity is defined as a ratio of the angular deflection of the AFM cantilever and the photodiode voltage for which the lateral inverse sensitivity is $S_l = \theta_l/V_l$, and the normal photodiode inverse sensitivity is $S_n = \theta_n/V_n$. Here, θ_n and θ_l are the normal and lateral angular deflection of the cantilever, respectively; V_n and V_l are the corresponding photodiode voltage outputs.

The normal spring constant k_n connects the flexural deflection x_n due to an applied normal force F_n, which can be determined by [35]:

$$F_n = k_n x_n \tag{2.16}$$

So, based on the beam mechanics, applied normal force can be evaluated by:

$$F_n = \frac{2}{3} k_n l S_n V_n \tag{2.17}$$

where l is the effective length of the AFM cantilever. Therefore, the normal inverse sensitivity S_n of the photodiode can be determined from:

$$S_n = \frac{3}{2l}\frac{dx_n}{dV_n} \tag{2.18}$$

For a cantilever with a rectangular cross section, the lateral force is related to the measured photodiode voltage V_l from:

$$F_l = \frac{Gwt^3}{3l(h+t/2)} S_l V_l \tag{2.19}$$

where G is the cantilever shear modulus and h is the tip height. Based on the laser beam mechanism, the displacement of the laser spot on the photodiode is decided by the reflecting cantilever's angular deflection and the distance between the reflecting point and the spot position on the photodiode. Normally, the distance can be considered as a constant; so, based on the hypothesis that the photodiode is "rotationally symmetric" [29], the lateral sensitivity is assumed to be equal to the normal sensitivity. Thus, the lateral force conversion factor α_0 can be calculated from:

$$\alpha_0 = \frac{Ewt^3}{3l(h+t/2)} S_n \tag{2.20}$$

However, the hypothesis of photodiode symmetry is often not exactly the case, due to the possible asymmetry of the laser spot shape as well as the diffraction effects from the cantilever, and may induce errors. The beam mechanics method requires accurate knowledge of the cantilever's elastic modulus and dimensions. As discussed below, in our experiments, the high-resolution optical microscope is used to measure the cantilever's length, width and the tip height, and the force oscillation method is used to determine the thickness of the cantilever.

2.1.2.5 Results and Discussion

As described above, two loading methods were used in our experiments: top loading and side loading. In the first case, we used Eq. 2.13 to calculate the lateral force conversion factor. For side loading, considering the loading position and the direction of the loading force, therefore, Eq. 2.14 was used to calculate the lateral force conversion factor. The inverse sensitivity Sn of the photodiode described in Eqs. 2.16 - 2.18 was measured for each AFM cantilever by a linear fit on the plot of the photodiode voltage output versus displacement of the cantilevers. Here, we assumed that the photodiode inverse sensitivities S_l and S_n had the same value in our experiments.

The experimental results are summarized in table 2.1, which lists the dimensions and the mechanics of the AFM cantilevers. The dimensions, including length L, width w and tip height h, were measured using the optical microscope. The first flexure resonant frequencies f_0 were used to determine the thickness of the cantilevers from Eq. 2.9. The normal spring constant k_n and lateral spring constant k_l were calculated from Eq. 2.11. The last three columns list α_0 calculated from Eq. 2.20 based on the beam mechanics, and a measured using the proposed top loading and side loading methods, respectively.

Each cantilever was bent laterally by the piezoresistive cantilever ten times. For each time the lateral force conversion factor was calculated as outlined in Fig. 2.5(a-c), in which the symbol of the blue solid circles shows data obtained from the experiments using the top loading method. The straight red lines are the linear fit of the corresponding data using the least square method; their gradients were used to calculate a from Eq. 2.13. Then, a value of the lateral force conversion factors was averaged from the results of the ten experiment repeats. The lateral force conversion factor a0 was calculated based on the beam mechanics. Fig. 2.5(d) shows a comparison of the lateral force conversion factors obtained from each method. The top loading and side loading results are approximately the same; the values of the side loading method are just a little higher due to the friction effect not being factored into Eq. 2.14. But larger differences occur between the measured factor α and α_0 calculated based on the beam mechanics because of cantilever dimension errors and the photodiode's asymmetry sensitivity.

Both methods inevitably have some sources of error. For the proposed method, we need to take into account errors generated by the calibration of the piezoresistive cantilever as well as those from the lateral force calibration of AFM cantilevers, that is, variables S_p, V_p and V_l for the top loading method, while θ and l need to be added for the side loading method. The errors in these measurements of S_p, V_p, V_l, θ and l are of the order of 11%, 0.02V, 0.02V, 1° and 2%. Considering the error generated from δh, the maximum overall error for the calibration of the lateral conversion factor a using the proposed method is 12.4%; it largely depends on the uncertainty of S_p. If an absolute force standard is used to calibrate the piezoresistive force sensor, an error of less than 6% can be expected. Using Eq. 2.20, the method based on the beam mechanics may yield an overall error as high as 33% with uncertainties: l : 2%, w : 2%, h : 10%, t : 10%, S_n : 10%. For these cantilevers, the two methods actually yield similar results and errors of similar magnitude. However, as the geometry and

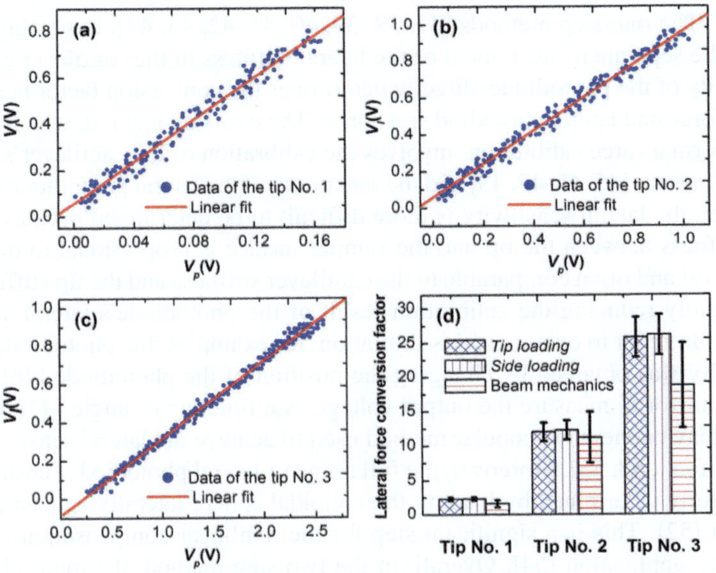

Fig. 2.5 Examples of voltage outputs of the photodiode plotted versus voltage output of the piezoresistive force sensor using cantilever No. 1, No. 2 and No. 3 respectively. The blue solid circles show the data obtained from the experiments using the top loading method. The straight red lines are the linear fit of the corresponding data using the least square method. (d) The lateral force conversion factors α averaged over ten experimental data repeats for each tip and the lateral force conversion factors α_0 are calculated from equation based on the cantilever mechanics.

material composition of the cantilevers become more complicated, the uncertainty of the beam mechanics method will increase significantly. Moreover, this calculation assumes that the photodiode's lateral sensitivity and normal sensitivity are the same, which may not be true. We will test this using a colloidal tip in our future work.

In contrast, the piezoresistive force sensor has several attractive features. The most significant fact is that it can provide a force standard for the direct calibration of the lateral force conversion factors without any knowledge of the photodiode, or cantilever shape, dimensions and physical properties, thereby overcoming almost all the difficulties in the calibration of the lateral force measurement.

2.1.3 *Optical Lever Calibration in Atomic Force Microscope with a Mechanical Lever*

A great deal of attention has been paid to techniques for the reliable and precise calibration of the lateral force application in atomic force microscopes (AFM) since the first friction force measurement with an AFM [36]. Two kinds of methods are commonly used in the lateral force calibration: a one-step method and a two-step method. The one-step methods [37, 38, 39, 40, 41, 42, 43, 44], bypassing difficulties in the separate measurement of the lateral stiffness of the cantilever and lateral sensitivity of the photodiode, directly determines the conversion factor between the lateral force and lateral photodiode response. The two-step method, which is similar to the normal force calibration, involves the calibration of the cantilever's torsional spring constant [45, 46, 47, 48] and the sensitivity of the lateral photodiode response. However, the lateral sensitivity is more difficult to be determined because the contact stiffness between the tip and the sample surface is proportional to the contact radius [49] and often comparable to the cantilever stiffness and the tip stiffness [50], significantly reducing the calibration result of the photodiode's lateral sensitivity [51, 52]. In order to overcome this limitation, for examples, the photodiode's lateral sensitivity was obtained by changing the position of the photodiode [46] or using a tilted mirror to measure the output voltage as a function of angle [47]. Colloidal probes may be the most popular method used to achieve the lateral sensitivity measurement [51, 52, 53]. Moreover, the full range of lateral photodiode sensitivity was successfully determined by loading the colloidal sphere laterally against a vertical sidewall [52]. This is a significant step for the nonlinear compensation in the lateral force application [54]. Overall, in the two-step method, the main obstacle in the lateral sensitivity calibration is the difficulty in determining the actual lateral deflection using the conventional displacement-voltage conversion between the tip displacement and photodiode output. Here, we present a new method to calibrate the lateral sensitivity of the optical lever using a small mechanical lever with a flexible hinge. This device can directly transfer the lever translation into the angular deflection of the any type of cantilever mounted on the mechanical lever using a simple geometric calculation.

In the alternative method for the calibration of the lateral sensitivity of the photodiode response, a mechanical lever, fabricated by the electric discharging machining (EDM) technique, is used as a so-called translator for translation-to-angle conversion. A lever with a dimension of 22mm $\times$ 8mm $\times$ 4mm was used in our experiments. Simulation and experimental results indicate that its dimensions can be in fact reduced much to match space limitations of AFM in actual use. A diagram of the experimental set-up is shown in Fig. 2.6(a)-(c).

In Fig. 2.6(a), the mechanical lever is fixed on the AFM stage and a testing cantilever is mounted on a holder attached to the upper beam of the lever, deliberately bringing the torsional axis of the cantilever and rotation axis of the lever in line. When the AFM stage moves vertically with a displacement Δz, it pushes the mechanical lever against a barrier located on the top surface of the upper beam with a distance of L from the center of the flexible hinge, converting the translation into a nanoscale rotation of the upper beam of the lever (In Fig. 2.6(b)). In this case, the accurate translation on the Z-axis of the AFM stage can be converted into a rotation on the cantilever, imitating a torsional deflection of the cantilever deduced by a torsional moment applied on it. In Fig. 2.6(c), a finite-element analysis of the displacement vector distribution of the upper beam shows that its rotation is nicely around the center of the flexible hinge when a displacement is applied on the lever's pan. This lever can accurately calibrate the lateral sensitivity of the photodiode

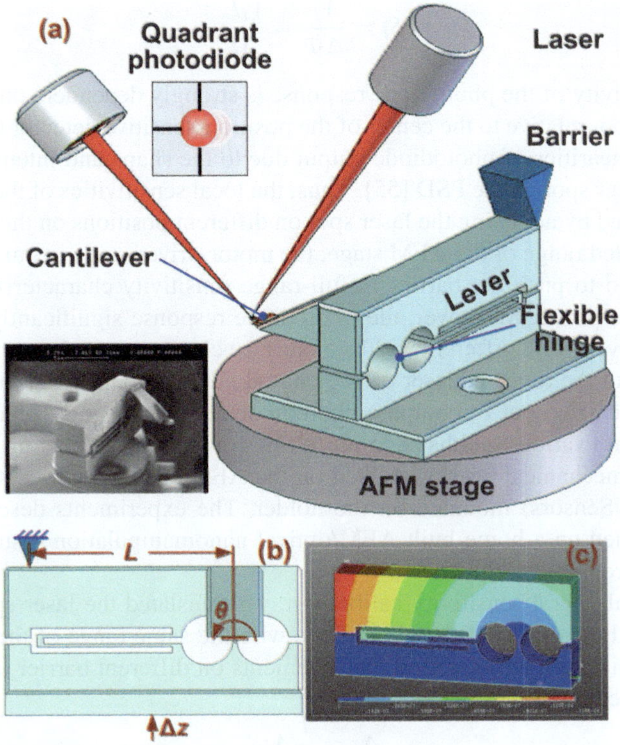

Fig. 2.6 Diagram of the experimental set-up. (a) A small mechanical lever with a flexible hinge. The inset shows a SEM image of a mechanical lever (15mm × 8mm × 4mm) with an attached cantilever. (b) The geometric transform between the AFM stage displacement Δz and a rotation angle θ of the cantilever. (c) Finite-element analysis of the displacement vector distribution of the upper beam.

response without any clearance or creep due to the flexible hinge used for kinematic transform. More importantly, this method can be used to directly calibrate the sensitivities of any type of cantilevers mounted on the mechanical lever instead of a tilted reflecting mirror [47] and without any changes [39] and load applied to the cantilevers, obtaining a nondestructive and more accurate conversion between the lateral angular deflection and the photodiode response.

As shown in Fig.1 (b), a geometric transform between the displacement Δz applied on the pan and the rotation angle $\Delta\Phi$ on the cantilever can be simplified as:

$$\Delta\Phi = \frac{\Delta z}{L} \tag{2.21}$$

where L is the distance between the barrier location and the center of the flexible hinge. Thus, when the voltage output V_l of the photodiode and the corresponding displacement Δz are known, the lateral angular sensitivity S_l of the photodiode can be easily obtained from the Eq. 2.21:

$$S_l = \frac{V_l}{\Delta\Phi} = \frac{V_l L}{\Delta z} \tag{2.22}$$

The sensitivity of the photodiode response is strongly dependent on the position of the laser spot relative to the center of the position-sensitive detector (PSD), introducing nonlinearities of photodiode output due to the shape and intensity distribution of the laser spot on the PSD [55]. Thus, the local sensitivities of the photodiode were calibrated by adjusting the laser spot on different positions on the photodiode. Due to a limited range of the AFM stage, the motor-driven stage in our experiments was employed to push the barrier in full-range sensitivity characterization of the photodiode response. Moreover, the photodiode response significantly differs between cantilevers with different widths or surface coatings, which strongly relate to the intensity of the reflected laser on the photodiode [52]. Therefore, further experiments were carried out to compare photodiode response using different types of cantilevers with various widths and reflectivities of their reflex coating.

After the mechanical lever was fixed on the AFM stage with a cantilever (ContAL, Budget Sensors) mounted on the holder. The experiments described below were performed on a home-built AFM/optical nanomanipulation system under an ambient environment in the air.

In the local lateral sensitivity calibration, we translated the laser spot symmetrically around the center of the photodiode with the same range of the photodiode outputs. In this case, the scanning displacements on different barrier loading positions were decided by:

$$\Delta z = \frac{l}{l_0}\Delta z_0 \tag{2.23}$$

where l_0 is a reference position of the barrier with a photodiode response of about $\pm$120mV deduced by the AFM stage displacement δz_0. In the experiments, $l_0 =$ 8mm was selected and $\delta z_0 = 4\mu$m was decided accordingly.

In the experiments, ten loading positions L from 8mm to 17mm with an interval of 1mm were used. Fig. 2.7 shows examples of the local sensitivity calibration results obtained on five positions from 8mm to 16mm with an interval of 2mm.

Open and closed symbols refer to the approaching and retracting data, respectively. The data show that very small hysteresis loops occur between the approaching and retracting plots due to the closed loop control of the Z nanostage as well as the non-clearance and non-creep flexible hinge. Slopes of each of the plots are the photodiode voltage outputs vs. displacements of the AFM stage, which can be easily converted into the lateral sensitivities by Eq. 2.23.

In Fig. 2.8(a), a full-range calibration result obtained on the loading position 15mm is plotted. Open and closed circles refer to the approaching and retracting data, respectively. The hysteresis loop of these plots is due to the backlash of the motor-driven stage. Diamond symbols refer to the lateral full-range sensitivity calculated by making the derivative of the approaching data. A nonlinear fit using the Gauss function indicates that the calibrated result agrees well with behavior predicated by Gaussian distribution of the laser spot positions on the photodiode. Another full-range sensitivity was also characterized by the local sensitivity calibration

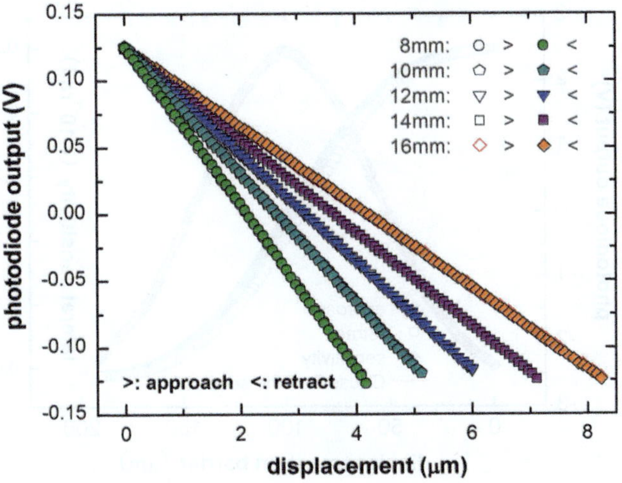

Fig. 2.7 Examples of the sensitivity calibration results acquired using the developed mechanical lever with different locating positions of the barrier.

method by translating the photodiode laterally from the left to the right-hand sides with an interval of 0.1V in photodiode output and repeating the local sensitivity calibration on each position, as shown in Fig. 2.8(b) by closed circles. For comparison, the full-range sensitivity obtained by moving barrier is also plotted here by the diamond symbols. Small differences between these two methods (max. difference on the photodiode center is about $0.011\text{V}/10^{-3}$ rad, approximately 2.2% of the corresponding sensitivity). This difference may be due to itself on the local sensitivity calibration, which average the photodiode responses on neighboring ranges of each spot position.

Four types of cantilevers, including the cantilever used in the former experiments (termed Tip 1 here) with the same reflex coating material (aluminum), were used in the further experiments of the local sensitivity comparison on the photodiode center. Each type of cantilever is from the same batch packed in the same box with 10 pieces. Fig. 2.9 shows a comparison of the average sensitivities of these four types of cantilevers, in which the error bars show an overall error of $\pm 5.6\%$ of the sensitivities, which is deduced from uncertainties: $L : 0.2\text{mm}, V_l : 0.015\text{V}, \Delta z : 5\text{nm}$. The results show that the sensitivity is much more dependant on the reflex coating than the width of the cantilever. Great divergence of the reflectivity occurs due to the different properties of the reflex coating even cantilevers have the same type of reflex coating and similar width. Therefore, it is necessary to re-calibrate the sensitivity when a different type of cantilever is used.

In summary, we have presented a new method to calibrate the lateral sensitivity of optical lever in the AFM using a mechanical lever with a flexible hinge. In the experiments, the small- and large-range scanning respectively were used to calibrate the local- and full-range sensitivities of the photodiode response, and the calibration

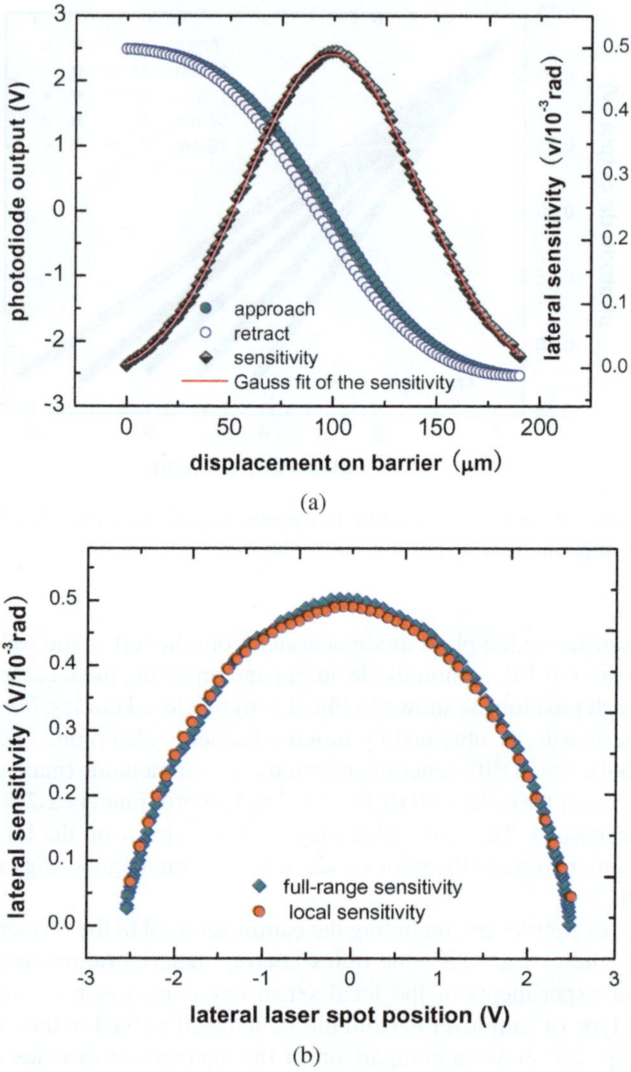

Fig. 2.8 Full-range calibration of the photodiode response. (a) Full-response plot of the photodiode vs. the barrier displacement on L =15mm. (b) Lateral sensitivity vs. spot position plot on the photodiode obtained by the local and full-range calibration methods.

results are accurate to ±5.6%. A method such of this may allow accurate, direct and nondestructive calibration of the lateral sensitivity of the optical lever in AFM without any modification to the actual AFM or the cantilevers, thereby enabling an accurate calibration of the lateral-force measurement in atomic force microscopy.

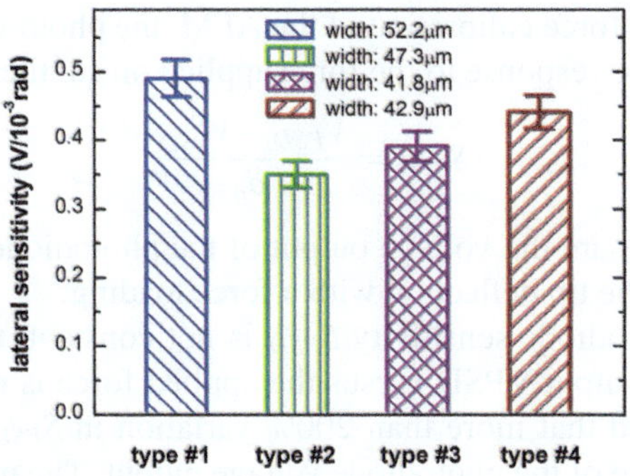

Fig. 2.9 Sensitivity calibration results using four types of cantilevers with the same aluminum coating and different widths.

2.1.4 Enhanced Accuracy of Force Application Using Nonlinear Calibration of Optical Levers

The atomic force microscope (AFM) has been widely used as a nano-effector with a function of force sensing to detect interaction forces between an AFM tip and a sample, thereby controlling the process of the nanomanipulation. However, both the extent and accuracy of force application are significantly limited by the nonlinearity of the commonly used optical lever with a nonlinear position-sensitive detector (PSD). In order to compensate the nonlinearity of the optical lever, a nonlinear calibration method is presented. This method applies the nonlinear curve fit to a full-range position-voltage response of the photodiode, obtaining a continuous function of its voltage-related sensitivity. Thus, Interaction forces can be defined as integrals of this sensitivity function between any two responses of photodiode voltage outputs, instead of rough transformation with a single conversion factor.

For this method, we will use the following results (see section 2.1.1).

Table 2.2 Calibration results of the naormal and lateral force application

$t(\mu m)$	$f_0(kHz)$	$k_n(N/m)$	$k_l(N/m)$	$\beta(\mu N/V)$	$\alpha(\mu N/V)$
2.14	13.30	0.24	74.90	0.47 ± 0.02	14.85 ± 1.66

2.1.4.1 Traditional Force Calibration

Various literatures analyzed and discussed the characterization of the sensitivity of the optical lever [56, 57, 58]. The sensitivity of the optical lever can be enhanced by increasing the intensity of the laser beam or by decreasing the beam divergence. Moreover, during the force calibration, the sensitivity of the optical lever has strong dependences on the position of the laser spot relative to the center of the PSD and geometry of the optical path [59, 57]. Main causes that introduce the nonlin-earities are the shape and intensity distribution of the laser spot on the PSD [58], which limit the range of real force application in AFM, especially when a very "soft" cantilever is used.

For the traditional force calibration of the AFM, the photo-diode sensitivity S_{PSD} is considered as linear response to the force applied on cantilever's tip by:

$$S_{PSD} = \frac{V_{PSD} - V_{PSD}^0}{\delta_p} \tag{2.24}$$

where V_{PSD}^0 and V_{PSD} are the voltage output of the photodiode before and after the force loading, δ_p is the tip deflection with a force loading.

Actually, the photodiode sensitivity S_{PSD} is not constant, that is the plot of the photodiode voltage output VPSD versus the applied force is nonlinear. In fact, our experiments indicated that more than 200% variation in S_{PSD} is a function of the range and initial value of the photodiode voltage output. The most important factors that introduce the nonlinearity are the shape and intensity distribution of the laser spot. If the spot is near the center of the photodiode, the response is linear. When the spot deviates from the center of the photodiode, the nonlinearity becomes more obvious. The force/displacement-voltage response of the AFM therefore should be accurately calibrated in the full range of the photodiode.

2.1.4.2 Nonlinear Compensation of the Force Application

For the convenience of normal and lateral calibration, an-gular sensitivity S_{PSD} is used in our experiments, which is in-dependent of the dimensions of the cantilevers and directly reveals the mechanism of the deflection measurement. The angular sensitivity S_{PSD} is defined as a ratio of the angular deflection of the cantilever and the photodiode voltage output. Thus, the normal photodiode sensitivity is $S_{PSD}^n = \frac{\delta V_{PSD}^n}{\theta_n}$ and the lateral sensitivity $S_{PSD}^l = \frac{\delta V_{PSD}^l}{\theta_l}$. Here, θ_n and θ_l are defined as the normal and lateral angular deflection of the cantilever.

The normal spring constant k_n connects the flexural deflection δ_n due to an applied normal force $F_n = k_n \delta_n$. So based on the beam mechanics, θ_n can be presented as:

$$\theta_n = \frac{3}{2L}\frac{F_n}{k_n} \tag{2.25}$$

where L is the effective length of the cantilever. Therefore, if we know the continuous function of V_{PSD}^n, so S_{PSD}^n can be determined by:

$$S_{PSD}^n = \frac{2L}{3}\frac{dV_{PSD}^n}{d\delta_n} \tag{2.26}$$

For a cantilever with rectangular cross section, the torsional angle θ_l related to the applied force F_l by:

$$\theta_l = \frac{3F_l L(h + t/2)}{Gwt^3} \tag{2.27}$$

where G is the shear modulus of the cantilever. Thus the lateral sensitivity S_{PSD}^l of the photodiode can be also determined by:

$$S_{PSD}^l = \frac{dV_{PSD}^l}{d\theta_l} \tag{2.28}$$

Continuous functions of the normal and lateral sensitivities can be determined by the calibration and nonlinear fit of the position-voltage curves. Thus, in the actual application, the angular deflection of the cantilever can be obtained by:

$$\theta_n = \int \frac{1}{S^n_{PSD}} dV^n_{PSD} \tag{2.29}$$

$$\theta_l = \int \frac{1}{S^l_{PSD}} dV^l_{PSD} \tag{2.30}$$

where the lower and upper limits are the initial and force deduced voltage outputs of the photodiode. Also the normal and lateral tip displacement can be calculated by:

$$\delta_n = \frac{2L}{3}\theta_n \tag{2.31}$$

$$\delta_l = (h+t/2)\theta_l \tag{2.32}$$

In the actual application, the whole nonlinear calibration protocol can be carried as follows:

1. Set the initial voltage output (without force loading) of the photodiode near the lower point by adjusting the position of reflecting laser spot.
2. Record original force/position-voltage responses by the normal and lateral force calibration.
3. Transform the force/position-voltage responses to vol-tage-angular sensitivity responses by 2.25 and 2.27 for normal and lateral cases, respectively.
4. Employ nonlinear fit voltage-angular sensitivity re-sponses to obtain continuous functions of V_{PSD} and then calculate the angular sensitivity S_{PSD} by 2.26 and 2.28.
5. Calculate the angular deflection on the AFM using 2.29 and 2.30 for normal and lateral force, respectively. Then the applied forces on the AFM tip can be easily obtained.

2.1.4.3 Experimental Results

The experiments described below were performed on an AFM based nanorobotic system. The voltage range of the position detector, unlike $\pm$10V of a common AFM, is $\pm$1.5V because electronics with a lower ratio of signal amplifier is used. Nonetheless, the general approach can be widely applicable and the only difference is just the calibrated conversion parameters described in table 2.2.

Inspired by the sigmoidal shape of the V_{PSD} versus θ curves presented in fig. 2.10, the method of Sigmoidal fit was employed to the normal and lateral voltage-angular sensitivity response (V_{PSD},θ) in the experiments. The common Dose response function was used in the Sigmoidal fit by:

$$V_{PSD} = A_1 + \frac{A_2 - A_1}{1+10^{(\theta_0-\theta)p}} \tag{2.33}$$

where A_1, A_2, p and θ_0 are the nonlinear fit parameters: lower limit, upper limit, slope and the value of θ as half value of V_{PSD}. All the responses are in the almost full range of the photodiode signal output.

The next step is to calculate the inverse angular sensitivities of the photodiode, which can be obtained as the derivative of the Sigmoidal fit described in Eq. (2.33):

$$S_{PSD}^{-1} = \frac{d\theta}{dV_{PSD}} = \frac{(1+\zeta)^2}{(A_2 - A_1)\zeta p \ln 10} \tag{2.34}$$

where $\zeta = -\frac{A_2 - V_{PSD}}{A_1 - V_{PSD}}$.

Thus we get a continuous function of the angular sensitivity S_{PSD} on the full range of photodiode voltage output V_{PSD}, rather than a single value. This function will be used to calculate the compensated normal and lateral angular deflections by Eq. (2.29) and Eq. (2.30) (rather than from Eq. (2.26) and Eq. (2.28), which assumes a linear transform between applied force and deflection with a single value of the sensitivity S_{PSD}), respectively. So a simple expression of the angular deflection between any two signal outputs V_{PSD}^0 and V_{PSD}^l can be obtained by:

$$\theta = -\left.\frac{\ln|\zeta|}{2.302585p}\right|_{V_{PSD}^0}^{V_{PSD}^l} \tag{2.35}$$

Voltage-angular deflection responses (V_{PSD},θ) of the normal and lateral cases used for the calculation of the photodiode sensitivities S_{PSD} were obtained by the real force calibrations with a "soft" cantilever with a normal spring constant of 0.24N/m. The Dose response function was used to fit the (V_{PSD},θ) responses and fitting parameters are shown in table 2.3, which would be used to calibrate the sensitivity S_{PSD} via Eq. (2.35).

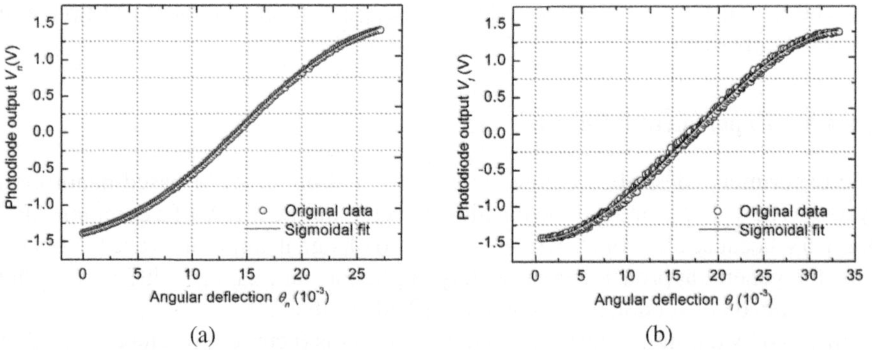

Fig. 2.10 The force calibration curves (V_{PSD},θ) and Sigmoidal fitting results using the Dose response function. (a) The normal V_{PSD} versus θ_n response. (b) The lateral V_{PSD} versus θ_l response. All the responses are in almost the full range of the photodiode signal output. The open circle symbol represents the original data calculated from the force calibration results and the red line is the Sigmoidal fitting results.

Table 2.3 Parameters of the sigmoid fit

Calibration Type	A_1	A_2	θ_0	p
Normal	-1.62358	1.76318	14.55008	0.07663
Lateral	-1.61967	1.71585	16.98019	0.06446

Sigmoidal fit results are shown in fig. 2.10, including the force calibration curves (V_{PSD},θ) (open circle) and fitting results using the Dose response function (red line). Fig. 2.10(a) shows the normal V_{PSD} versus θ_n response, and the lateral V_{PSD} versus θ_n response is presented in Fig. 2.10(b). All the responses are in the almost 95% full range of the photodiode signal output (± 1.43V). For the lateral calibration, the angular deflection is calculated via Eq. (2.28) with the readout of the piezoresistive force sensor.

In order to further verify the proposed method for the non-linearity compensation, an apparent sensitivity compensation experiment was evaluated using the response of the normal inverse S_{PSD} versus the normal voltage V_{PS}. Fig. 2.11(a) shows that more than 200% variation in normal S_{PSD} is the function in the full range of the photodiode voltage output. The fitting sensitivity (red line) generated from the Sigmoidal fit is in accordance with the shape of the real sensitivity curve. The blue straight line, obtained from a linear fit of the bottom on the real sensitivity curve, indicates that an inverse sensitivity 6.92^{-3} rad/V is the minimum value of this curve, presenting the highest sensitivity when the laser spot is near the center of the photo-diode.

For easier representation the results of the sensitivity com-pensation, the ratio of the real and the fitting sensitivity was multiplied by the minimum inverse sensitivity in fig. 2.11(a), giving an apparent compensated sensitivity as shown in fig. 2.11(b). The slope of the linear fit of the compensated sensitivity is 0.087 (red line), resulting in a variation of 3.6% in contrast with more than 200% before the compensation. The range of the force measurement was extended from 36% to 95% of the full range (within 5% minimum value of the sensitivity, the photodiode signal is in -0.5V to $+0.58$V, which represents 36% of the full range of ± 1.5V), and the corresponding force application range improved from 0.25μN to 0.69μN of the cantilever with a spring constant of 0.24N/m.

Fig. 2.11(c) shows the further comparison of the positions cal-culated from the traditional and the proposed method. The diamond symbol shows a nonlinear relationship between the calculated positions by traditional method Ztra versus real posi-tion Zreal recorded by the AFM stage, resulting in an overall position error $\Delta Z = 0.434$μm (28.9% of the total displacement in the full range of the photodiode). The symbol of the circles displays an approximately straight line of compensated position Z_{com} with a gradient of 0.9996. Note that both plots have a same value of the linear fit near the center of the photodiode, where the position difference keeps constant due to the linear sensitivity in this area. The experiments results indicated that an excellent nonlinear fit obtained by the proposed method.

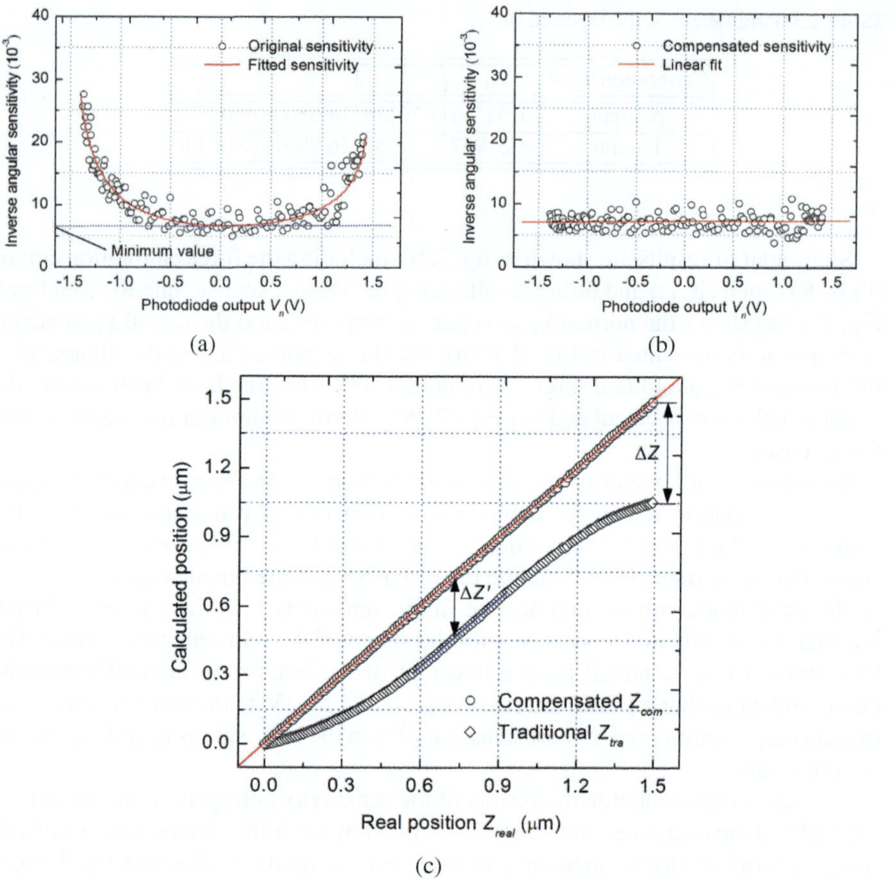

Fig. 2.11 (a) The normal inverse sensitivity versus normal voltage as the slope $d\theta/dV_{PSD}$ of the normal (V_{PSD},θ_n) response shown in Fig. 2.10(a). (b) Compensated normal inverse sensitivity using Sigmoidal fit of data shown in Fig. 2.11(a). The corresponding linear fit (the slope is 0.087) shows the nonlinearity is reduced from more than 200% to 3.6%. (c) Compensated position Zcom and traditional calibrated Ztra versus real position Zreal recorded by the AFM stage. The slope of the linear fit of the compensated position curve is 0.9996.

2.2 Cross-Talk Compensation in Atomic Force Microscopy

In a typical commercial AFM, the optical detection system simply consists of a laser beam that hits the top of the cantilever and reflects back onto a four quadrant position sensitive photodetector (PSPD). Note that the optical system does not detect the tip deflection, but rather measures the inclination of the cantilever near its free end. Assuming that the AFM tip base perfectly coincides with the shear center of the cantilever and perfectly aligned optical system, a vertical force registers only a normal bending signal that can be measured by the illumination difference of the collected laser beam between the upper and lower quadrants of the PSPD ((A+B)-(C+D), as shown in Fig. 1(a) and conventionally called V_{A-B} hereafter in this work). Similarly, applying a lateral force on the tip, perpendicular to the long axis of the micro cantilever, results in a torsional deflection and a lateral bending of the AFM cantilever. Since, generally, the lateral stiffness of the beam is much larger than the torsional stiffness, measured lateral signal response from the left and right quadrants of the PSPD ((A+C)-(B+D), as shown in Fig. 1(a) and conventionally called V_{LFM} in this work) is mainly due to the twisting of the cantilever. The longitudinal force component, perpendicular to the short axis of the AFM cantilever, contributes to the normal bending, which can be measured similar to the vertical force. Note that, due to this effect, it is very difficult to decouple the vertical and longitudinal forces acting on the AFM tip simultaneously.

Since response signals of the optical detection system are voltages (V), a crucial step for a reliable force measurement by using an AFM is to convert these signals to force values. Dominant sources of error are the normal and torsional stiffness calibration as well as the normal and lateral sensitivity calibration. Calibration methods of normal stiffness and sensitivity yield results well below acceptable error limits, while lateral stiffness and sensitivity calibration tehniques are not yet standardized [63, 64, 65, 66].

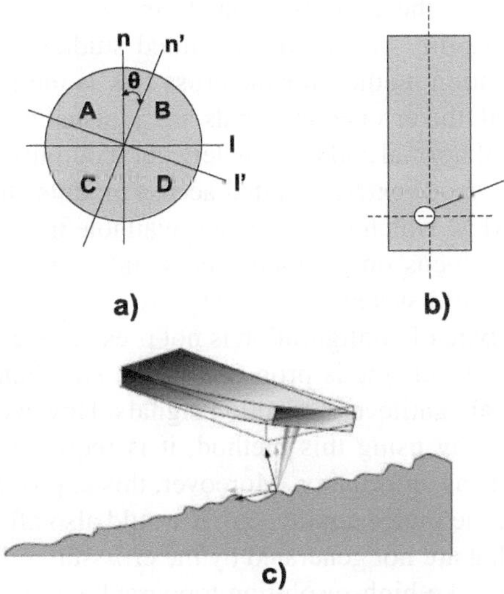

Fig. 2.12 a) False signal measurement possibility due to the inherent misalignment of the photodetector. Same effect would be detected during scanning if the AFM probe is misaligned due to mounting errors at the cantilever holder. b) The shear center of the AFM probe may not coincide with the AFM probe. c) The possible mechanical cross-talk behavior due to the physical interaction between the AFM tip and tilted substrate.

Among different techniques used in AFM, lateral force microscopy (LFM) has a practical importance due to its high sensitivity in the lateral direction. However, this technique suffers from the aforementioned errors as well as the cross-talk of the lateral and normal signals, which may result in a wrong measurement of the forces. In a typical friction experiment using LFM, the AFM probe is moved relative to the sample and normal and lateral deflections of the cantilever are measured. Several different physical reasons for the cross-talk have been identified in literature such as geometric [67], mechanical [68, 69] and electronic reasons [70] as shown in Fig. 2.12. Geometric cross-talk is mainly due to a rotational misalignment of the PSPD, but other effects such as the mounting error of the AFM cantilever to the probe holder also contribute to the cross-talk. Mechanical cross-talk is due to two factors. One of them is the misalignment of the shear center of the cantilever to the geometric center of the cantilever. This is more pronounced for particle attached AFM probes since it is difficult to glue the particle exactly at the center of the cantilever. The other factor is only apparent in some AFM systems, where slope correction does not actually tilt the substrate (or the head), but transforms the motion axes of the head, so that it moves according to the sample slope. The cantilever base, however, is not parallel to the substrate causing normal and frictional forces on the surface not to coincide with the normal and lateral forces measured. Electronic cross-talk is the result of magnetic interference of the signals and can only be prevented by the manufacturer [70].

Although cross-talk of the aforementioned signals is a well known problem which restricts the capability of the AFM, limited studies are attempted to solve it. Electronic compensation method for the cross-talk is the most widely used one [67, 70]. In this method, the erroneous signals are adjusted by adding and subtracting different fractions of normal and torsion deflection outputs while the AFM is out of contact. However, it needs external signal access modules to reach the input and output signals of the AFM which is not always available in all commercial AFM's. Moreover, these works focus on geometric cross-talk only. Geometric cross-talk may be removed if the AFM system has a rotary two-dimensional stage as proposed in [71]. However, this type of configuration is not present in commercial AFM systems. In [69], a feedback scheme is proposed which simultaneously regulates the normal as well as lateral cantilever deflection signals. However, in order to correct the cross-talked signals by using this method, it is required to modify the AFM probe holder which acts as an actuator. Moreover, this approach may give satisfactory results to improve the image quality, but, it would also affect the real signals of the physical systems that are not generated by the cross-talk.

AFM is designed to take high-resolution topographical images of substrates. It works by scanning over the substrate in a raster pattern while keeping the normal deflection of the cantilever constant with a feedback controller. The vertical positions of the cantilever base during this force-controlled motion are collected into a image matrix and displayed as topography images.

While this idea of scanning is very powerful in giving very detailed images of the substrate, it inherently assumes that V_{A-B} changes only due to vertical forces. This assumption is not true if there is cross-talk between the deflection signals, in which case, the achieved topographical images will have frictional artifacts.

One possible artifact due to cross-talk is, then, lower (or higher) topography measurements than the actual value on the sides of the image for small scan sizes. When the tip starts its motion in the lateral axis, it will be in the static friction regime. As it moves, the friction force increases until it reaches the maximum static friction value, at which point it starts sliding over the substrate. This increase in the friction force will have its effect on V_{A-B}, which in turn causes the cantilever base to move and hence, give erroneous topography data. Another artifact occurs if the substrate has zones of different friction factors. Again, due to cross-talk effects these regions will look higher or lower than their surroundings, which may not actually be correct.

Nevertheless, to be able to use the AFM for reliable force feedback for more than one dimension for any AFM application, one of the most important problems is the cross-talk effects between the measured force values. It is simply not acceptable to experience an unreal lateral force while the tip is vertically pressed onto the sample or unreal normal force while one is doing a lateral force measurement due to the cross-talking of signals. Thus, this paper focuses on the cross-talk compensation of force measurements in the AFM systems without adding additional complexity to the AFM setup.

2.2.1 Cross-Talk Compensation Procedure

Until now, effects of cross-talk in real experiments are discussed. In the ideal case, the coupling behavior of the aforementioned signals should be most apparent if the AFM tip is stuck to the surface similar to the initial deflection of the AFM tip in the static phase of the friction loop. A commercially available AFM (Autoprobe CP-II; Veeco, Santa Barbara, CA, USA) and two types of AFM cantilevers, contact and non-contact (PPP-CONTR and Pointprobe, NanoSensors, Neuchatel, Switzerland), are used for the experiments. In order to fix the AFM probe to the substrate, we first spin coated ultraviolet (UV) light curable glue (Loctite 3761, Henkel, Cleveland, OH, USA) on a glass substrate. UV glue selected as a high viscosity grade to ensure no penetration of the fluid to the AFM cantilever base. The final thickness of the film glue after spin coating is approximately 2 μm which is much more less than the height value of the AFM tip.

AFM cantilever is carefully approached to the glass surface and the glue is cured by UV light exposure using a UV curing light gun, bonding the AFM tip strongly to the substrate. Note that, although this is a destructive technique, it is expected to give valuable information about the coupling of the signals. Here, we checked the behavior of the signals for all possible scanning directions.

In Figs. 2.13-2.15, the scanner is moved in three directions that is lateral, longitudinal and vertical directions, respectively, with a small amount (*i.e.* 0.1 μm) while detecting the corresponding signals. For the lateral movement, since the AFM tip is fixed on the substrate, it is expected to detect a linear change in the LFM signal and no change in the A-B signal. However, as can be seen in the figure, the vertical signal is giving a slope showing a coupling behavior of the signals. Note that, the LFM

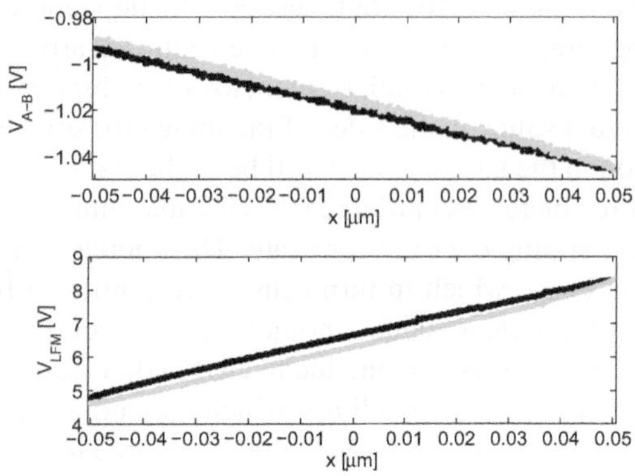

Fig. 2.13 Normal (A-B) and lateral (LFM) deflection signals for a forward (black) and backward (gray) substrate motion in the lateral (x) direction for a fixed AFM probe-substrate assembly.

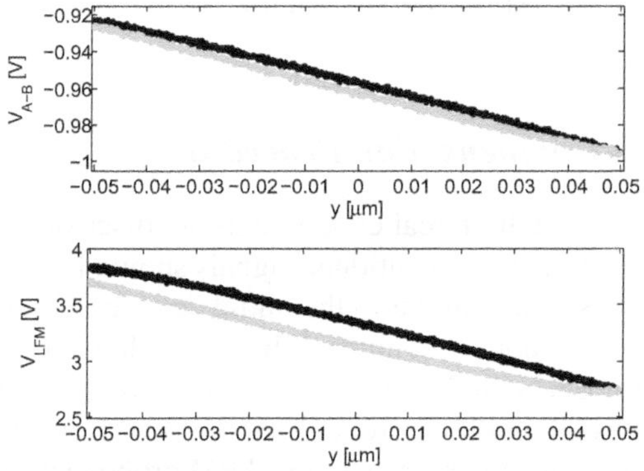

Fig. 2.14 Normal (A-B) and lateral (LFM) force signals for a forward (black) and backward (gray) substrate motion in the longitudinal (y) direction for a fixed AFM probe-substrate assembly.

signal is showing a hysteresis between the forward and backward motions which is attributed to a visco-elastic behavior of the adhesive bonding. Same behavior is observed for the other movements.

To measure and compensate for the cross-talk effect on the two voltage signals (normal (A-B) and lateral (LFM) deflection signals), we devised a simple strategy. A slope-corrected flat surface would exert a constant normal force for lateral motions and no lateral force for vertical motions if there was no cross-talk effect. Exploiting this fact obviously gives a quantitative measurement of cross-talk between the aforementioned signals. That is, the observed change in a signal without any physical reason is the result of only the cross-talk and this change can, then, be used to

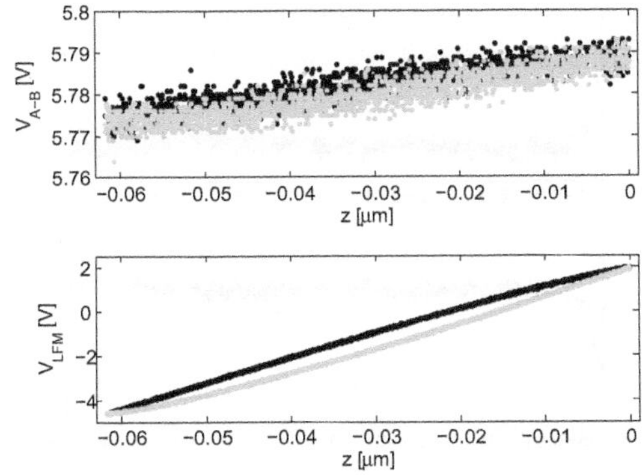

Fig. 2.15 Normal (A-B) and lateral (LFM) force signals for an approach (black) and retract (gray) cycle in the vertical (z) direction (*i.e.* force-distance curve) for a fixed AFM probe-substrate assembly.

measure and remove this effect. Note that, a significant amount of thermal drift in the normal (z) direction would affect the normal deflection signal causing it not to remain constant. On the other hand, thermal drift effects are more pronounced at the initial stages of experimentation and settles down after a proper waiting time. Hence, we wait for 30 minutes before starting experiments. Moreover, since drift velocities are reduced (less than 10 nm/min) after the mentioned waiting time, we conduct our experiments in high substrate velocities in order to further eliminate the thermal drift effect.

The experimental procedure to calibrate cross-talk, thus, uses a flat, smooth and rigid surface (*e.g.* freshly cleaved mica or single-crystal silicon). The probe is first approached to this surface, and slope correction is performed using small line scans (about 0.1 μm) since the surface should at least be locally flat in this region of interest. After slope correction, force feedback control (servo) of the AFM is turned off and a lateral friction loop is performed. A typical result of this motion on a single-crystal silicon wafer is given in Fig. 2.16. Note that even when the servo is off, the normal deflection of the cantilever should remain the same as no vertical motion is made. Then a vertical approach-retract cycle is performed to the surface with minimal amount of elastic indentation. Resulting voltage curves for normal and lateral deflections for this motion are given in Fig. 2.17.

A four-quadrant PSPD uses the normal (α) and lateral (β) angles of a rectangular cantilever at the point where the laser reflects off, to determine the normal and lateral voltage signals, respectively. These angles are directly proportional to the applied forces assuming small deflections as:

$$\begin{pmatrix} \alpha \\ \beta \end{pmatrix} = C \begin{bmatrix} 0 & \frac{3L_t}{L^2} & \frac{3}{2L} \\ \frac{2L_t}{L^2} & 0 & 0 \end{bmatrix} \begin{pmatrix} F_x \\ F_y \\ F_z \end{pmatrix}, \tag{2.36}$$

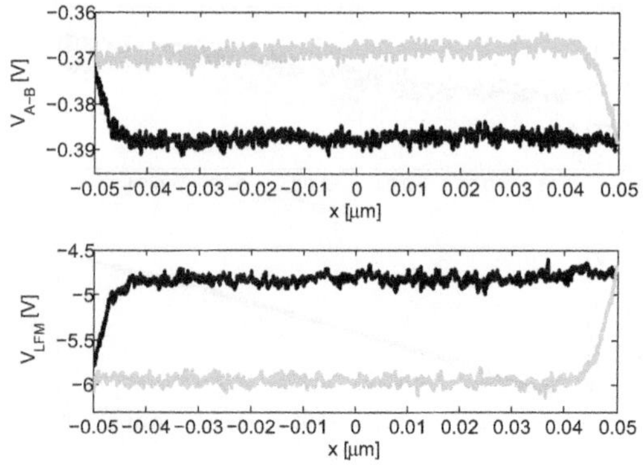

Fig. 2.16 Normal (A-B) and lateral (LFM) force signals for a forward (black) and backward (gray) line motion in the lateral (x) direction (*i.e.* friction loop) on a single-crystal silicon wafer.

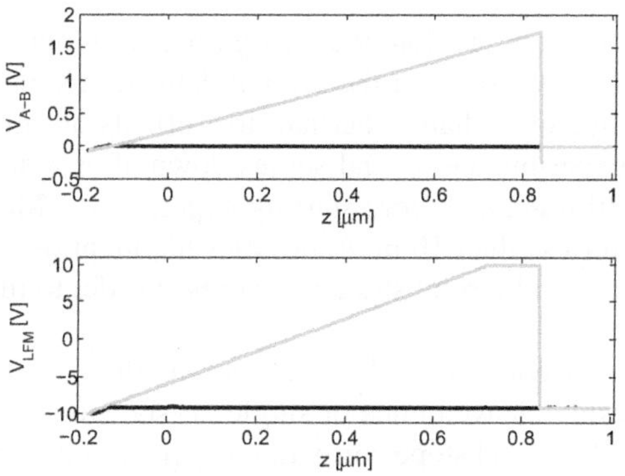

Fig. 2.17 Normal (A-B) and lateral (LFM) force signals for an approach (black) and retract (gray) cycle in the vertical (z) direction (*i.e.* force-distance curve) on a single-crystal silicon wafer. Note that cross-talk is so strong that the LFM signal saturates at the retract phase.

where $C = \frac{L^3}{3EI}$, L is the cantilever length, L_t is the tip height, E is the Young's modulus, and I is the moment of inertia, and F_x, F_y and F_z are the lateral, normal and axial forces, respectively.

Note that, forces are not readily available before the calibration of the probe stiffnesses. Moreover, the two voltage signals are typically scaled differently as the lateral twist of the cantilever is much smaller than its normal bending.

If the only source of cross-talk is geometric [70], cross-talk would simply be a rotational transformation between the measured and actual angles where θ is the misalignment of the optical detection system. Therefore, to correctly apply the rotational transformation, angles must be extracted from the voltage signals. Assuming there is no longitudinal force ($F_y = 0$) and small deflections, the relation between the angles to the voltage signals is given as:

$$\begin{pmatrix} \alpha^m \\ \beta^m \end{pmatrix} = \begin{bmatrix} \frac{3}{2L} & 0 \\ 0 & \frac{1}{L_t} \end{bmatrix} \begin{pmatrix} \delta_z \\ \delta_x \end{pmatrix} = \begin{bmatrix} \frac{3}{2Ls_n} & 0 \\ 0 & \frac{1}{L_t s_l} \end{bmatrix} \begin{pmatrix} V_{A-B} \\ V_{LFM} \end{pmatrix}, \tag{2.37}$$

where α^m and β^m are the measured values of the bending and twisting angles of the cantilever, δ_z and δ_x are the normal and lateral deflections of the tip, and s_n and s_l are the normal and lateral sensitivities of the AFM, respectively.

The actual bending (α^a) and twisting (β^a) angles of the cantilever are, then

$$\begin{pmatrix} \alpha^a \\ \beta^a \end{pmatrix} = \begin{bmatrix} cos(\theta) & -sin(\theta) \\ sin(\theta) & cos(\theta) \end{bmatrix} \begin{pmatrix} \alpha^m \\ \beta^m \end{pmatrix}. \tag{2.38}$$

PSPD angle of misalignment (θ) is the main unknown in these equations. Cantilever length (L) can be measured by a calibrated optical microscope, sensitivities can be calibrated using basic routines [76, 75]. Measuring the tip height (L_t) is the most difficult one, which can ideally be measured under a scanning electron microscope (SEM). However, another possibility is to take it as another unknown parameter to calibrate from the two experiments made.

Nevertheless, this approach is incomplete since it makes the assumption of only geometric cross-talk. The effect of other kinds of cross-talk may not simply be described by a rotational transformation. Therefore, a more general approach is to use an affine transformation between the measured and actual voltages such as:

$$\begin{pmatrix} V^a_{A-B} \\ V^a_{LFM} \end{pmatrix} = \underbrace{\begin{bmatrix} a_{11} & a_{12} \\ a_{21} & a_{22} \end{bmatrix}}_{\mathbf{A}} \begin{pmatrix} V^m_{A-B} \\ V^m_{LFM} \end{pmatrix}. \tag{2.39}$$

Using the two experimental results, relations between the elements of each row in $\mathbf{A} \in \Re^{2\times 2}$ can be found. Explicitly, for a lateral motion on a surface, forward and backward traces of the normal deflection signal should follow the same curve. Therefore, the actual difference between the mean values of V^a_{A-B} should be zero between forward ($\overline{V}^a_{A-B_f}$) and backward ($\overline{V}^a_{A-B_b}$) traces:

$$\Delta V^a_{A-B} = \overline{V}^a_{A-B_f} - \overline{V}^a_{A-B_b} = 0 \tag{2.40}$$

and using (2.39) and (2.40),

$$\Delta V^a_{A-B} = a_{11}\Delta V^m_{A-B} + a_{12}\Delta V^m_{LFM} = 0, \tag{2.41}$$

which gives

$$\frac{a_{11}}{a_{12}} = -\frac{\Delta V^m_{LFM}}{\Delta V^m_{A-B}}, \tag{2.42}$$

for the first experiment that is previously described.

Similarly, the slope of the actual lateral deflection signal V^a_{LFM} should again be equal to zero for a vertical approach retract cycle, which means

$$\frac{dV^a_{LFM}}{dz} = a_{21}\frac{dV^m_{A-B}}{dz} + a_{22}\frac{dV^m_{LFM}}{dz} = 0 \tag{2.43}$$

and hence

$$\frac{a_{21}}{a_{22}} = -\frac{dV^m_{LFM}}{dV^m_{A-B}}, \tag{2.44}$$

in the second experiment. This ratio can be simply found by fitting a line to the V_{LFM} vs. V_{A-B} curve, shown in Fig. 2.18, during the retract phase of the vertical motion. In fact, the linearity of this curve can be interpreted as an indication of cross-talk.

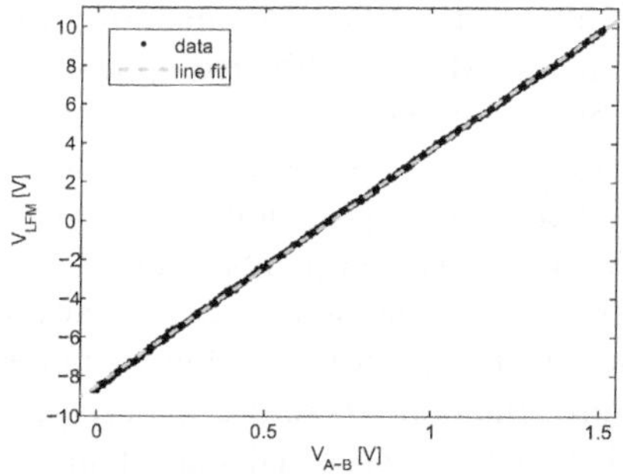

Fig. 2.18 Lateral (LFM) vs. normal (A-B) deflection signals during the retract phase of the vertical motion before saturation.

Equations (2.42) and (2.44), by themselves, are not enough to calculate **A**. Since vertical and lateral deflection signals are voltage signals, whose sensitivities should be calibrated to achieve the actual forces, without loss of generality, another constraint can be arbitrarily selected as

$$a_{i1}^2 + a_{i2}^2 = 1, \tag{2.45}$$

for $i = 1,2$. Here, each row of **A** is treated as a vector, whose dot product with the measured voltage vector gives the actual voltage in each axis. This selection ensures that each row vector in the transformation matrix is a unit vector (normalized) and hence do not cause unnecessary scaling of the measured signals.

Using (2.45) with (2.42) and (2.44) allows us to calculate the transformation matrix **A**. Using the calculated transformation to the measured voltage signals, normal and lateral deflection curves, compensated for cross-talk are given in Figs. 2.19 and 2.20.

To investigate the error associated with the proposed method, we performed an error propagation analysis for the friction loop data shown in Fig. 5. Using the standard deviations of the two measured signals as a degree of uncertainty along with equations (6) and (7), the propagated amount of uncertainty in the calculated ratio a_{11}/a_{12}, comes out to be 11.7% for this particular system configuration and experimental conditions. This value of uncertainty would be further decreased if one uses an AFM with a higher signal to noise ratio.

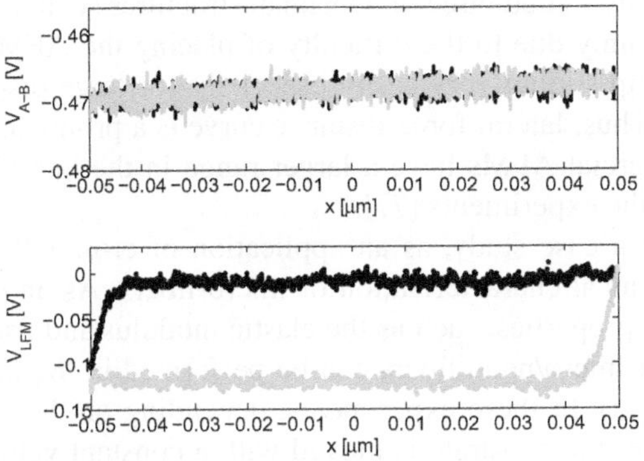

Fig. 2.19 Normal (A-B) and lateral (LFM) force signals for a forward (black) and backward (gray) line motion in the lateral (x) direction (*i.e.* friction loop) on a single-crystal silicon wafer after cross-talk compensation.

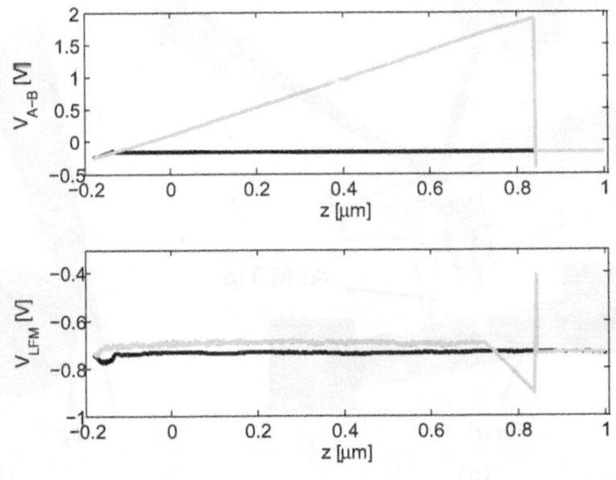

Fig. 2.20 Normal (A-B) and lateral (LFM) force signals for an approach (black) and retract (gray) cycle in the vertical (z) direction (*i.e.* force-distance curve) on a single-crystal silicon wafer after cross-talk compensation. Note that the saturation in the measured LFM signal causes an unrealistic change in the corrected signal.

2.2.2 A Case Study for the Cross-Talk Compensation

Vertical AFM force-distance curve is a fundamental tool to study surface interactions. It is a conventional tool to measure the adhesion and elastic interactions. However, when large displacements are needed in the vertical direction, AFM suffers from the limited movement range of the piezo tube. Also, elastic or plastic response measurements for suspended three dimensional structures such as cylindrical fibers are complicated mainly due to the difficulty of placing the AFM probe tip on the sample and AFM tip sliding very easily especially for large displacements in the vertical direction. Thus, lateral force-distance curve is a promising technique since most of the commercial AFMs have a larger range in the lateral direction giving more flexibility in the experiments [77].

What follows is a case study, as an application of cross-talk compensation in AFM-based mechanical characterization of micro-fibers. As an example, measuring the mechanical properties, such as the elastic modulus and fracture strength, of suspended polymer micro/nano fibers can be performed in the lateral direction as shown in Fig. 2.21(a). In these experiments, depending on the AFM's configuration, either the tip or the substrate is moved with a constant velocity in the lateral (x) direction until the tip comes into contact with the suspended fiber and deflects it. In the whole process, force feedback loop is disabled to have a constant height movement of the AFM tip.

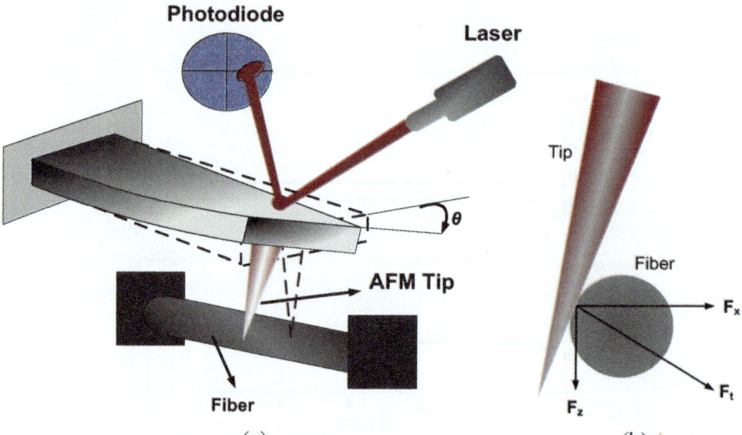

(a) (b)

Fig. 2.21 a) AFM tip is used to characterize the mechanical behavior of a polymer micro-fiber by pushing it laterally in a perfectly aligned optical system. b) The AFM tip is making an ellipsoidal contact with the fiber rather than a point contact resulting in a total force vector that is a combination of lateral and normal forces.

While studies assume that the AFM tip makes a fixed point contact with the cylindrical fiber at its centerline, the geometry of the tip will not allow the forces exerted on the fiber to be only in the lateral direction, especially for micro-fibers that have a relatively large diameter in comparison to nano-fibers. Typically, AFM tips have a conical or pyramidal shape which causes a certain angle of contact with the fiber and the normal direction of the area of contact determines the force direction as shown in Fig. 2.21(b). Thus, the micro-fiber does not only deflects in the lateral direction but also in the normal direction.

This means that, normal forces on the cantilever should be accounted for, since otherwise the force measurement would be an underestimate. Nevertheless, as mentioned before, when two forces need to be measured simultaneously with an AFM, cross-talk effects must be removed to have a reliable measurement.

Based on the discussion above, even after cross-talk correction, the two signals should resemble each other on such experiments. Therefore, to ensure that the reason of this similarity is physical, we exploited the fact that the normal force on the fiber has the same direction regardless of whether the micro-fiber is pushed from the left or right.

As the effects of cross-talk should not change sign due to motion direction, two experiments are made to get the elastic response of poly-(methyl methacrylate) (PMMA) micro-fibers, when deflected laterally by a small distance with the AFM tip. Single micro-fibers are produced using probe based drawing with a sharp needle [78]. In this process, the polymer is continuously pumped inside a glass tube and the sharp end of the glass tube is dipped and retracted on the substrate and placed to a predetermined location. The solvent evaporates during the manipulation path leaving a solid and cylindrical suspended single fiber across previously produced trenches using lithography techniques. After locating the fibers on the trenches, the orientation and boundary conditions of each fiber on the trenches is checked using an optical microscope equipped with a 100 $\times$ magnification. In order to ensure fixed boundary conditions at the end of the fibers, a small droplet of UV glue (Loctite 3761, Henkel,Cleveland, OH, USA) is applied at the end of the fibers. Resulting normal and lateral deflection signals corrected for cross-talk are given in Figs. 2.22 and 2.23 when the tip is approached to the fiber from the left and right, respectively.

From these results one can see that the behavior of the normal deflection signal does not change due to the direction of tip motion and hence has physical reasons. Measurement of forces from these signals is beyond the scope of this paper. However, after the calibration of the normal and lateral stiffness values of the AFM cantilever, and normal and lateral sensitivity values of the cross-talk corrected signals one can obtain the actual force exerted on the micro-fiber as the vector sum of the two force components. Thus, the effects of the cross-talk can be eliminated and the resulting force on the manipulated system is ensured to originate from physical reasons only.

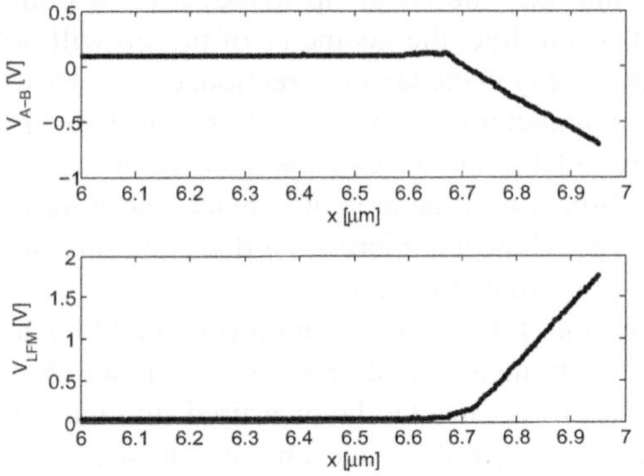

Fig. 2.22 Compensated normal (A-B) and lateral (LFM) force signals for a lateral fiber breaking experiment to characterize the mechanical properties of a PMMA micro-fiber when tip is approaching to the fiber from left.

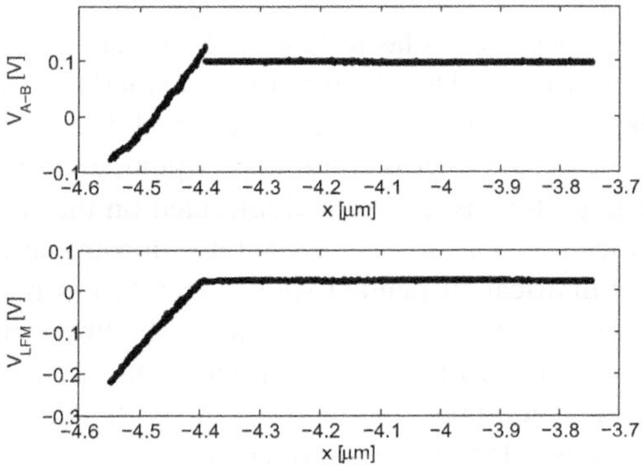

Fig. 2.23 Compensated normal (A-B) and lateral (LFM) force signals for a lateral fiber breaking experiment to characterize the mechanical properties of a PMMA micro-fiber when tip is approaching to the fiber from right.

2.3 Thermal Drift Compensation in AFM Based Nanomanipulation

Spatial uncertainties in an AFM system are partially caused by hysteresis, creep and other nonlinearities of the piezo scanning stage. These effects can mostly be reduced by measuring the actual displacement of the scanning unit using position sensors and feeding these data back to operate the scanner in a closed loop. However, position sensors are afflicted with noise and for small scan areas (in our setup $1 \times 1\ \mu m^2$), this often leads to oscillations in the closed-loop control. There exist other approaches that avoid these problems by modeling the scanning stage characteristics and by applying feedforward strategies [79, 80, 81].

More critical and less straightforward to counteract are spatial uncertainties that are induced by thermal drift. Even small changes in temperature cause all AFM components to vary slightly in size (due to thermal expansion and contraction) which result in an unknown, time-variant displacement between AFM probe and sample. This motion is generally very slow, but can obviously be detrimental to the success of nanomanipulation in the long run. By operating the AFM under homogeneous environmental conditions, the effect of thermal drift can be reduced, but even in highly temperature stable conditions thermal drift is still observable and amounts from 0.01 to 0.2 nm/s as reported in [82]. Especially when dealing with objects in the order of magnitude of a few nanometers (as shown in Fig. 2.24), the effect of thermal drift becomes a crucial issue for the success of manipulation (i.e. pushing a certain object).

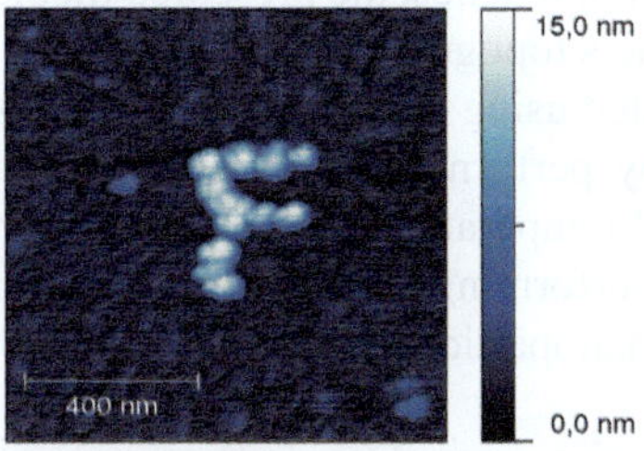

Fig. 2.24 Topography image recorded in non-contact mode (scan area $1.0 \times 1.0\ \mu m^2$) showing 15 nm sized Au nanoparticles on mica (Z axis is scaled up). The nanoparticles were manipulated with the AFM tip (pushing in contact mode) to form the letter 'F'. When dealing with objects of this order of magnitude or smaller, spatial displacements due to thermal drift can affect the manipulation and have therefore be compensated for.

The thermal drift-induced displacement between the AFM probe and the sample surface is not directly observable due to the lack of real-time capable visual feedback. Therefore, the only possibility to gain information about spatial drift inside the AFM is to use the AFM tip as a sensor itself.

Some commercial AFM software products provide functions to compensate for the effect of thermal drift. However, these functions are usually based on the cross-correlation of successively acquired topography images. This method can obviously only give correct results under the assumption that drift behaves linearly which has proven not to be the case, especially for inhomogeneous ambient conditions. Additionally, the AFM tip is occupied for minutes during image acquisition and cannot be used simultaneously for manipulation. Hence, this technique is not applicable for nanomanipulation. Other approaches use smaller, local scans to calculate drift, but also these occupy the AFM for periods of time in which manipulation is not possible [83].

More sophisticated approaches exist that try to track certain features on the sample (i.e. the center of a nanoparticle [84] or the highest point of a nanoobject [85]). Even though these techniques are able to measure drift reliably with update rates in the order of seconds, knowledge about the sample (i.e. the shape of a nanoobject) is necessary.

While the first approach is slow but independent of the sample's topography, the latter is faster but requires certain knowledge about the sample surface. This article presents a novel method for the estimation of drift, the key idea of which is to reduce the drift estimation to a localization problem as known from mobile robotics. To account for the uncertainties that arise from noisy and faulty sensor data – which occur in AFM applications – a particle filter based state estimation algorithm was implemented to measure the lateral displacement between AFM probe and sample. Even though there is thermal drift orthogonal to the sample plane (Z direction), this drift can be neglected for manipulation, because the height of the tip in relation to the sample can be obtained by measuring the deflection of the cantilever.

In contrast to existing approaches, the presented method does not depend on certain features in the sample's topography. To demonstrate this versatility and also its robustness, it was evaluated using a highly unstructured surface (a silicon substrate coated with gold) and by performing long-term drift measurements (up to 17 h) with externally-triggered temperature changes. Additionally, the algorithm's efficiency was analyzed by performing experiments on a mica substrate with randomly distributed 100 nm gold nanoparticles.

2.3.1 Drift Tracking with Bayesian Filtering

For the manipulation of objects in the order of magnitude of a few nanometers, lateral displacements between AFM probe and sample caused by thermal drift have been identified as a crucial error source. As mentioned in the introduction, the developed algorithm's main intention is to allow for robust drift measurements even for samples with unknown topography. Moreover, drift should also be estimated reliably in the case of only sparsely structured samples. Another requirement is that the AFM tip can only be used for the drift estimation temporarily, since during the manipulation experiments at which the proposed algorithm is aimed, the AFM tip is mainly used as an endeffector.

Therefore, instead of scanning an area or tracking certain features on the sample to measure drift (as in existing approaches), the developed algorithm periodically records short height profiles (as depicted in Figure 2.25) that can be recorded in a very small time-frame (0.1 s). These scanned lines are used as sensor data and are used to gain information about the drift-induced lateral displacement between tip and sample. This is performed by a comparison of these height profiles with line scans that are extrapolated from a previously recorded topography image. However, even though it seems obvious that the drift-induced displacement cannot be extracted directly from a single height profile and a reference image, the developed algorithm use these data to update its estimation of the current system state, namely drift. Due to a low signal-to-noise ratio and sporadic faulty measurements, the recorded line scans only represent the height profile of the sample roughly. Even if two height profiles are recorded immediately after each other at the same position (aside from drift, but this can be neglected for small time frames), they differ slightly due to noise(refer to Fig. 2.25). Additionally, depending on the type of sample used, these line scans may also contain little information. To account for these perceptional uncertainties, a probabilistic approach which is known as Monte Carlo Localization (MCL) is used as a foundation for the drift estimation algorithm:

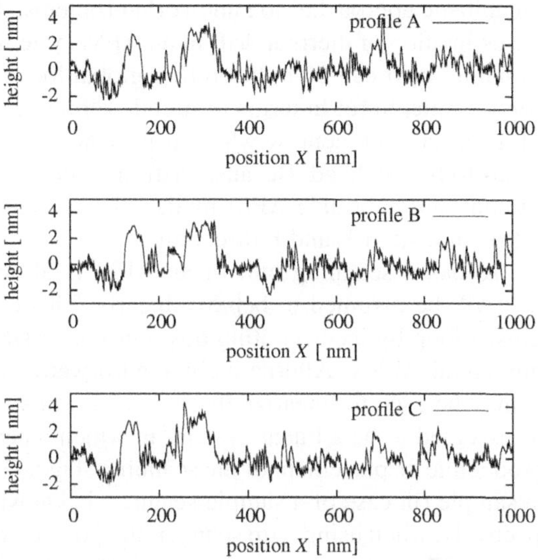

Fig. 2.25 Height profiles utilized as sensor data for the drift tracking algorithm. Profiles are recorded at the same position immediately after each other with tip velocities of 10 μm/s. Because drift can be neglected in this small time frame, the deviation observable in the depicted line scans indicate the low signal-to-noise ratio of this kind of measurement.

Since several years, MCL has successfully been applied in the domain of mobile, autonomous, macroscale robotics [86]. It is part of the superordinate concept known as *probabilistic robotics*, whose key idea is to explicitly deal with the uncertainties that exist in a robotic system using the calculus of probability theory. Instead of using a singular value as a guess for the system's state, information is represented by a probability density over the state space. In this way, uncertainties can explicitly be incorporated in the description of the system state. It has been proven that these kinds of algorithms are robust against noisy and faulty sensor data and that they perform well even when the system's behavior can only be poorly modeled [87]. This is mostly the case for mobile, robotic systems, because uncertainties exist in the execution of actions as well as in the perception of the environment. If nanorobotic systems, especially AFM systems, are compared to the previously mentioned mobile, autonomous macrorobotics, it can be stated that both systems have some common characteristics. First, also in atomic force microscopy, the environment is only partially observable (e.g. tip-convolution) and perception is tainted with uncertainty. Second, the effect of actions (e.g. in case of tip-based manipulations) are difficult to predict and thermal drift often results in positioning uncertainties.

Due to those similarities, it would stand to reason that algorithms used in macrorobotics to overcome the perceptional and action uncertainties can also be successfully applied in the domain of nanorobotic systems. As a proof of concept for the adaptability of probabilistic approaches to nanoworld problems, the MCL approach was adapted for the estimation of thermal drift in an AFM system.

The task of estimating drift can be reduced to a global localization problem as known from the robotics literature: In terms of mobile robotics, the sample surface can be considered as an environment in which a moving robot, in the discussed case the AFM tip, has to be localized. Because drift in Z direction is negligible for constant-force and constant-amplitude AFM modes, the presented algorithm confines on the estimation of drift in X and Y direction.

Although the piezo-based scanning unit of an AFM is afflicted by creep and hysteresis effects, it will be assumed to behave linearly. This can be achieved by operating it in a closed loop by incorporating position sensor signals, which is the case for most commercial AFMs. Alternatively, feedforward strategies (like proposed in [80, 81]) may be used to linearize the scanning unit's movement. Hence, the scanner can be assumed to be a linear system, in which a certain control input will result in a reproducible displacement of the scanning stage. Generally speaking, the position of the sample (in case of a sample scanner) is known in its own frame of reference. Respectively, when using a tip scanner, the position of the tip is known in its frame of reference. The only unknown quantity is though the displacement between the tip's frame of reference and the frame of reference of the sample, which is caused by thermal drift.

Because the AFM's primary function is imaging a sample's topography, the environment of this localization problem can be defined as well-known. Even though it seems obvious that the initial drift can be defined as zero in relation to a reference image recorded immediately before, this only gives a rough approximation because of drift that is present even during image acquisition. Using a pessimistic estimate of this drift, this initial drift can be constrained to a certain area inside the state space, but within this area the position is unknown. Therefore, drift estimation can be seen as a global localization problem as defined in [87].

Different probabilistic algorithms exist that are able to solve global localization problems. All are based on Bayesian filtering for state estimation, and differ only in the implementation and the way the probability distributions describing the system state are represented. Here, a particle filter based approach was chosen, because it outperforms other approaches in reference to efficiency, resolution and robustness. Additionally, since the height profiles selected as sensor data are uninformative, it is advantageous to be able to track multiple hypotheses on the system state, particularly at the beginning of the drift estimation. This is the case for particle filters, since they can handle arbitrary probability densities. A detailed introduction and comparison of probabilistic localization algorithms can be found in [87]. However, since the proposed algorithm is founded on Bayes filters, a short introduction will be given.

2.3.1.1 Bayes Filter

According to existing literature [87], knowledge about the system state at time t is denoted by the belief density $Bel(s_t)$, in which s_t represents a random variable over the state space. For each possible system state, a probability is given by this density function. To update this belief, the following recursive formula is applied which is known as Bayes filter:

$$Bel(s_t) = p(o_t|s_t) \int p(s_t|s_{t-1}, a_{t-1}) Bel(s_{t-1}) ds_{t-1} \qquad (2.46)$$

Given a distribution $Bel(s_{t-1})$ of the system state at time $t-1$, a new distribution for the state at time t is predicted by incorporating information about the action that was performed since $t-1$. Because this action, in our case the drift induced lateral displacement $a_{t-1} = (x_{t-1}, y_{t-1})$, is not known exactly, it is also given by a density function

$$p(s_t|s_{t-1}, a_{t-1}). \qquad (2.47)$$

This so called *motion model* calculates the conditional probability for a system state s_t if the previous state was s_{t-1} and an action a_{t-1} was performed. In the AFM drift context, this action can be considered as the drift since $t-1$, which is unknown in the beginning. The resulting distribution $\int p(s_t|s_{t-1}, a_{t-1}) Bel(s_{t-1}) ds_{t-1}$ is called *prediction*. To compensate for the uncertainties that were added by this prediction step, the resulting distribution has to be corrected by applying the *sensor model*

$$p(o_t|s_t). \qquad (2.48)$$

Given a certain observation o_t, in the presented approach one or multiple height profiles, the predicted system state s_t is corrected.

2.3.1.2 Particle Filter Implementation

Because the state space of this application is continuous, a direct implementation of Bayes filter as denoted in equation (2.46) is not feasible. Instead, a particle filter based approach was used in which the belief $Bel(s_t)$ is approximated by a set of M particles

$$\chi_t := s_t^{[1]}, s_t^{[2]}, \ldots, s_t^{[M]}.$$

The likelihood for a certain particle $s_t^{[m]}$ to be included in this particle set should approximately be proportional to the probability given by $Bel(s_t)$. According to the Bayes filter, the following recursive procedure is used to update the particle filter:

2.3.1.3 Prediction Step

In the prediction step, a new distribution for the current system state at time t is proposed using the motion model, which is given by the state transition probability $p(s_t|s_{t-1},a_{t-1})$. In case of drift estimation, this probability density expresses the expected drift since the last update of the filter. For each particle $s_{t-1}^{[m]}$ of the particle set χ_{t-1}, a new hypothetical state $\overline{s}_t^{[m]}$ is proposed and added to a temporary particle set $\overline{\chi}_t$.

2.3.1.4 Measurement Update

To correct for the uncertainties added in the prediction step the sensor model $p(o_t|s_t)$ is applied. Each particle $\overline{s}_t^{[m]}$ is evaluated against some observation o_t by calculating the importance factor $w_t^{[m]} = p(o_t|\overline{s}_t^{[m]})$. This importance factor then represents the probability that the current observation was made at the hypothetical state (drift position) of the corresponding particle. This way, each single hypothesis (represented by a particle) is evaluated by the sensor model.

2.3.1.5 Resampling

A new particle set χ_t is generated by resampling from the temporary set $\overline{\chi}_t$. First, all importance factors are scaled by a constant factor so that $\sum_{m=1}^{M} w_t^{[m]} = 1$. Particles are then drawn out of the temporary particle set $\overline{\chi}_t$ with the probability given by their normalized weights $w_t^{[m]}$. The higher a particle's weight, the higher its probability to be chosen and propagated to the posterior particle set. This way, particles with high weights survive and populate while others are eliminated. After this resampling step, the particles within the newly created particle set χ_t are distributed approximately to $Bel(s_t)$ as required by the definition. This procedure is also known as *importance sampling*.

2.3.1.6 Drift Estimation

In case of drift estimation, the demanded system state is the lateral displacement (aside from intended movements) of the AFM probe in relation to the sample

$$s = (x, y).$$

To implement the particle filter algorithm as desribed above, a sensor and a motion model have to be defined:

2.3.1.7 Observations and Sensor Model

To construct a suitable sensor model for the height profiles (like depicted in Fig. 2.25) utilized as observations, it is necessary to compare a measured height profile with the profile that is expected at a certain position. Such a height profile is extracted from the previously acquired reference image. To compare these two line scans, the summation of squared differences of all data points was selected as a measure for deviation. Experimental data are used to obtain the distribution of these expected deviations for two line scans taken at the same position, as shown in Fig. 2.26. Together with the extraction of height profiles from the reference image, this distribution was chosen as a sensor model.

Because the acquisition of the reference image takes some minutes, drift in Z direction can lead to unpredictable distortions of the reference image. If X is assumed to be the fast scanning direction, height profiles in Y direction extracted from the reference image may not reflect the true height profile due to Z drift, especially when imaging very flat surfaces. To avoid this problem of Z drift, only horizontal line scans are considered as observations.

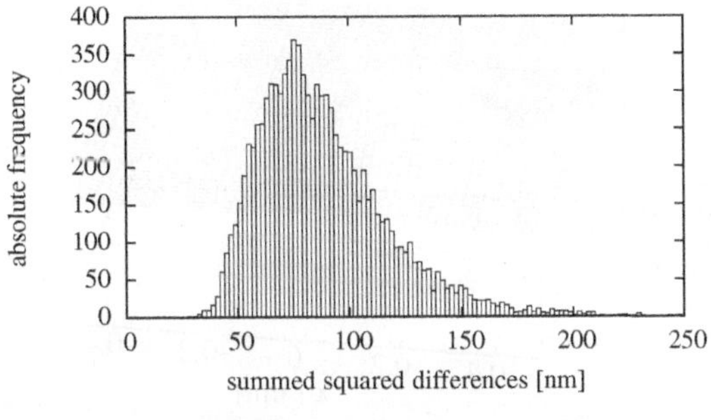

Fig. 2.26 Histogram displaying the absolute frequencies of the sum of squared differences calculated from height profiles successively recorded at the same position. 10000 height profiles of 1 μm in length were measured, each with a velocity of 10 μm/s. The sum of squared differences was obtained by comparing each two consecutive profiles.

2.3.1.8 Motion Model

During the first iterations of the particle filter, the particle set has not yet converged to a singular cluster indicating that the actual drift position and the resulting drift velocity cannot be determined reliably. As long as the particles form multiple clusters (representing different hypotheses of the system state), the following strategy is pursued: Because the current drift is unknown, a pessimistic approximation for the expected drift is used for the prediction of the new system state probability density. This pessimistic approximation assumes a maximum drift velocity, within these limits the drift velocities are assumed to be equiprobable and thus a uniform distribution was chosen as the motion model. For simplicity, this uniform distribution is limited to a rectangular area. After a few iterations when the particles have converged to a single cluster indicating an unimodal probability distribution and thus only one probable drift state, the motion model is switched. First, the latest drift measurements are analyzed and the expected drift vector v_d is calculated (by using a weighted moving average). This calculated drift is then utilized to predict the new drift state. Because the current drift velocity may change (e.g. due to temperature changes), a Gaussian distribution is added to the the expected drift based on the historical drift data to account for these uncertainties. An example of such a motion model is depicted in Fig. 2.27. The application of this motion model results in a slightly blurred belief distribution which comprises the uncertainty that was added due to the occurred drift since the last filter update.

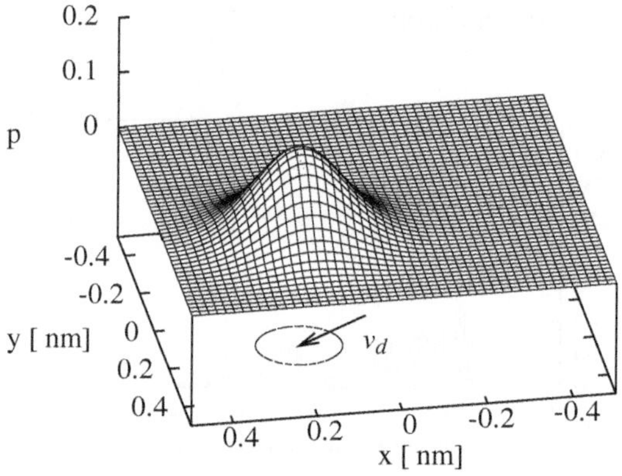

Fig. 2.27 Example of a probability density function used as the motion model after convergence of the particle filter. Based on recent drift measurements, the expected drift vector v_d since the last update of the particle is calculated. To incorporate uncertainties in drift velocity and direction, a Gaussian distribution is added to this drift vector.

The method for estimating the drift values, is as follows: First, a high resolution topography image of the sample is recorded, which will be denoted as the *reference image* in the following. Second, to learn the parameters of the sensor model, a defined number of height profiles are recorded successively at the same position. The parameters of these line scans are set to exactly the same values as for the height profiles that are recorded later to update the particle filter. Moreover, parameters like tip velocity and P and I gain of the Z-position control loop must also be the same as for the reference image. Otherwise, deviations between the height profiles extracted from the reference image and the profiles recorded as observations would affect the performance of the algorithm. The sum of squared differences of all successive height profiles are analyzed (like in Fig. 2.26) and the parameters for a distribution to fit these values approximately are calculated. As previously mentioned, the initial drift is not known exactly due to drift during the acquisition of the reference image. To compensate for this uncertainty, the particle filter is initialized by distributing all particles uniformly in a rectangular search area.

When new sensor data are available (in form of height profiles), the particle filter is updated by first applying the motion model to each particle to account for the uncertainties due to drift since the last update.

To evaluate this predicted drift state, the current observation has to be compared to the observation that is expected at each particle's position. Therefore, height profiles (with same parameters like the current observation) are sampled from the reference image. These expected observations are then evaluated against the measured observation o_t by applying the sensor model. After resampling, the particles are distributed according to the actual belief of the drift position.

To extract a singular value for drift estimation, X and Y of all particles are averaged. Additionally, to measure the quality of estimation, the variance of the particle distribution is analyzed. The more the particles are concentrated at a specific point, the more the variance decreases. If the variance falls below a previously defined threshold, the drift estimation is assumed to be correct.

If the particle set is out of the bounds of the reference image, height profiles from the reference image can obviously not be sampled and the drift estimation algorithm has to be restarted. Otherwise, the algorithm continues by acquiring a new observation and updating the particle filter again.

2.3.1.9 Drift Compensation Procedure

To use the results of the drift estimation algorithm for successful nanomanipulation, we need to define a procedure that allows for the compensation of drift over a long enough time interval that allows us to finish manipulation operations before another step of drift estimation can be executed. That is, until now, we have used consecutive line scans to estimate drift continuously. This method, however, exclusively works on drift estimation leaving no time for any other operations to run in parallel.

Focusing on nanoparticle pushing, let us define $T \in \Re^+$ as the time required for a single particle pushing operation to be finalized. This value, then, is the minimum period that successive line scans for drift estimation can be made in. For this period while drift estimation is not running, drift values should be updated using some extrapolation of the previous values. Since drift is a very smooth function in time, which can be modeled as linear function for small time intervals, and for simplicity, we are using a linear fit to the previous $N \in Z^+$ estimations of the particle filter with respect to time, to measure the velocity of drift and use a linear extrapolation to update drift values.

However, for the drift velocities to give reasonable estimates initially, the drift estimation algorithm should have converged. Therefore, to enable the drift estimation to converge initially, we allow it to run for a time period ($T_{\mathrm{init}} \in \Re^+$) before starting any manipulation tasks. The complete flowchart of the drift compensation procedure is depicted in Fig. 2.28.

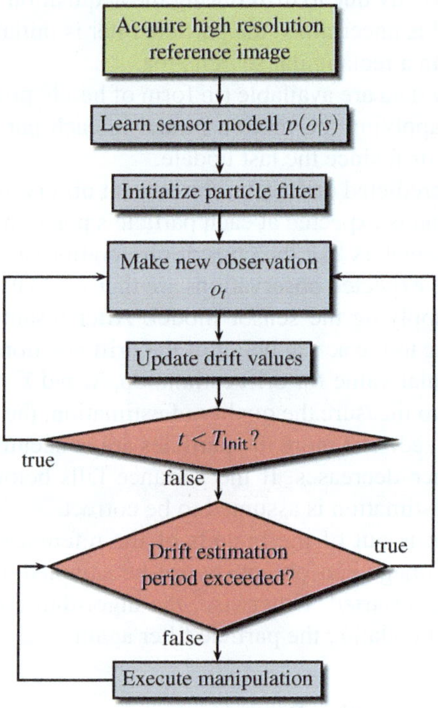

Fig. 2.28 Flowchart of the drift compensation algorithm.

2.3.1.10 Experimental Setup

For experimental validation of the described algorithm, two different setups were used. The first setup consists of an commercial AFM (Veeco, Autoprobe M5), which is accessed by a Windows 95-based PC (AFM PC) with an AFM cantilever (Veeco, RTESPA, k =40 N/m). A client-server program is created on the AFM PC, which allows an external PC to connect to the AFM through TCP/IP Ethernet (the Veeco SPMAPI is used to interface with the AFM). The main control PC uses real-time Linux (RTAI 3.3, Ubuntu Linux 2.6.15) and an interface program written in C++. It interfaces with the AFM PC through a direct Ethernet connection. A 3-DOF piezo-electric nanopositioning stage (PI (Physik Instrumente) P-753, 12 μm range, <1 nm precision) is used as a custom positioner in the AFM, which allows for a faster control bandwidth compared to actuating the AFM's nanopositioner through TCP/IP. The three axes of the piezo stage are controlled through its dedicated controller (PI (Physik Instrumente) E-612) by the main control PC using three D/A outputs (Adlink PCI-6208). Positions are read from the amplifier using an A/D converter (National Instruments PCI-6024E). In addition, the AFM's normal-deflection signal (A-B voltage signal) is read directly from the AFM with the A/D converter and the topography signal is streamed from the AFM PC over the ethernet connection. The total control bandwidth is 1 kHz, which is limited by the data acquisition system and the processor speed.

For operation, a sample is placed on the custom nanopositioner and the AFM probe is automatically brought into contact with the sample. By default, the AFM servos the Z position of the probe to maintain a constant force. For our experiments, this Z-servo controller is enabled and the topography values are recorded while the sample is moved in the horizontal plane yielding a custom rasterized scan of the surface topography. Before beginning the experiments, we wait for ten minutes for the system to stabilize.

As a second setup, a common AFM manufacturered by Nanotec (Spain) with the provided software WSxM [88] is utilized. Using the lithography language that is part of WSxM, it is possible to assemble basic AFM operations (parameter settings, drawing lines, etc.) with some control structures to a complex program, which is then uploaded and executed on the DSP that controls the AFM. The drift tracking algorithm was implemented in C++/QT as a standalone application that is executed on a dedicated machine. The height profiles were directly written to the file system by the lithography module of the AFM software. Therefore, the drift tracking implementation has a mechanism to monitor the file system to realize when new sensor data from the AFM arrives.

2.3.2 Experimental Results

2.3.2.1 Drift Measurements on Gold Coated Silicon Substrate

The performance of the drift tracking algorithm was analyzed by using several types of samples ranging from a well-structured AFM calibration grid and DVD/CD surfaces to gold nanoparticles. However, because the algorithm was mainly designed to work even with almost unstructured sample surfaces, the described results will first be confined to experiments conducted with a gold coated silicon substrate utilized as sample. Roughness parameters of this sample were determined to $R_{RMS} = 0.765$ nm and $R_a = 0.592$ nm. The results were obtained using the described Nanotec AFM system.

Fig. 2.29 Topography images ($10 \times 10\ \mu m^2$) of the Au/Si surface before (top left, bottom left (3D view, Z scaled)) and after (top right) the drift estimation (17 hours later). To compare the results of the drift estimation algorithm, the upper topography images were cross-correlated and the maximum of this correlation was determined to be at $X = 247$ nm and $Y = -220$ nm, which is mostly identical ($\Delta X = 6$ nm, $\Delta Y = 6$ nm) to the results obtained by the particle filter algorithm.

As a reference image, the topography of a $10 \times 10\ \mu m^2$ scan area was recorded using 512×512 pixels (depicted in Fig. 2.29, top left). The algorithm was started with 2000 particles uniformly distributed over an area of $800 \times 800\ nm^2$ to account for the initial uncertainty. Observations were acquired every 10 s in all experiments.

While first experiments were conducted in contact mode using single height profiles at the center position each 2 µm in length, it turned out that these kind of observations are not sufficient to track drift reliably. Even though the particles concentrated quickly, the estimated Y component of the drift became unstable after few iterations of the particle filter. As an improvement, three horizontal line scans (1 µm length, vertical distance of 10 nm) were obtained as observations, which were combined by the sensor model multiplicatively. This way, only particles that show good correspondence with all three height profiles get high importance factors and therefore populate. Experiments have shown that the particle set converges more quickly than with the previous single line observations and that drift could be measured successfully. However, after about one hour, it could be observed that the drift estimation became unstable. Topography images taken after the drift estimation have shown that the sample was damaged due to the high impact of the tip in contact mode.

To avoid damage of the sample, the experiments were repeated in non-contact mode (with modulated excitation frequency and phase locked loop). To obtain high quality topography images, cantilevers with a sharp tip (typical tip radius of 10 nm) were used[1]. As depicted in Fig. 2.30, drift could be measured over a period of several hours (17 h). Additionally, to prove the robustness of the algorithm and to show the influence of temperature changes to drift, the temperature inside the AFM was changed during the experiment. At time $t = 4.73$ h, the ventilation in the laboratory was switched off, which implies a faster increase of temperature of 1 °C within 90 min, when ventilation was turned on again at time $t = 6.23$ h.

A second image was recorded immediately after the drift estimation experiment (Fig. 2.29, top right) and was cross-correlated with the reference image to validate the results of the algorithm (see Fig. 2.29, bottom right). Aside from a deviation of $\Delta X = 6$ nm and $\Delta Y = 6$ nm, the maximum of this cross-correlation was identical to the result obtained from the particle filter drift estimation algorithm. The results have also shown, that the presented algorithm works reliably even for samples that are very poorly structured as the Au/Si substrate used in the described experiments.

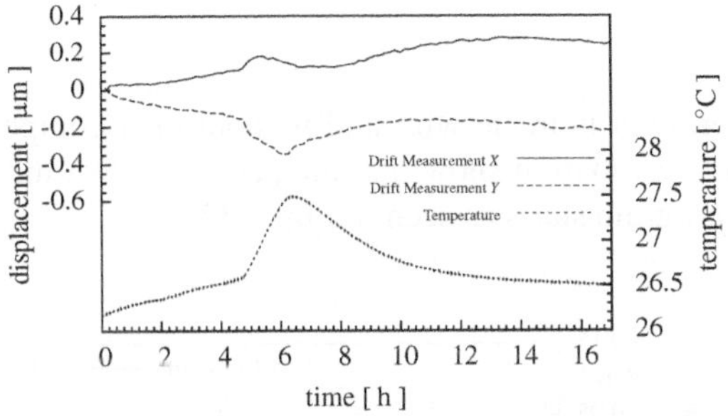

Fig. 2.30 Measured X and Y component of drift and temperature inside the AFM over a period of 17 h with the gold coated silicon substrate used as a sample.

2.3.2.2 Drift Compensation on Gold Nanoparticles

To further validate the efficiency of the drift compensation algorithm, experiments were conducted also on a nanoparticle substrate (using the customized Autoprobe M5 setup). The substrate contains 100 nm gold nanoparticles on poly-l-lysine (PLL) on a mica surface. Sample preparation is made following the procedure defined by Baur *et al.* [89]. To avoid any unintended movement of nanoparticles, the AFM was operated in non-contact mode for the following experiments. No particle pushing is made, to better demonstrate the effect of drift compensation on a possible nanomanipulation task. Rather, we made line scans over the same nanoparticle for an extended time period. After an initial AFM scan, the user is asked to select separate drift and manipulation regions as seen in Fig. 2.31. This is done to ensure that

[1] Type: N1B, manufacturer: NIIFP, Moscow.

no changes are made to the drift estimation region. The manipulation region is selected such that it contains a single particle over which line scans will be made. If the compensation procedure works, these line scans should give the same results regardless of time.

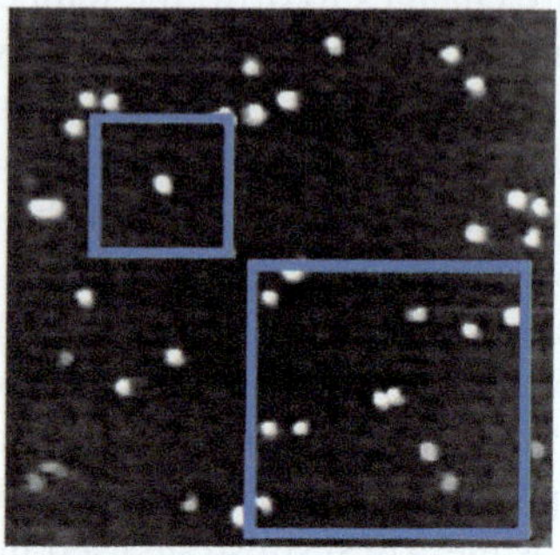

Fig. 2.31 Initial $10 \times 10\ \mu m^2$ AFM scan where the user chooses separate regions for drift estimation (large square) and particle manipulation (small square).

The same experiment is made with and without drift compensation. Line scan results without drift estimation show that the particle is lost merely after about 10 minutes after the program starts as seen in Fig. 2.32.

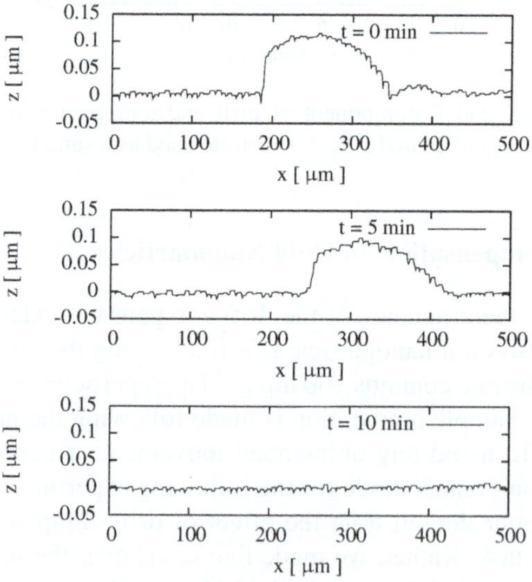

Fig. 2.32 Consecutive line scans on the same position without drift compensation.

When drift compensation is turned on, the results are improved as the line scans give the same profile, even 60 minutes after the experiment starts as seen in Fig. 2.33. Moreover, before and after images of the same $5 \times 5\ \mu m^2$ area in Fig. 2.34 show that drift is, in fact, being effectively compensated for by the algorithm. Drift values given by the particle filter over 60 minutes are plotted in Fig. 2.35.

Note that, 60 minutes is not, by any means, a limit of the drift compensation algorithm, but rather the time the experiment was stopped. The parameters for the algorithm are defined as $T_{\mathrm{init}} = 5$ min, $T = 2$ min and $N = 10$. These parameters and the corresponding results mean that manipulation tasks can start after only 5 minutes of initial estimation period and can run afterwards as if no drift exists.

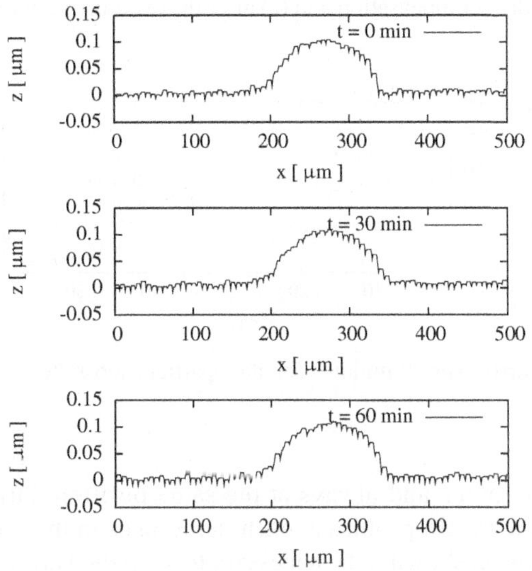

Fig. 2.33 Consecutive line scans on the same position with drift compensation. At time $t =$ 30 min the estimated drift amounts to $X = -112$ nm and $Y = 237$ nm. At time $t = 60$ min drift is estimated to $X = -168$ nm and $Y = 190$ nm.

2.3.2.3 Drift Distortions in Reference Images

In section 2.3.2.1 it was mentioned that the horizontal height profiles that were used as sensor data were recorded at the center of the reference image. In theory, however, it would not make a difference at which vertical position these profiles are recorded. This is true under the assumption that the reference image is free of drift artefacts. But, due to the drift presence during image acquisition and scanning times of a few minutes, a time variant displacement between tip and sample leads to a distortion in the reference image. This makes it necessary that the height profiles are recorded

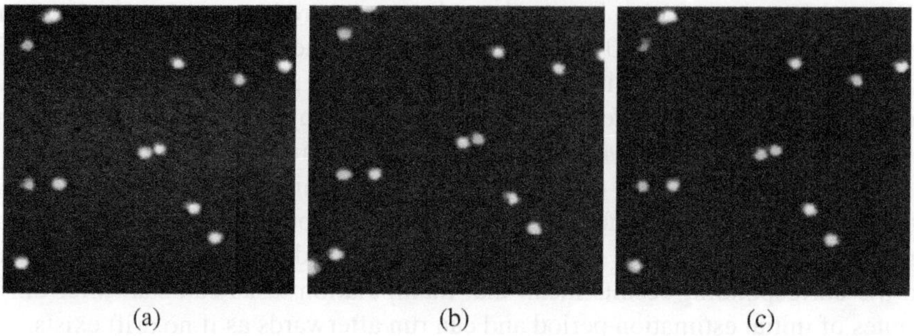

Fig. 2.34 AFM images of the same $5 \times 5\ \mu m^2$ area (a) before the experiment, (b) after the experiment, without drift compensation and (c) after the experiment, with drift compensation.

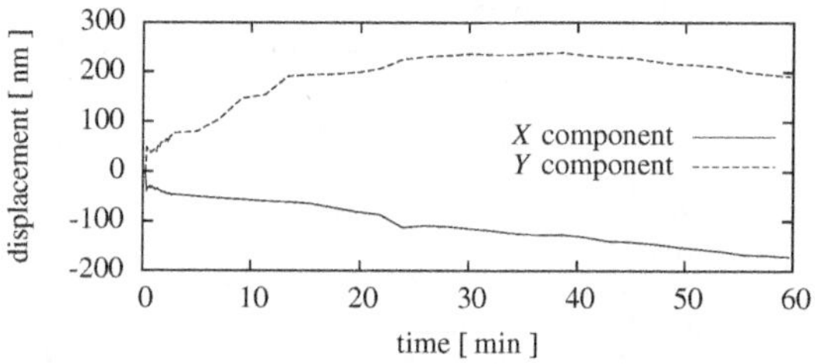

Fig. 2.35 Measured drift over 60 minutes (on nanoparticle substrate).

in the center of the image and always at the same position during an experiment. When using multiple height profiles at each iteration as in the previously described experiments, it has been shown to be necessary to keep the vertical distance between these height profiles small in order to minimize the negative effect the distortion of the reference images might have. To avoid the negative effect of drift induced distortions in reference images, it is planned to integrate an algorithm for obtaining drift-free images. This could be realized in accordance to the method presented by Lapshin in [90]. Moreover, the acquisition of drift-corrected images is also helpful for the automation of manipulation experiments in order to perform accurate path planning.

The presented results have also shown that probablistic algorithms, as known from the macrorobotics context, can successfully be applied in the domain of nanorobotics. Especially the uncertainties that greatly reduce the reliablity in nanomanipulation enable a wide range of speculative applications for probabilistic robotics. These applications of course not only cover localization, but may also include planning and control of nanomanipulation tasks.

2.4 Conclusion

In order to obtain highly-precise force detection and extend the force application range of the AFM based nanomanipulation system, the normal and lateral force applications need to be accurately calibrated and the corresponding nonlinear sensitivities well compensated. For the calibration of the lateral force, a robotics based method, making use of an accurately calibrated piezo-force sensor composed of a tippless piezoresistive cantilever and corresponding electronics, is proposed to determine the lateral force conversion factor. This method may be used to directly calibrate factor between the lateral force and the photodiode signal for cantilevers with a wide range of spring constant, regardless of their size, shape, material or coating effect and any knowledge of the optical lever. A practicable approach is also shown to compensate the sensitive nonlinearity of photodiode by calculating the cantilever deflection using the nonlinear fit of the sensitivity, which was achieved from the Sigmoidal fit of the normal and lateral force-voltage curves, thereby extending the effective force application range of the optical lever. The experimental results demonstrated that the sensitivity error of normal responses could be reduced from more than 200% to 3.6% and the range of the force application was extended.

In a second part, calibration and correction of cross-talk in atomic force microscopy (AFM) is demonstrated. Several reasons and effects of this inherent problem on experimental results are discussed. We propose a general procedure that can be used on most AFM systems to compensate for cross-talk on the cantilever bending and twisting signals. The method utilizes two initial experiments on a flat surface to achieve an affine transformation between the measured signals and the actual signals. Using this transformation directly on the voltage signals allows us to remove the detrimental effects of cross-talk on AFM based force measurement experiments. The achieved transformation matrix can be turned into a simple circuit and applied on-line, by users who have access to the raw signals in the AFM head. As a case study, a lateral deflection based mechanical characterization test for a poly-(methyl methacrylate) micro-fiber that is suspended on a trench is investigated in terms of the effectiveness of the cross-talk compensation.

One critical issue for the automation of AFM-based nanomanipulation is recognized to be the compensation of thermal drift. As a precondition for successful compensation, measuring drift reliably is a necessity. In the last part, a versatile drift estimation algorithm based on particle filter localization was presented. In contrast to other state-of-the-art drift estimation approaches, the developed algorithm is not based on the tracking of certain features. Instead, arbitrary height profiles are used as sensor data as an input for particle filter-based state estimation. These sparse sensor data, in combination with the power of probabilistic approaches, lead to a sample independent drift estimation algorithm. The results have shown that measuring drift with this algorithm is even possible with nearly featureless samples, like the Au-coated Si-surface used in the experiments. Other approaches, that measure drift by means of object tracking, would probably fail for these kind of samples.

References

1. Binning, G., Quate, C.F., Gerber, C.: Phys. Rev. Lett. 56, 930 (1986) 31
2. Meyer, G., Amer, N.M.: Appl. Phys. Lett. 57, 2089 (1990) 32
3. Albrecht, T.R., Akamine, S., Carver, T.E., Quate, C.F.: J. Vac. Sci. Technol. A 8, 3386 (1990) 32
4. Sarid, D., Elings, V.: J. Vac. Sci. Technol. B 9, 431 (1991) 32
5. Neumeister, J.M., Ducker, W.A.: Rev. Sci. Instrum. 65, 2527 (1994) 32
6. Hazel, J.L., Tsukruk, V.V.: J. Tribol. 120, 814 (1998) 32
7. Hazel, J.L., Tsukruk, V.V.: Thin Solid Films 339, 249 (1999) 32
8. Cleveland, J.P., Manne, S., Bocek, D., Hansma, P.K.: Rev. Sci. Instrum. 64, 403 (1993) 32, 36
9. Ruan, J., Bhushan, B.: ASME J. Tribol. 116, 378 (1994) 32
10. Sader, J.E., Chon, J.W.M., Mulvaney, P.: Rev. Sci. Instrum. 70, 3967 (1999) 32
11. Cumpson, P.J., Zhdan, P., Hedley, J.: Ultramicroscopy, 100, 241 (2004) 32
12. Aksu, S.B., Turner, J.A.: Rev. Sci. Instrum. 78, 043704 (2007) 32
13. Golovko, D.S., Haschke, T., Wiechert, W., Bonaccurso, E.: Rev. Sci. Instrum. 78, 043705 (2007) 32
14. Ohler, B.: Rev. Sci. Instrum. 78, 063701 (2007) 32
15. Gates, R.S., Reitsma, M.G.: Rev. Sci. Instrum. 78, 086101 (2007) 32, 36
16. Langlois, E.D., Shaw, G.A., Kramar, J.A., Pratt, J.R., Hurley, D.C.: Rev. Sci. Instrum. 78, 093705 (2007) 32, 36
17. Carpick, R.W., Ogletree, D.F., Salmeron, M.: Appl. Phys. Lett. 70, 1548 (1997) 34, 40
18. Lantz, M.A., O'shea, S.J., Hoole, A.C.F., Welland, M.E.: Appl. Phys. Lett. 70, 970 (1997) 34, 40
19. Cain, R.G., Biggs, S., Page, N.W.: J. colloid Interf. Sci. 227, 55 (2000)
20. Cannara, R.J., Eglin, M., Carpick, R.W.: Rev. Sci. Instrum. 77, 053701 (2006) 34
21. Liu, E., Blanpain, B., Celis, J.P.: Wear 192, 141 (1996) 34
22. Ecke, S., Raiteri, R., Bonaccurso, E., Reiner, C., Deiseroth, H.J., Butt, H.J.: Rev. Sci. Instrum. 72, 4164 (2001) 34
23. Wang, F., Zhao, X.Z.: Rev. Sci. Instrum. 78, 043701 (2007) 34
24. Ogletree, D.F., Carpick, R.W., Salmeron, M.: Rev. Sci. Instrum. 67, 3298 (1996) 34
25. Varenberg, M., Etsion, I., Halperin, G.: Rev. Sci. Instrum. 74, 3362 (2003) 34
26. Asay, D.B., Kim, S.H.: Rev. Sci. Instrum. 77, 043903 (2006) 34
27. Feiler, A., Attard, P., Larson, I.: Rev. Sci. Instrum. 71, 2746 (2000) 34
28. Morel, N., Ramonda, M., Tordjeman, P.: Appl. Phys. Lett. 86, 163103 (2005) 34
29. Liu, W.H., Bonin, K., Guthold, M.: Rev. Sci. Instrum. 78, 063707 (2007) 34, 41
30. Cumpson, P.J., Hedley, J., Clifford, C.A.:J. Vac. Sci. Technol. B 23, 1992 (2005) 34
31. Jeon, S., Braiman, Y., Thundat, T.: Appl. Phys. Lett. 84, 1795 (2004) 34
32. Li, Q., Kim, K.S., Rydberg, A.: Rev. Sci. Instrum. 77, 065105 (2006) 34
33. Behrens, I., Doering, L., Peiner, E.: J. Micromech. Microeng. 13, S171 (2003) 35
34. Mendels, D.A., Lowe, M., Cuenat, A., Cain, M.G., Vallejo, E., Ellis, D., Mendels, F.: J. Micromech. Microeng. 16, 1720 (2006)
35. Green, C.P., Lioe, H., Cleveland, J.P., Proksch, R., Mulvaney, P., Sader, J.E.: Rev. Sci. Instrum. 75, 1988 (2004) 41
36. Mate, C.M., McClelland, G.M., Chiang, S., Erlandsson, R.: Phys. Rev. Lett. 59, 1942 (1987) 43
37. Ogletree, D.F., Carpick, R.W., Salmeron, M.: Rev. Sci. Instrum. 67, 3298 (1996) 44
38. Varenberg, M., Etsion, I., Halperin, G.: Rev. Sci. Instrum. 74, 3362 (2003) 44
39. Feiler, A., Attard, P., Larson, I.: Rev. Sci. Instrum. 71, 2746 (2000) 44, 45

40. Morel, N., Ramonda, M., Tordjeman, P.: Appl. Phys. Lett. 86, 163103 (2005) 44
41. Liu, W.H., Bonin, K., Guthold, M.: Rev. Sci. Instrum. 78, 063707 (2007) 44
42. Cumpson, P.J., Hedley, J., Clifford, C.A.: J. Vac. Sci. Technol. B 23, 1992 (2005) 44
43. Li, Q., Kim, K.S., Rydberg, A.: Rev. Sci. Instrum. 77, 065105 (2006) 44
44. Xie, H., Vitard, J., Haliyo, S., Régnier, S., Boukallel, M.: Rev. Sci. Instrum. 79, 033708 (2008) 44
45. Neumeister, J., Ducker, W.A.: Rev. Sci. Instrum. 65, 2527 (1994) 44
46. Liu, E., Blanpain, B., Celis, J.P.: Wear 192, 141 (1996) 44
47. Bogdanovic, G., Meurk, A., Rutland, M.W.: Colloids Surf. B 19, 397 (2000) 44, 45
48. Jeon, S., Braiman, Y., Thundat, T.: Appl. Phys. Lett. 84. 1795 (2004) 44
49. Carpick, R.W., Ogletree, D.F., Salmeron, M.: Appl. Phys. Lett. 70, 1548 (1997) 44
50. Lantz, M.A., O'shea, S.J., Hoole, A.C.F., Welland, M.E.: Appl. Phys. Lett. 70, 970 (1997) 44
51. Cain, R.G., Biggs, S., Page, N.W.: J. Colloid Interf. Sci. 227, 55 (2000) 44
52. Cannara, R.J., Eglin, M., Carpick, R.W.: Rev. Sci. Instrum. 77, 053701 (2006) 44, 46
53. Ecke, S., Raiteri, R., Bonaccurso, E., Reiner, C., Deiseroth, H.J., Butt, H.J.: Rev. Sci. Instrum. 72, 4164 (2001) 44
54. Xie, H., Vitard, J., Haliyo, S., Régnier, S.: IEEE Sens J. 8, 1478 (2008) 44
55. Schaffer, T.E., Hansma, P.R.: J. Appl. Phys. 84, 4661 (1998) 46
56. Schaffer, T.E., Hansma, P.R.: Characterization and optimization of the detection sensitivity of an atomic force microscope for small cantilevers. J. Appl. Phys. 84(9), 4661–4666 (1998) 49
57. Beaulieu, L.Y., Godin, M., Laroche, O., Tabard-Cossa, V., Grtter, P.: Calibrating laser beam deflection systems for use in atomic force microscopes and cantilever sensors. Appl. Phys. Lett. 88(8), 083108 (2006) 49
58. Schaffer, T.E., Hansma, P.R.: Characterization and optimization of the detection sensitivity of an atomic force microscope for small cantilevers. J. Appl. Phys. 84(9), 4661–4666 (1998) 49
59. Cannara, R.J., Eglin, M., Carpick, R.W.: Lateral force calibration in atomic force microscopy: A new lateral force calibration method and general guidelines for optimization. Rev. Sci. Instrum. 77(5), 053701 (2006) 49
60. Sümer, B., Sitti, M.: J. Adh. Scie. Techn. 22, 481 (2008) 185
61. Cannara, R.J., Brukman, M.J., Cimatu, K., Sumant, A.V., Baldelli, S., Carpick, R.W.: Science 318, 5851 (2007)
62. Sun, Y., Akhremitchev, B., Walker, G.: Langmuir 20, 5837 (2004)
63. Ogletree, D.F., Carpick, R.W., Salmeron, M.: Revi. Sci. Instrum. 67, 3298 (1996) 55
64. Feiler, A., Attard, P., Larson, I.: Revi. Sci. Instrum. 71, 2746 (2000) 55
65. Li, Q., Kim, K.-S., Rydberg, A.: Revi. Sci. Instrum. 77, 065105 (2006) 55
66. Liu, W., Bonin, K., Guthold, M.: Revi. Sci. Instrum. 78, 063707 (2007) 55
67. Varenberg, M., Etsion, I., Halperin, G.: Revi. Sci. Instrum. 74, 3569 (2003) 56
68. Piner, R., Ruoff, R.S.: Revi. Sci. Instrum. 73, 3392 (2002) 56
69. Shegaonkar, A.C., Salapaka, S.M.: Revi. Sci. Instrum. 78, 103706 (2007) 56
70. Hoffmann, A., Jungk, T., Soergel, E.: Revi. Sci. Instrum. 78, 016101 (2007) 56, 60
71. Fujisawa, S., Ogiso, H.: Revi. Sci. Instrum. 74, 5115 (2003) 56
72. Sitti, M., Hashimoto, H.: IEEE/ASME Trans. Mechatronics 8, 287 (2003)
73. Vogl, W., Ma, B., Sitti, M.: IEEE Trans. on Nanotechnology 5, 397 (2006)
74. Onal, C.D., Sitti, M.: Int. J. Robotics Research (in press)
75. Pettersson, T., Nordgren, N., Rutland, M.W., Feiler, A.: Revi. Sci. Instrum. 78, 093702 (2007) 61
76. Cannara, R.J., Brukman, M.J., Carpick, R.W.: Revi. Sci. Instrum. 76, 053706 (2006) 61

77. Bulgarevich, D.S., Mitsui, K., Arakawa, H.: J. Phys.: Conf. Ser. 61, 170 (2007) 64
78. Nain, A.S., Wong, J.C., Amon, C., Sitti, M.: Appl. Phys. Lett. 89, 183105 (2006) 65
79. Mokaberi, B., Requicha, A.A.G.: IEEE Transactions on Automation Science and Engineering (2007) 67
80. Schitter, G., Allgöwer, F., Stemmer, A.: Nanotechnology 15(1), 108 (2004) 67, 70
81. Li, Y., Bechhoefer, J.: Review of Scientific Instruments 78(1), 013702 (2007) 67, 70
82. Mokaberi, B., Yun, J., Wang, M., Requicha, A.A.G.: Proceedings on IEEE International Conference on Robotics & Automation, ICRA 2007 (2007) 67
83. Chen, H., Xi, N., Li, G.: IEEE Transactions on Automation Science and Engineering 3, 208 (2006) 67
84. Mokaberi, B., Requicha, A.A.G.: IEEE Transactions on Automation Science and Engineering 3, 199 (2006), doi: 10.1109/TASE.2006.875534 68
85. Tranvouez, E., Boer-Duchemin, E., Comtet, G., Dujardin, G.: Rev. Sci. Instrum. 78, 115103 (2007) 68
86. Fox, D., Burgard, W., Dellaert, F., Thrun, S.: Proceedings of the Sixteenth National Conference on Artificial Intelligence (AAAI 1999) (1999) 69
87. Thrun, S., Burgard, W., Fox, D.: Probabilistic robotics. MIT Press, Cambridge (2005) 69, 70, 71
88. Horcas, I., Fernández, R., Gómez-Rodríguez, J.M., Colchero, J., Gómez-Herrero, J., Baro, A.M.: Review of Scientific Instruments 78(1), 013705 (2007) 77
89. Baur, C., Gazen, B.C., Koel, B., Ramachandran, T.R., Requicha, A.A.G., Zini, L.: Journal of Vacuum Science & Technology B 15(4), 1577 (1997) 79
90. Lapshin, R.V.: Measurement Science and Technology 18(3), 907 (2007), `http://stacks.iop.org/0957-0233/18/907` 82

Chapter 3
Nanomechanics of AFM Based Nanomanipulation

Nanotechnology, which aims at the ideal miniaturization of devices and machines down to atomic and molecular sizes has been a recent hot topic as a promising high technology for the forthcoming century. By precise control of atoms, molecules, or nano scale objects, new sensors and man-made materials, tera-byte capacity memories, micro scale robots/machines, DNA computers, quantum devices, micro scale distributed intelligence system devices with integrated sensors, actuators and communication tools, etc., would be possible within the near future.

When downscaled, the volumetric forces (e.g., the gravity) tend to decrease faster than other kinds of forces, such as the capillary forces or the viscous force. Altough these forces still exist at the macroscopic scale, they are often negligible (and neglected) in macroscopic assembly. A reduced system is consequently brought face to face with the relative increase of these so-called surface forces. As they decrease slower than the weight, there is always a cutoff size below which these forces disturb the handling task because they make the microcomponent stick to the tip of the gripper (the weight can no longer overcome them and ensure release). The first part of this chapter gives a review of existing models used by other authors aiming at modeling micromanipulation tasks.

Two possible nano-tribological characterization methods are proposed in the second part of this chapter: (1) Sliding the micro/nano-object on the substrate while it is attached to an AFM probe, (2) Nano-robotic pushing of the micro/nano-object with the sharp tip of an AFM probe. Modeling of these methods are realized and experiments are conducted for the latter method using a piezoresistive AFM probe as a 1-D force sensor and nano-manipulator. In the experiments, 500 nm radius gold-coated latex particles are pushed on a silicon substrate. Preliminary results show that different frictional behavior such as sliding, rolling, and rotation could be observed, and shear stresses and frictional behavior could be estimated using these techniques at the nano-scale.

In this chapter, in order to understand robotic grasping at the nanoscale, contact mechanics between nano grippers and nano samples is studied. Hertz and Johnson-Kendall-Roberts (JKR) models were introduced to simulate elastic contacts between various profiles of flat surface, sphere and cylinder for different types of nano

H. Xie et al.: Atomic Force Microscopy Based Nanorobotics, STAR 71, pp. 87–143.
springerlink.com

samples and nano grippers. Analyses and evaluation instances indicate that friction forces, commonly used in macro grasping to overcome gravity, at the nanoscale are often insufficient to overcome the relatively strong adhesion forces when picking up the nano sample deposited on a substrate because of the small grasping contact area. For stable nanoscale grasping, two-finger grippers are proposed, and a non-parallel gripper with a "V" configuration was demonstrated to have better grasping capabilities than a parallel gripper. To achieve robotic nanoscale grasping, a nano gripper constructed from two individually actuated and sensed microcantilevers is presented. Experimental results for the pick-and-place manipulation of silicon nanowires validate the theoretical analyses and capabilities of the proposed nano gripper.

3.1 The Physics of the Micro/Nanoworld

Classical adhesion models [105, 84, 135] usually proposed to study adhesion in micromanipulation or atomic forces microscopy (AFM) are based on the elastic deformation of two antagonist solids (microcomponent/gripper in micromanipulation, cantilever tip/substrate in AFM). The theoretical background proposed in this part aims at detailing:

1. every phenomenon leading to a force interaction: capillary forces, electrostatic forces (in both liquid and air environment, but restricted to conductive materials), van der Waals forces and contact forces ;
2. the influence of surface science concepts such as topography, deformation and wettability.

These elements constitute a basis to model adhesion without using empirical global energy parameters such as surface energies.

When dealing with stiction and adhesion problems in micromanipulation, it is often referred to a list of many concepts (van der Waals interaction, capillary force, adhesion, pull-off...) which sometimes can recover one another. [118] have proposed to sort out these forces by making the distinction whether there is contact or not. When there is no physical contact between two solids, the forces in action are called distance or surface forces (according to the scientific literature in this domain [76, 86, 144], these latter are electrostatic, van der Waals and capillary forces). When both solids contact one another, there is deformation and adhesion forces through the surfaces in contact. In this case, the authors considered contact forces and adhesion or pull-off forces. Electrostatic or capillary effects can be added, but van der Waals forces are non considered anymore, because they thought are already involved in the pull-off term.

Before going through these models, let us mention that many (attractive) effects contribute to adhesion. Based on [122], we propose the schematic forces summary presented in Table 3.1.

Table 3.1 Forces summary and their interaction distances

INTERACTION DISTANCE	PREDOMINANT FORCE
Up to infinite range	Gravity
From a few nm up to 1 mm	Capillary forces
> 0.3 nm	Coulomb (electrostatic) forces
0.3 nm $<$ separation distance $<$ 100 nm	Lifshitz - van der Waals
< 0.3 nm	Molecular interactions
$0.1 - 0.2$ nm	Chemical interactions

Additional effects turn out to be also of importance: let us cite the Casimir effect that will not be detailed in this contribution. We refer to [110].

It seems however that capillary effect dominates all the microworld from a few nm up to the tenth of mm. van der Waals effects turn out to compete with capillary effect but only within the nano range up to a few tens of nm. We therefore mainly focus on both effects together with the electrostatic adhesion which comes from either from the intense electrostatic fields coming from microrobotic actuation - they can be avoided using thermal actuation - or from the moderate effect of contact potentials.

3.1.1 van der Waals Forces

The so-called van der Waals forces are often cited in papers dealing with micromanipulation and microassembly, probably because the founding papers of these bibliography reviews [76, 86] present these forces next to the capillary and the electrostatic forces as being of the utmost importance in the sticking of microparts. Other authors [71] prefer to neglect these forces because they are of a smaller order The reasons for this opposition do not seem to be clear, all the more so since some authors propose to use it as a suitable gripping principle [67, 87]. The will to clarify this situation is a first reason to study van der Waals forces.

A good and very didactic introduction to the subject can be found in [102] while a more exhaustive description of the van der Waals (VDW) forces is proposed by [65, 90, 103]. Let us now have a look on the ways to compute the van der Waals interaction between two macroscopic bodies: the first one is known as the microscopic or Hamaker approach and the second one is called the macroscopic or Lifshitz approach.

From a strictly theoretical point of view, the van der Waals forces are non-additive, non-isotropic and retardated. However, [128] proposed a straight and powerful way to establish the potential interaction by assuming a pairwise additivity of the interactions. Moreover, this approach does not consider the retardation effect. The results are therefore limited to separation distances between an upper limit of about $5 - 10$nm (because we neglect the retardation effect) and a lower limit of about one intermolecular distance assumes that $l \ll r$. This lower boundary is reinforced by the value of the equilibrium distance (about $0.1 - 0.2$ nm) arising from the Lennard-Jones potential: for separation distances smaller than $0.1 - 0.2$nm, very strong repulsive forces occur that can no longer be neglected). This lower limit is sometimes called the van der Waals radius [102]. We should keep in mind that even with these restrictions, the results are not exactly correct for the interaction of solids and liquids because of the pairwise summation assumption. However, [103] and [146] consider that these approximations are useful in several applications.

The Lifshitz method, also called macroscopic approach, consists in considering the two interacting objects as continuous media with a dielectric response to electromagnetic fields. The dispersion forces are then considered the mutual interaction of dipoles oscillating at a given frequency. When the separation distance becomes bigger than a cut-off length depending on this frequency and the light speed, the attraction tends to decrease because the propagation time becomes of the same order as the oscillation period of the dipoles, the field emitted by one dipole interacting with another dipole with a different phase. This effect has first been pointed out by Casimir and Polder [79] and computed by Lifhitz using the quantum field theory [127]. Although this approach is of the greatest complexity, similar results can be obtained by using the Hamaker's results, on the condition to replace the Hamaker constant by a pseudo - constant involving more parameters. This method is out of our scope, which is to roughly evaluate the importance of the van der Waals forces in microassembly and to investigate the influence of geometry, roughness and orientation on the manipulation of microcomponents. We will therefore limit ourselves to the Hamaker method, despite its limitations. The interested reader will find further information about the Lifshitz approach in [65], chapter VI, and in [103].

Table 3.2 Comparison between the approximations from the literature (D, separation distance and R, the sphere radius). Note that the sign '-' of the forces has been omitted: they must be considered attractive.

Object 1	Object 2	Expression	Reference
Plane	Plane//	$W \approx -\frac{A}{12\pi D^2}$; $F \approx \frac{A}{6\pi D^3}$ (by surface unit)	[65, 103, 152]
Cylinder	Cylinder //	$W \approx \frac{AL}{12\sqrt{2}D^{\frac{3}{2}}}(\frac{R_1R_2}{R_1+R_2})^{\frac{1}{2}}$; $F \approx \frac{AL}{8\sqrt{2}D^{\frac{5}{2}}}(\frac{R_1R_2}{R_1+R_2})^{\frac{1}{2}}$ (L, cylinders length; R_i, cylinders radii)	[103]
Cylinder	Cylinder ⊥	$W \approx -\frac{A\sqrt{R_1R_2}}{6D}$; $F \approx \frac{A\sqrt{R_1R_2}}{6D^2}$	[65, 103]
Sphere	Plane	$W \approx -\frac{AR}{6D}$; $F \approx \frac{AR}{6D^2}$	[65, 152]
Sphere	Sphere	$W \approx -\frac{AR}{6D}$; $F \approx \frac{AR}{6D^2}$ (including conical and spherical asperities)	[65, 103, 152]

Based on the Hamaker method, there exist models: (1) without roughness nor orientation [65, 103]; (2) with roughness but without orientation [66, 118, 153]:

(3) without roughness but with orientation [87, 118]. Note that we have not found any description of a configuration including both roughness and orientation. Ideally, these forces should be computed again taking into account the mechanical deformations at contact. To close this section, let us recall some useful references: [118, 102, 88, 65, 103, 76].

3.1.2 Capillary Forces

Capillary forces between two solids arise from the presence of a liquid meniscus between both solids. The presence of this liquid is due either to the user - who puts liquid to provoke an adhesion force, for example to pick up a component - or due to the condensation of the surrounding humidity - either spontaneously due to environmental conditions or due to the cooling of a gripper for example [82]. On a more general way, these forces arise from the surface tension of the interface between two media: water-air, water-oil, oil-air...Therefore, they are also called surface tension forces or surface tension effects. They are of the utmost importance in the microworld because they clearly dominate all the other effects but maybe in some cases at a few nanometer scale the van der Waals forces with which they compete on a balanced manner.

The key concepts to the understanding and the modeling of capillary forces are the *surface energy*, *surface tension*, the *contact angles* and *wettability* together with the *Young-Dupr equation*, the *pressure drop* across the interface described by the so-called *Laplace* equation, the *curvature* of a surface in the 3D space. When a droplet is posed on a solid substrate (see figure 3.1), the liquid spreads out and we can distinguish three phases (vapor, liquid, solid) separated by three interfaces that join one another at the triple line, also called contact line.

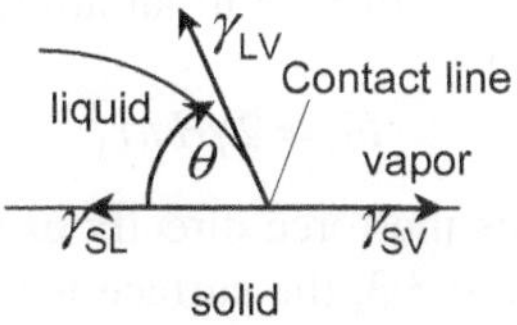

Fig. 3.1 Illustration of the Young-Dupré equation

At this triple line, the liquid-vapor interface makes an angle θ with the substrate. If the contact line is at equilibrium θ is called the static contact angle, which is linked to the interfacial tensions by the Young-Dupré equation [65, 103]:

$$\gamma_{LV} \cos\theta + \gamma_{SL} = \gamma_{SV} \tag{3.1}$$

Due to the surface tension, there exists a pressure difference across the interface between a liquid and a gas. In the case of a soap bubble for example, the pressure inside the bubble is bigger, to compensate the outside pressure and to overcome the tension effect. In a more general case, the pressure difference is linked to the curvature of the interface according to the Laplace equation [65]:

$$2\gamma H = 2\gamma(\frac{1}{R_1} + \frac{1}{R_2}) = p_{\mathrm{in}} - p_{\mathrm{out}} \tag{3.2}$$

where H is the mean curvature and R_1 and R_2 two principal curvature radii.

Models of capillary forces found in the literature are usually valid only at equilibrium. Let us now consider two solids linked by a liquid bridge[1], also called meniscus (figure 3.2(a)). In order to link this to the general frame of micromanipulation, let us call the upper solid the 'tool' or the 'gripper' (it will be used as a gripper) and the lower one as the object (it will be used as micropart or microcomponent). Since axial symmetry is assumed, it can be seen in figure 3.2(a) that the contact line between the meniscus and the object (the gripper) is a circle with a radius r_1 (r_2). The pressure inside the meniscus is denoted by p_{in} and that outside the meniscus by p_{out}. θ_1 is the contact angle between the object and the meniscus and θ_2 is the angle between the gripper and the meniscus. z represents the separation distance (also called the gap) between the component and the gripper. h is called the immersion height. At its neck, the principal curvature radii are ρ' (in a plane perpendicular to the z axis, i.e. parallel to the component) and ρ (in the plane rz).

The object is submitted to the 'Laplace' force, arising from the pressure difference $p_{\mathrm{in}} - p_{\mathrm{out}}$, and to the 'tension' force, directly exerted by the surface tension. In what follows, we will consider that these two forces constitute what we will call the capillary force[2]. The 'Laplace force' is due to the Laplace pressure difference that acts over an area πr_1^2 (see figure 3.2(b)) and can be attractive or repulsive according to the sign of the pressure difference, i.e. according to the sign of the mean curvature: a concave meniscus will lead to an attractive force while a convex one will induce a repulsive force.

$$F_{\mathrm{L}} = 2\gamma H \pi r_1^2 \tag{3.3}$$

The 'tension force' implies the force directly exerted by the liquid on the solid surface. As illustrated in figure 3.3, the surface tension γ acting along the contact circle must be projected on the vertical direction, leading to:

$$F_{\mathrm{T}} = 2\pi r_1 \gamma \sin(\theta_1 + \phi_1) \tag{3.4}$$

[1] The presented configuration is axially symmetric, to introduce the capillary force from a 'mechanical' point of view, i.e. using concepts like pressure or tensions. In a more general case, the configuration is not axially symmetric and an energetic approach has to be implemented, see therefore [115].

[2] [130] uses the terms 'capillary' force for the term arising from the pressure difference and 'interfacial tension force' for that exerted by the surface tension.

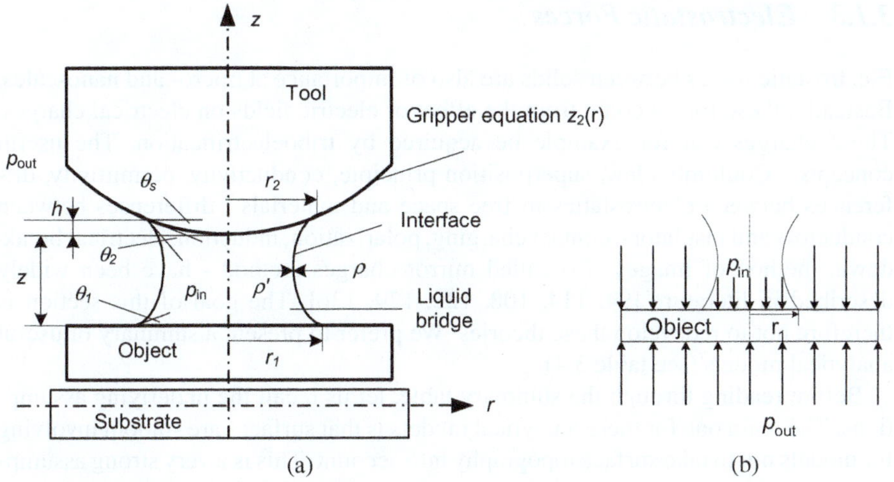

Fig. 3.2 (a) Effects of a liquid bridge linking two solid objects (b) Origin of the Laplace force: attractive case

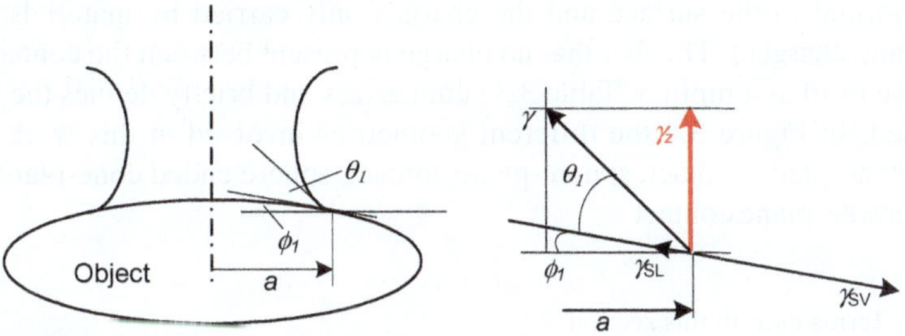

Fig. 3.3 Origin of the tension force and detail

Therefore, the capillary force is given by:

$$F_C = F_T + F_L = 2\pi r_1 \gamma \sin(\theta_1 + \phi_1) + 2\gamma H \pi r_1^2 \quad (3.5)$$

On a more general way - for example in the case of non axially symmetric geometries -, the force is computed from the derivation of the surface energy. ϕ_1 denotes the slope of the component at he location of the contact line: it will be often considered equal to zero. This leads to the well-known approximation [103]:

$$F_{max} = -4\pi R\gamma \cos\theta \quad (3.6)$$

If $z \neq 0$ (the gap between the sphere and the plane), the total capillary force can be rewritten as follows:

$$F = -\frac{4\pi R\gamma \cos\theta}{\frac{z}{h} + 1} \quad (3.7)$$

Additional information can be found in [115, 113, 103].

3.1.3 Electrostatic Forces

Electrostatic forces between solids are also of importance at micro- and nanoscales. Basically, these forces come from the effect of electric fields on electrical charges. These charges can for example be acquired by triboelectrification. The useful concepts - Coulomb's law, superposition principle, conductivity, permittivity, differences between electrostatics in free space and materials , differences between conductors and insulators, contact charging, polarization, induction, electrical breakdown, method of images also called mirror charges method - have been widely described in literature [94, 114, 108, 122, 129, 136]. The goal of this section is therefore not to redevelop these theories. We prefer to present a summary of useful analytical models (see table 3.4).

Before reading through the summary table, let us recall the underlying assumptions. The main one for these analytical models is that surfaces are smooth involving for models not to take surface topography into account. This is a very strong assumption since, no matter how carefully or expensively a surface is manufactured, it can never be perfectly smooth. The second assumption defines materials as conductive involving that the potential is uniformly distributed along the surface, the electric field is normal to the surface and the charges only carried by materials surfaces (no volumic charges). The fact that no charge is present between the contacting objects is the third assumption. Table 3.3 summarizes and briefly defines the different terms used. In Figure 3.4 the different geometries involved in this work are presented, plane-plane contact, sphere-plane contact, sphere ended cone-plane contact and hyperbole-plane contact.

Table 3.3 Terms used in this section

Term	Definition	Units	Usual Values
ε_0	Free space permittivity	$C^{-2}N^{-1}m^{-2}$	8.85^{-12}
R	Sphere radius	m	10 nm to 100 μm
z	Separation distance	m	1 nm to 100 μm
V	Potential difference	V	0.5 to 20 V
θ	Cone half aperture angle	rad	til 10°
L	Length of tip	m	10 μm to 500 μm
A	Area of contact	m^2	
r_{max}	Maximum distance to the axis	m	
δ	Truncated cone height	m	
l	Plane width	m	
W	Electrostatic energy	J	
C	Capacitance	F	

Plane-plane and sphere-plane models are the most encountered in the literature. The expressions have been derived from the electrostatic energy $W(d)$.

$$F_{elec}(z) = -\frac{\partial W(z)}{\partial z} = -\frac{1}{2}\frac{\partial C}{\partial z}V^2 \tag{3.8}$$

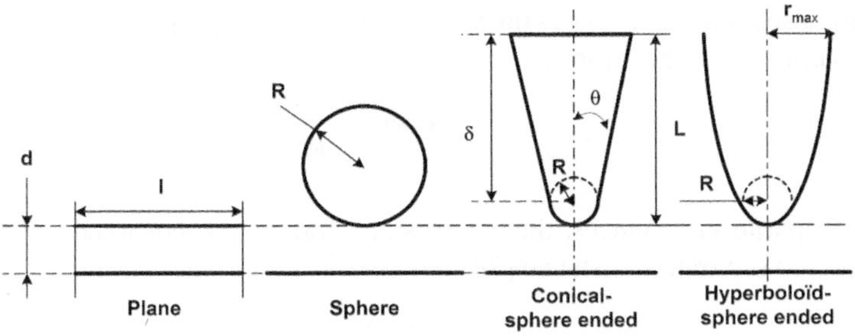

Fig. 3.4 Representation of the involved geometries

The simple case [86] is the plane-plane contact where two smooth planar surfaces are brought into contact. The surface of contact has an area A and the capacitance is obtained from the well-known plane capacitor case.

$$C(d) = \frac{\varepsilon_0 A}{z}$$

$$F_{plane} = \frac{\varepsilon_0 V^2 A}{2z^2} \tag{3.9}$$

This model gives the electrostatic pressure and knowing the area of the surface the electrostatic force can be deduced. The experience shows however that it is very difficult to determine the area of contact in real configurations. The planar model is thus very restricted in terms of applications. Moreover studied objects are rarely totally flat. In application it may thus be used at very close separation distances between object when the contact can be estimated by flat surfaces.

The sphere models have been developed for more complex shapes and longer separation distances. Many author such as ref. [114] have been using them when studying the adhesion phenomenon disturbing micromanipulations. These models give an estimation of the electrostatic forces for the contact between a conductive sphere and a conductive plane. As the previous model, they are derived from the electrostatic interaction energy, and the capacitance between a sphere and a plane is given by the following expression.

$$C_{sphere} = 4\pi\varepsilon_0 R \sinh(\alpha) \sum_{n=1}^{\infty} (\sinh n\alpha)^{-1}$$

with $\alpha = \cosh^{-1}((R+z)/R)$. It is usual in literature to analyze the contact between tip and surface in atomic force microscopy as a sphere above a conducting plane [76]. The developed expressions depend on the separation distance and more precisely on the ratio between the sphere radius R and the separation distance z. Three models have been developed from the general expression given by ref. [85], depending on the separation distance range.

For small separation distances the electrostatic force is proportional to the inverse of the separation distance: [86, 76, 70, 78]:

$$F_{sphere1} = \frac{\pi \varepsilon_0 R V^2}{z} \quad R \gg z \tag{3.10}$$

For large separation distances, the electrostatic force is proportional to the inverse of the squared separation distance [78, 101]:

$$F_{sphere2} = \frac{\pi \varepsilon_0 R^2 V^2}{z^2} \quad R \ll z \tag{3.11}$$

For all separation distances, [101, 77] a general expression has been developed from Equation 3.10 and Equation 3.11:

$$F_{sphere3} = \pi \varepsilon_0 \frac{R^2 V^2}{z(z+R)} \tag{3.12}$$

These models are restricted in their applicable separation distances. They are often used to get a quantitave assesment of the electrostatic forcebetween the probe and the substrate in scanning probe microscopy. A review of analytical models is given in the table 3.4.

As indicated in the underlying assumptions of these analytical models, surface roughness is not taken into account. [126, 125, 124] have clearly shown the reduction of electrostatic forces in presence of (even very small) surface roughness. Indeed, a first comparison between simulation results [124] and literature results [147] is shown in figure 3.5(a). The first observation is that even though results are in good correlation, simulated forces are stronger in smooth configuration than what was experimentally obtained. Moreover the difference between experimental results and simulations increases when the separation distance decreases. This observation was attributed to the fact that even though the spot of contact has been chosen to be smooth, it can never be perfectly smooth. Even a very small roughness may influence the results at such small separation distances. [126] introduced roughness with the generation of a fractal surface using fractal parameters $D = 1.55$ and $G = 1.5 \times 10^{-12}$ for the planar contacting surface in order to have a maximum asperity peaks of 0.8 nm and an average roughness of 0.3 nm (which is often assumed to be negligible). The first observation is that even a roughness as small as this one is influencing the results from simulations, decreasing the electrostatic forces. This is specially true when the tip gets closer to the surface. The influence of surface roughness is also more important at higher applied voltages. The results from simulations including roughness are closer to the experimental measures. Hao et al. [93] also measured electrostatic forces for a sphere-ended conical tip. The tip radius is 270 nm and the half aperture angle is 5°. Figure 3.5(b) shows experimental results obtained compared with simulations for different roughness parameters. Conclusions are identical than what was observed for ref. [147].

Table 3.4 Review of analytical models

Contact type	Expression	Ref.
Plane-plane	$F_{plane} = \frac{\varepsilon_0 V^2 A}{2z^2}$	[86]
Sphere-plane	$F_{sphere1} = \frac{\pi\varepsilon_0 R V^2}{z}$ for $R \gg z$	[86, 76, 70, 78]
Sphere-plane	$F_{sphere2} = \frac{\pi\varepsilon_0 R^2 V^2}{z^2}$ for $R \ll z$	[78, 101]
Sphere-plane	$F_{sphere3} = \pi\varepsilon_0 \frac{R^2 U^2}{z(z+R)}$ for $R \ll z \ll L$	[101, 77]
Cylinder-plane	$F_{cyl}(N/m) = \frac{\pi\varepsilon_0\varepsilon_R\sqrt{R}V^2}{2\sqrt{2}z^{3/2}} = \frac{\pi\varepsilon_0\varepsilon_R\lambda V^2}{4\sqrt{2}\pi\sqrt{A}z^{3/2}}$	[151]
Conical tip (charged line)	$F_{ch} \cong \frac{\lambda_0^2}{4\pi\varepsilon_0} ln\left(\frac{L}{4z}\right)$ for $R \ll z$ with $\lambda_0 = 4\pi\varepsilon_0 V\left[\ln\left(\frac{1+\cos\theta}{1-\cos\theta}\right)\right]^{-1}$	[93]
Conical tip (asympt. model)	$F_{as} = \pi\varepsilon_0 V^2\left[\frac{R^2(1-\sin\theta)}{z[z+R(1-\sin\theta)]} \cdots + k^2\left(\ln\frac{L}{z+R(1-\sin\theta)} - 1 + \frac{R\cos^2\theta\sin\theta}{z+R(1-\sin\theta)}\right)\right]$ with $k^2 = 1/[\ln(\tan(\theta/2))]^2$	[101]
Hyperb. tip	$F_{hyp1} = \pi\varepsilon_0 V^2 k^2\left[ln\left(1+\frac{L}{R}\right) - \frac{(z-R/tan\theta^2)L}{z(L+z)}\right]$ with $k^2 = 1/[\ln(\tan(\theta/2))]^2$	[120]
Hyperb. tip	$F_{hyp2} = 4\pi\varepsilon_0 V^2 \frac{ln\left[1+\left(\frac{r_{max}}{R}\right)^2\left(1+\frac{R}{z}\right)\right]}{ln^2\left(\frac{1+\eta_{tip}}{1-\eta_{tip}}\right)}$ with $\eta_{tip} = \sqrt{\frac{z}{z+R}}$	[143, 142]

As a conclusion, we have shown in this section advanced results concerning electrostatic forces at the nano- and microscales.

3.1.4 Elastic Contact Mechanics

This subsection reminds the Hertz contact theory and the related adhesion models [99, 105, 135, 84]. In case of a sphere (radius R) on a planar surface, pull-off force is approximately given by JKR (for the lower boundary) or DMT (for the higher boundary) contact models [109], [84].

$$\frac{3}{2}\pi RW \leq F_{\text{pull-off}} \leq 2\pi RW \tag{3.13}$$

where W is the work of adhesion between the two medium.

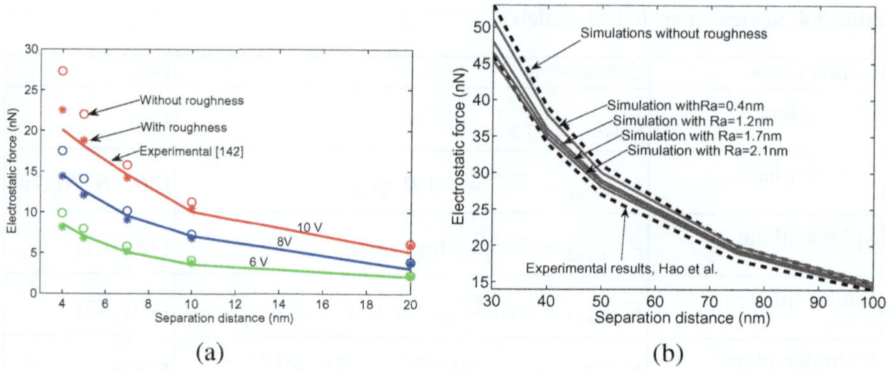

Fig. 3.5 (a) Comparison with experimental measures from [147], who measured electrostatic forces for a sphere-ended conical tip of radius 40 nm and half aperture angle 10° for different voltages. The characteristics of the tip were found using SEM images. The experimental results are compared with simulations, first without and then with roughness parameters (fractal representation) (b) Electrostatic normal force (nN) versus separation distance (nm) for a sphere-ended conical tip of radius 270 nm and half aperture angle θ=5°. Plot shows experimental results obtained by Hao et al. [93], simulations results without roughness parameters and simulation results including different average roughness

According to [135], λ coefficient can be used to choose the most appropriate contact model for a given case. This coefficient is expressed for an interface between two bodies 1 and 2 with:

$$\lambda = 2\sigma_0 \left(\frac{R}{\pi W_{12} K^2} \right)^{\frac{1}{3}} \tag{3.14}$$

where K is the equivalent elastic modulus, calculated using the Poisson's ratios μ and Young's modulus E.

$$K = \frac{4}{3} \left(\frac{1-\mu_1^2}{E_1} + \frac{1-\mu_2^2}{E_2} \right) \tag{3.15}$$

W_{12} is expressed as $W_{12} = \gamma_1 + \gamma_2 - \gamma_{12} = 2\sqrt{\gamma_1 \gamma_2}$ with γ_{12} interfacial energy, γ_1 and γ_2 surface energy of the object, substrate or tip [150].
Using λ, pull-off force can be estimated with:

$$\lambda < 0.1 \Longrightarrow \text{DMT model} \quad P = 2\pi R W_{12}$$
$$\lambda > 5 \Longrightarrow \text{JKR model} \quad P = \frac{3}{2}\pi R W_{12}$$
$$0.1 < \lambda < 5 \Longrightarrow \text{Dugdale model}$$
$$P = \left(\frac{7}{4} - \frac{1}{4} \frac{4.04\lambda^{\frac{1}{4}} - 1}{4.04\lambda^{\frac{1}{4}} + 1} \right) \pi W_{12} R \tag{3.16}$$

When two media are in contact, the surface energy W_{12} is equal to:

$$W_{12} \simeq 2\sqrt{\gamma_1 \gamma_2} \quad (3.17)$$

with γ_i the surface energy of the body i.

From the previous formulas the energy W_{132} required to separate two media 1 and 2 immersed in a medium 3 is given by:

$$W_{132} = W_{12} + W_{33} - W_{13} - W_{23} = \gamma_{13} + \gamma_{23} - \gamma_{12}$$

Nevertheless, these models - which are widely used to interpret AFM measurements or to design grippers and microtools - rely on elastic deformation assumption, which could be no longer valid at scales smaller than 1 μm.

3.2 Nanomechanics of Contact Pushing or Pulling Using One Probe

The nano-scale friction differs from the macro-scale friction and henceforth objects are almost wearless, adhesional friction dominates at low loads [11], and friction becomes an intrinsic property of the particular interface [12]. There has been many works on nano-tribology of AFM probe tips on different surfaces [13], [14]. In these studies, AFM tip is contacted and moved on a surface and frictional forces are measured by torsional bending of the probe. However, these studies are limited to specific tip materials, and cannot characterize different type of motions of objects on substrates such as rolling, rotation, etc. Therefore, this part contributes by introducing an autonomous nano-robotic manipulation system for nanometer-sized object frictional parameter and behavior characterization in any environment and any mode of motion. Two possible characterization methods are proposed: (1) Sliding the micro/nano-object on the substrate while it is bonded to an AFM probe, (2) Nano-robotic pushing of the micro/nano-object on the substrate with the sharp tip of an AFM probe. Quasi-static motion equations of these methods are derived, and experiments are conducted for the latter method using a piezoresistive AFM probe as a 1-D force sensor and nano-manipulator. In the experiments, 500 nm radius gold-coated latex particles are pushed on a silicon substrate, and frictional parameters and behavior are estimated using the proposed models and experimental pushing force data.

The section is organized as follows. In the next subsection, the main concept of the nano-tribological characterization system is defined with assumptions, and sliding, rolling, and spinning frictional models are given. In subsection 2, two possible methods are introduced, and the automatic pushing control scheme for the second method is explained. Subsection 3 describes the latex particle pushing experiments and results.

3.2.1 Problem Definition

The aim is to develop methods for measuring the friction between a micro/nano-object and the substrate while observing the object motion behavior, e.g. sliding, stick-slip, rolling, and spinning. As a possible method, the object-substrate motion scheme in Figure 3.6 is proposed.

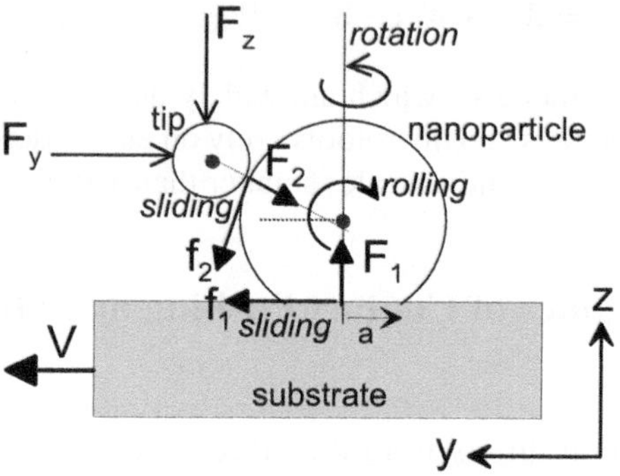

Fig. 3.6 AFM tip-driven motions of a nano-particle on the moving substrate: sliding, stick-slip, rolling, sticking, and rotation.

Here, the micro/nano-object is fixed to an AFM probe base or pushed by an AFM probe tip, and, by moving the substrate with a controlled constant speed, F_y and F_z forces are measured in real-time to predict the object motion behavior and object-substrate frictional parameters. Here, the following assumptions are made:

- Friction is taken to be mainly adhesional without wear and ploughing effects by assuming low load, smooth and unreactive surfaces, and high adhesional forces.
- The substrate is moving with a constant speed V to neglect the inertial effects of the substrate and the positioning stage. Also, V is chosen to be high so that the stick-slip behavior is not observed.
- Micro/nano-particles with radius R are selected as the object. However, the same characterization methods can be directly applied to any micro/nano-object such as carbon nano-tubes, nano-wires, nano-crystals, DNA, RNA, cell, etc. by generalizing the contact mechanics models.
- The AFM probe tip shape is assumed to be spherical with radius R_t. R_t is selected to be much smaller than R in order to have relatively lower tip-particle adhesion than the particle-substrate interface during their detachment. Thus, the tip-particle sticking problem after pushing is minimized.
- The tip and substrate are chosen to be the same material for simplicity and less uncertainty.
- The substrate is selected as a conductor or semiconductor material, and it is electrically grounded for preventing the electrostatic adhesion accumulation due to triboelectrification.

- Ambient environmental conditions are assumed with $40-60\%$ relative humidity and $23\ ^oC$ temperature. Therefore, a water layer could exist on the surface depending on their hydrophilicity and humidity level which would change the tribological behavior [15]. However, the same method is valid for any other environment such as vacuum or liquid.
- All surfaces are assumed to be not contaminated and very smooth. Thus, the roughness effect on adhesion and friction is neglected.

During the AFM probe tip based pushing method, possible particle motions are sticking, sliding, stick-slip, rotation, and rolling. Therefore, nano-scale models for the sliding and rotational friction forces are required. Sliding is the most general case in AFM based lateral frictional force microscopy (stick-slip is also observed frequently in atomic scale imaging and slow speeds [14]). Since the friction is mainly adhesional with the given assumptions, sliding friction becomes as [21], [22]

$$\begin{aligned} f_1 &= \tau A \\ f_2 &= \tau A_2 \end{aligned} \tag{3.18}$$

where τ is the shear stress of the particle and substrate contact and the tip and particle contact points, and $A = \pi a^2$ and $A_2 = \pi a_2^2$ are the real contact areas and a and a_2 are the real contact radii for the particle-substrate and tip-particle interfaces respectively. In general, τ is a function of contact area and pressure [23] and the first-order approximation is given by

$$\tau = \tau_0 + c(P + P_0) \tag{3.19}$$

where $P = L/A$ is the pressure for a given contact area A and a normal load L, P_0 is the capillary pressure in the case of a liquid layer, and c is the proportionality constant. For a bulk material, τ is expected to be $\tau \approx G/29$ [22] theoretically, where G is the shear modulus.

Rolling has been addressed in a few nano-related works. Butt *et al.* [16] modeled the rolling friction force f_r between two identical micro-particles as

$$f_r = 6\pi\gamma_p d \tag{3.20}$$

where γ_p is the particle surface energy and d is the critical rolling displacement. Note that $\sigma \leq d \leq a$ is assumed, where σ is the mean distance between neighboring atoms and a is the contact radius. Another group [17] has reported the rolling friction moment model based upon Amonton's law, where the rolling friction moment M_r is given by

$$M_r = \mu_r(L + F_a) \tag{3.21}$$

with F_a is the adhesion force, and μ_r is the rotational friction coefficient. However, since the adhesion based friction is assumed to be dominant, the particle-substrate rolling friction is modeled as

$$f_1^{\psi} = \tau_{\psi} A \tag{3.22}$$

where τ_ψ is the rotational friction coefficient. For most materials, f_1^ψ is much less than f_1. If this would be the case at the nano-scale, then mostly rolling would start before sliding depending on the applied pushing load.

Comparing the relative magnitudes of above frictional forces,

$$f_1 > f_2$$
$$f_1 > f_1^\psi$$
$$f_2 > f_2^\psi \tag{3.23}$$

assuming $R_t < R$, i.e. $A_2 < A$, and $\tau_\psi < \tau$. Thus, depending on F_y, firstly rolling or sliding of the tip-particle interface, and then sliding of the particle-substrate interface would be expected.

3.2.2 *Methods*

For a particle-substrate interface friction characterization, the following two methods are proposed in this chapter:

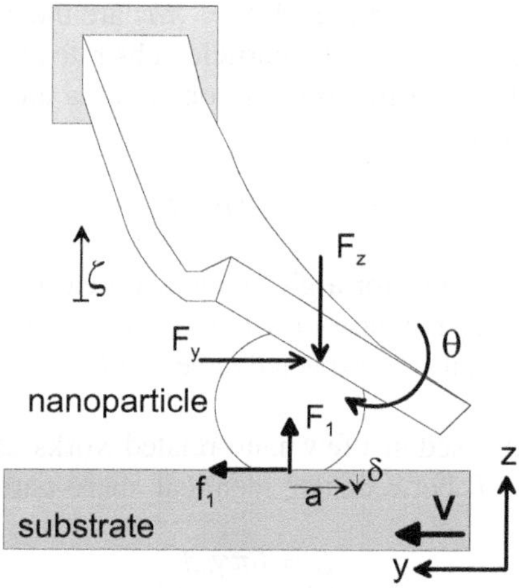

Fig. 3.7 The sliding motion of a nano-particle, which is bonded to the rectangular AFM probe (front view).

3.2.2.1 Method I

Let a nano-particle be attached to the bottom end of a rectangular AFM probe as shown in Figure 3.7. The AFM probe attached with a micro-particle was originally investigated for measuring the pull-off force against the particle-substrate adhesion with well-defined geometries [18]. Adapting it to nano-tribological measurements, the particle that is attached to the AFM probe base is contacted and moved on the substrate laterally in the y-direction with a constant speed V, and the torsional deflection is measured to compute the *direct* particle-substrate friction f_1. This method can be applied to frictional characterization of micron or 100s of nanometer size of particles where the object and substrate geometries are well defined. However, for 10s of nanometer sizes of particles, attaching a single nano-particle and sliding control could be very challenging, and this method allows only the characterization of the sliding and stick-slip frictional behaviors since the particle is glued to the nano-probe base.

For this method, the normal deflection $\Delta\zeta$ and torsional twisting $\Delta\theta$ from the initial positions are measured using an optical detection system. Since the substrate moves along the y-axis only, $F_x = 0$ here. Thus,

$$\Delta\zeta = \frac{F_z\cos\alpha}{k_z} - \frac{F_z\sin\alpha}{k_{xz}} \qquad \Delta\theta = S_\theta\frac{F_y}{k_y^\theta} \qquad k_z = \frac{Ewt^3}{4l^3}$$
$$k_{xz} = \frac{2L_s}{3}k_z \qquad k_y^\theta = \frac{Gwt^3}{3lH^2} \approx \frac{L_s^2}{2}k_z \qquad k_y^b = \frac{Etw^3}{4l^3} = \left(\frac{w}{t}\right)^2 k_z \tag{3.24}$$

where k_z, k_{xz}, k_y^θ, and k_y^b are the spring constants for bending due to the normal force, bending due to the lateral force moment, twisting due to the lateral force, and lateral bending due to the lateral force, respectively. Also, $E = 2(1+\nu)G$, G, $\nu \approx 0.33$, l, w, t, $L_s = l/H$, and H are the Young's modulus, shear modulus, Poisson's ratio, length, width, thickness, structural constant, and tip height of the beam, respectively, S_θ is the sensor coefficient for the twisting measurement, and α is the beam tilt angle from the base guaranteeing the point contact of the particle with the substrate.

F_y twists the probe but also deflects it laterally. However, since F_y is to be measured from $\Delta\theta$ twisting only, $k_y^b \gg k_y^\theta$ condition should be held for possible measurement errors. This means that following condition should be held for the probe:

$$wH \gg lt\,. \tag{3.25}$$

Then, by keeping k_z in a reasonable range, t and l should be minimized while w and H ($= 2R$ in this case) are maximized for this method.

Assuming a quasi-static slow motion of the particle on the substrate, at each instant:

$$f_1 = F_y\,. \tag{3.26}$$

Thus, by measuring $\Delta\theta$, f_1 is computed. On the other hand, $\Delta\zeta$ measurement gives normal load F_z. F_z would deform the particle with a contact area $A = \pi a^2$ and indentation depth δ using the Johnson-Kendall-Roberts (JKR) model [19] such that

$$a^3 = R/K\left(F_z + 3\pi R\Delta\gamma + \sqrt{6\pi R\Delta\gamma F_z + (3\pi R\Delta\gamma)^2}\right)$$
$$\delta = a^2/R - 2/3\sqrt{3\pi\Delta\gamma a/K}\,. \tag{3.27}$$

Here, $1/K = (3/4)[(1-\nu_p^2)/E_p + (1-\nu_s^2)/E_s]$ is the reduced elastic modulus for the particle-substrate system, E_p and E_s are the Young moduli, and ν_p and ν_s are the Poisson's coefficients of the particle and substrate, respectively, R is the particle radius, and $\Delta\gamma = \gamma_p + \gamma_s - \gamma_{ps} \approx 2\sqrt{\gamma_p\gamma_s}$ [20], γ_p, γ_s, and γ_{ps} are the particle, substrate, and particle-substrate interface surface energies, respectively.

Measuring $\Delta\theta$ and $\Delta\zeta$ and computing F_y and F_z, f_1, A and δ are determined. During moving the substrate (or the base of the beam) with a constant speed V, sliding or stick-slip motion is possible. Assuming only sliding would be considered with a high V, sliding starts if

$$f_1^* = F_y^* = \tau^s A \tag{3.28}$$

Thus, the static shear strength τ^s could be measured from F_y^*. In the kinetic friction region, $f_1 = \tau A$, and by measuring f_1 and F_z at this region, τ could be also predicted.

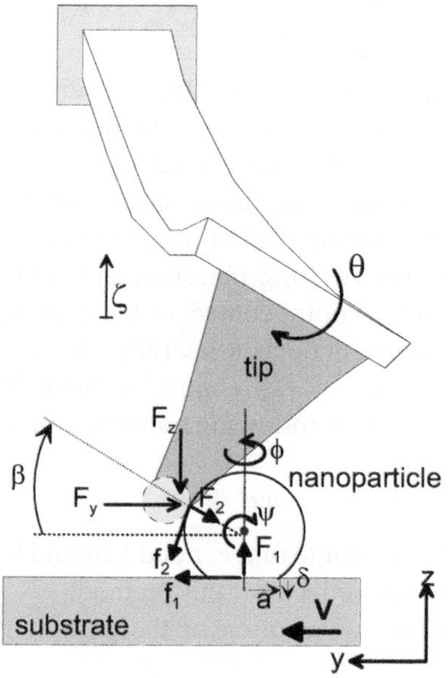

Fig. 3.8 Nano-tribological measurement technique by controlled AFM probe tip pushing of the micro/nano-object using 2-D force sensing for direct friction measurement.

3.2.2.2 Method II

A micro/nano-particle can be pushed on the substrate using a sharp AFM probe tip with a constant speed as in Figure 3.6. This could be the most *general* method for any material with any size and geometry for any mode of motion (sliding, rolling, rotation, etc.). Depending on the AFM probe deflection measurement technique, this method can be applied in two ways.

3.2.2.3 2-D Force Sensing AFM Probe

In this method, the particle is pushed in the *y-direction* as shown in Figure 3.8, and F_y and F_z are measured simultaneously from the $\Delta\theta$ and $\Delta\zeta$ deflection data.

In this case, beam deflection equations are same as Eq.(3.24). Quasi-static equilibrium equations are:

$$\begin{aligned} f_1 &= F_y = F_2\cos\beta - f_2\sin\beta \\ f_2 &= F_z\cos\beta - F_y\sin\beta \\ F_z &= F_1 \end{aligned} \tag{3.29}$$

F_z deforms the particle as in Eq. (3.27). As possible modes of motion:

- *Sticking:* If $F_y = f_1 < \tau^s A$ and $f_2 < \tau^s A_2$, the particle would stick to the substrate and tip.
- *Sliding:* If $F_y = f_1 \geq \tau^s A$ and $f_2 < \tau^s A_2$, the particle would slide while stuck to the tip.
- *Rolling:* If $(f_1 - f_2)R \geq f_1^{\psi}$ and $f_2 \geq \tau^s A_2$, i.e. $f_2 = \tau A_2$, the particle would start to roll while sliding from the tip. The same equality can be written as

$$\left[F_z\cos\beta - (1+\sin\beta)F_y\right]R \geq \tau_\psi A\,. \tag{3.30}$$

- *Rotation (Spinning):* If there is an offset of x_0 along the x-axis, a spinning could occur along the z-axis when $F_y x_0 \geq \tau_\phi A$ where τ_ϕ is the rotational friction coefficient.

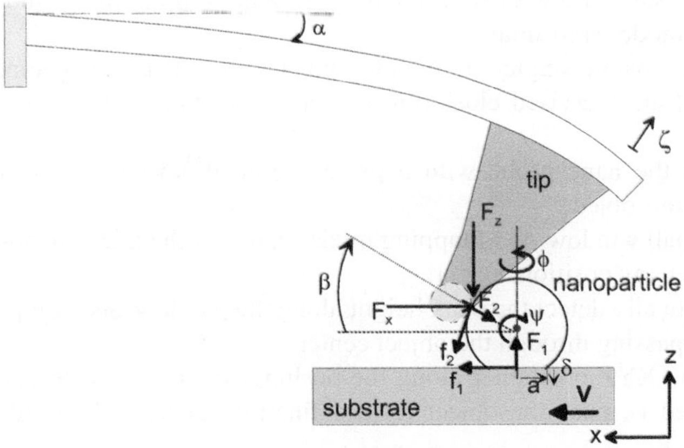

Fig. 3.9 Nano-tribological measurement technique by controlled AFM probe tip pushing using 1-D force sensing for indirect measurement.

3.2.2.4 1-D Force Sensing AFM Probe

Using 1-D optical or piezoresistive deflection data, only $\Delta\zeta$ is measured as shown in Figure 3.9. This kind of simple setup would indirectly measure the frictional force F_x by pushing the particle in the *x-direction*.

The normal deflection of the probe is given by

$$\Delta\zeta = \frac{F_z\cos\alpha - F_x\sin\alpha}{k_z} - \frac{F_z\sin\alpha + F_x\cos\alpha}{k_{xz}} \tag{3.31}$$

Here, $\Delta\zeta$ is the only measured parameter which depends both on F_x and F_z. Assuming F_z is relatively very small with respect to F_x by setting $\beta \approx 0$, and the bending due to F_x is maximized by selecting α tilt angle large and an AFM probe with a large tip height and short probe length, F_x can be extracted from the deflection data.

2-D force sensing case sticking, sliding, rolling, and rotation behaviors are the same in this method by just replacing $F_y = F_x$.

3.2.3 Automatic Pushing Scheme

This tribological characterization system is supposed to operate automatically for future industrial applications. The aim is to scatter the micro/nano-objects to be characterized on a substrate in a semi-fixed way, and then automatically (or by a user-defined interface) find a single object, and push and record the pushing force data. The data is then analyzed using the given models. The basic pushing control steps [21] for a micro/nano-object is as follows:

1. Scan the substrate with semi-fixed micro/nano-objects on it, and get the 3-D tapping mode AFM image.
2. Detect a separate single micro/nano-object with an expected geometry and size using an unsupervised clustering object segmentation algorithm proposed in [24].
3. Position the nano-probe with a pre-determined XYZ distance to the single micro/nano-object.
4. Get a small window AFM tapping mode scan over the micro/nano-object again to correct any positional error.
5. Automatically detect the peak height along the window, and compute the pushing line passing through the object center.
6. Move the XYZ positioner along the pushing-line in 1-D with constant height and speed V, push the object for a defined distance, and record the pushing force data.

3.2.4 Experiments

Using a custom-made AFM system [25] with a piezoresistive AFM nano-probe (ThermoMicroscopes Inc., non-contact piezolever), gold-coated latex particles (JEOL Inc.) are pushed on a silicon substrate using Method II with 1-D force sensing approach and automatic pushing control with a user interface [4]. Experiments are realized in laboratory environment with 23 ^oC temperature and $50-60\%$ relative humidity. Gold-coated latex particles with 500 nm radii are pushed with the probe automatically, and their top-view high-resolution ($\times 5000$ magnification and 90 nm/pixel) optical microscope (Olympus Inc.) images are used for calculating their positions before and after the pushing operation. A closed-loop piezoelectric XYZ positioner with 15-20 nm accuracy enables the constant speed and height trajectory along the pushing line. The speed of the stage is set to 1.6 μm/s for all experiments. Cantilever deflection is measured through a Wheatstone bridge which gives the voltage difference due to the resistance change of the cantilever by the applied tip forces. Calibrating the normal bending sensitivity of the probe S_z as 34.4 nm/V using a hard silicon surface, $\Delta\zeta$ is computed at each instant.

Piezoresistive nano-probe parameters are given as: the tip radius of $R_t \approx 30$ nm, the length of $l = 155$ μm, the width of $w = 50$ μm, the thickness of $t = 3$ μm, the tip height of $H = 3$ μm, $L_s = l/H = 51.7$, normal bending stiffness of $k_z = 8$ N/m, and the measured resonant frequency of 135.3 KHz. Using these parameter values, $k_{xz} = 276$ N/m, $k_y^\theta = 11225$ N/m, and $k_y^b = 2222$ N/m are computed. Moreover, $\alpha = 15^o$ and $\beta \approx 0^o$ are taken. Measured $\Delta\zeta$ corresponds to

$$\Delta\zeta = 0.12F_z - 0.036F_x \,. \tag{3.32}$$

For calculating the contact area, $\Delta\gamma = 0.248$ J/m^2 is computed for SiO_2 substrate (few nanometers thick natural SiO_2 layer exists on silicon substrate in ambient conditions) and the gold layer of the particle with a possible water layer in between. Here, $\gamma_{SiO_2} = 160$ mJ/m^2 [26], $\gamma_{Au} = 1.5$ J/m^2 and $\gamma_{H_2O} = 73$ mJ/m^2 are taken. Since $\beta \approx 0^o$, $F_z = f_2$ and A_2 is the tip-particle contact area due to the contact load F_x. For no contact load case ($F_x = 0$), $A_2 \approx 1.5 \times 10^{-16}$ m^2 using Eq. (3.27) with $E_p = 3.8$ GPa, $E_{SiO_2} = 73$ GPa, $\nu_p = 0.4$, $\nu_{SiO_2} = 0.17$, and $R = 500$ nm. On the other hand, the particle-substrate contact area A is determined by F_z, and $A = 109 \times 10^{-16}$ m^2 for $F_z = 0$. Then, from Eq. (3.32), $F_x \approx -30\Delta\zeta$. Using S_z, $\Delta\zeta$ voltage data is transformed to nN frictional forces by

$$f_1 = -1032\Delta\zeta(V)\; nN \,. \tag{3.33}$$

During the pushing experiments, the following modes of motion are observed:

- *Sliding:* In Figure 3.10, it can be seen that the particle is pushed on the substrate for around 4 μm distance. At first, the particle starts to roll, and after some distance it started to slide with a large peak. The large peak corresponds to the static sliding friction, and after this peak, kinetic friction is observed with a smaller friction. From the peak, around 4.04 V is measured. Using Eq. (3.33), $f_1^* = 4.17$ μN is computed. Since $F_z \approx 0$, $A \approx 109 \times 10^{-16}$ m^2 gives the static

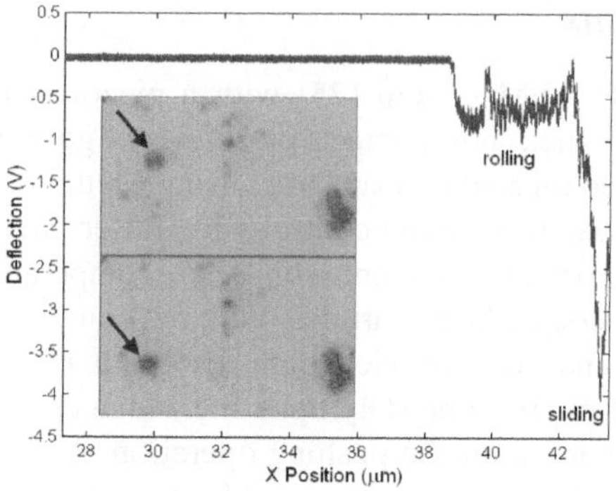

Fig. 3.10 Initially rolling and then sliding behaviors for a 1 μm gold-coated latex particle pushing case where the particle is pushed around 4 μm: Inner images are the before (upper image) and after (lower image) the pushing operation top-view optical microscope images, and the arrow indicates the pushed particle (deflection (V) $\times$ 1032 = nanoforce (nN) in the y-axis).

shear strength as $\tau^s = 382$ MPa. In the kinetic friction region, $f_1 = 2.58$ μN, $\tau = 237$ MPa is held. Using the theoretical shear modulus and Poisson's ratio values for the SiO_2 and gold interfaces, $\tau \approx G/29 \approx 311$ MPa is computed using $G = [(2 - \nu_{Au})/G_{Au} + (2 - \nu_{SiO_2})/G_{SiO_2}]^{-1} = 9$ GPa for $G_{SiO_2} = 31.4$ GPa, $G_{Au} = 30$ GPa, and $\nu_{Au} = 0.42$. For a thick water layer case between the particle and substrate, $\tau = 144$ MPa [27]. Thus, measured shear strength values are close to the theoretical values.

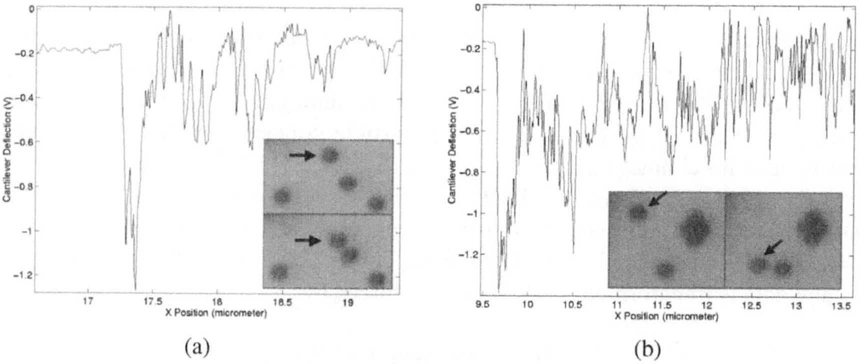

Fig. 3.11 (a) Rolling case where the particle is pushed around 1 μm as shown in two high-resolution optical microscope images (before and after pushing images) (b) Rolling case where the particle is pushed around 3μm.

- *Rolling:* Rolling behavior could be observed in Figure 3.10, 3.11, and 3.11 by the periodic oscillation behavior in the force deflection data. After an initial static frictional phase, cantilever deflects almost with a periodic motion of 500 nm intervals which is the radius of the particle. In this region, $\Delta\zeta = 0.7$ V is approximated in average, and this corresponds to $f_1^{\psi} = 722$ nN frictional force. Using $A \approx 109 \times 10^{-16}$ m^2, $\tau_{\psi} = f_1^{\psi} R/A = 33$ N/m rolling friction coefficient is predicted.

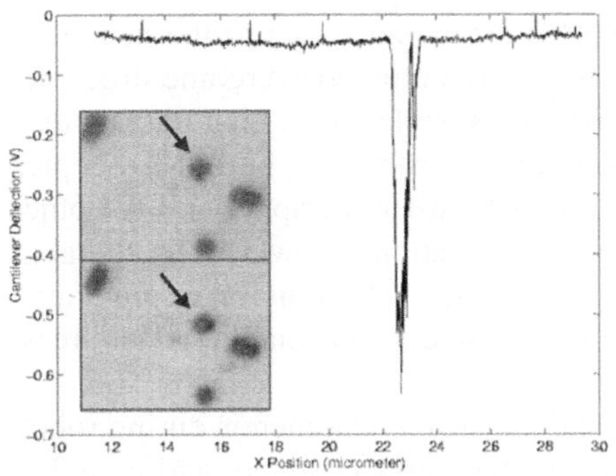

Fig. 3.12 Rotation case where the particle is rotated around 90° around the vertical z-axis (a small black dot on the pushed particle in the inner optical microscope images and no positional change of the particle are the proofs to show the approximate 90° spinning behavior).

- *Rotation:* Rotation of the particle occurred when there is a significant x_0 offset during pushing. In Figure 3.12, optical microscope image shows a 90° particle rotation around the z-axis taking the small black dot attached on the particle as the reference. From the deflection data, after the second peak, the tip loses contact with the particle after around 500 nm displacement. The rotational friction is predicted from the second peak as $f_1 = 155$ nN which gives $\tau_{\phi} = f_1 x_0/A \approx 4$ N/m rotational friction coefficient with $x_0 \approx 300$ nm.

By pushing micro/nano-objects or attaching them on the nano-probe, frictional forces can be directly or indirectly estimated, and sliding, spinning, or rolling frictional parameters can be estimated in any environment. Gold-coated latex particles with 500 nm radius are pushed on a silicon substrate in the experiments. Preliminary results show that pushing operation can result in sliding, rolling, or rotation behavior depending on the particle-substrate frictional properties, and sliding and rolling frictional coefficients are predicted for gold-silicon interfaces. From the pushing data of the 500 nm radius gold-coated latex particles on a silicon substrate with a natural silicon oxide layer, static, kinetic, rolling, and spinning shear stresses are estimated as 382 MPa, 237 MPa, 33 N/m, and 4 N/m, respectively. Thus, if there is an offset x_0 from the pushing direction and particle center, the particle would start to spin firstly, and the tip and particle would lose contact after a very short time. In order to prevent this, $x_0 = 0$ should be precisely controlled. Besides of spinning, rolling or sliding starts first depending on the particle radius, τ^s, and τ_{ψ}. For measured τ^s and τ_{ψ} values, if the particle radius is smaller than around 86 nm ($R < \tau_{\psi}/\tau^s$ assuming $f_2 \approx 0$), the particle would start to slide first and then roll. However, since $R = 500$ nm in the experiments, particles first started to roll and then slide. On the other hand, sliding and rolling could be observed simultaneously if the friction is higher than the static sliding and rolling frictions.

3.3 Nanomechanics of Pick-and-Place Manipulation Using Two Probes

To achieve robotic nanoscale grasping, the main difficulties are fabricating very sharp end-effectors with a size comparable to the nano object to be manipulated and with enough grasping force output to overcome strong adhesion forces, coupled with force sensing capabilities while controlling interactions between the nano object and the end-effectors or the substrate. For macroscopic, even for microscopic applications, friction forces between a gripper and the objects is generally strong enough for stable grasping operations. However, frictional grasping does not work in this case because contact-induced friction forces are often less than the relatively large adhesion forces attributable to the small contact areas that are favorable for nano object release.

In order to understand physical phenomenon during robotic nanoscale grasping, various contact profiles of flat surface, sphere and cylinder for different types of fingers and nano samples are analyzed based on the Hertz model [47]. Grasping capabilities of nano grippers with single- and two-finger configurations are discussed. As an example, we present a prototype of two-tip AFM-based nano robotic system, in which two collaborative cantilevers with protruding tips are used to construct a nanotip gripper. In our approach, interactions between the nano gripper and the nano object during grasping are analyzed and simulated theoretically. We have used the developed nanotip gripper to build a 3-D nanowire crossbar.

This section is organized as follows. Subsection 2 describes nanoscale grasping problems. modelling of interactive forces and contact mechanics during grasping is presented in subsection 3. Nanoscale grasping with different grippers is analyzed in subsection 4. In subsection 5, we show a prototype of the nanotip gripper and a grasping result for a nanowire crossbar.

3.3.1 Problem Definition

3.3.1.1 Challenges

Modelling and simulation of contact mechanics between the nano object, gripper and a substrate is required to understand nanoscale grasping interactions, thereby providing a theoretical estimation of grasping forces, release adhesion forces and maximum contact stress for reliable pickup and smooth release operations, as well as protecting the gripper and the nano object from damage. Here, research on the following aspects associated with nanoscale grasping should be addressed:

- adhesion forces : while adhesion is clearly of interest for a wide range from macroscopic applications to the microscopic scale, its importance obviously becomes dominant at the nanoscale. The effects of adhesion during a grasping operation are friction and interfacial wear as well as a contribution to pickup and release operations. For example, adhesion forces between a gripper and a nano object can be used to counteract substrate adhesion. On the other hand, control and reduction of the adhesion between the gripper and the nano object is essential for the release operation. Therefore, studying the adhesion during nanoscale grasping is crucial to developing proper grasping tools and schemes.

- nanoscale friction : commonly used in macroscopic grasping to overcome gravity, friction forces can be also used to break adhesion in nanoscale applications. Therefore, understanding and accurately estimating friction forces is important for a reliable nanoscale grasping operation. However, the empirical da Vinci-Amontons laws, common knowledge in macroscopic friction as an outcome of a collective action of a number of asperities, are no longer suitable for nanoscale contact as it is regarded as a single-asperity contact where the friction is dependent on the contact area and Young's modulus of the contacting interface [48, 49].
- contact stress : to protect the brittle gripper and the nano object from damage, it is essential to predicate the maximum stress at the contact area and maintain it below the contact yield stress. If the adhesion forces, external load, contact area and the stress distribution are known, the contact stress can be accurately estimated.
- grasping : to pick up the nano object, the grasping force must be greater than the adhesion forces applied on the nano object by the substrate. The grasping force may be comprised of adhesion forces, friction forces, or other interactive forces. To obtain an adequate and stable grasping force, some basic problems need to be identified, such as, interaction analysis, proper gripper configuration design, and choice of effective grasping strategies and techniques.
- release : to release the nano object in its target position, interactive forces on the releasing direction applied from the gripper should be less than the adhesion forces from the substrate. Solutions are needed to reduce the effects of the adhesion forces from the gripper, such as, a method to decrease the contact area, surface modifications, and external actions with special manipulation schemes.

3.3.1.2 Contact Configurations

Figure 3.13 shows four configurations of nanoscale grasping configurations using two types of two-finger grippers with cylindrical and rectangular fingers. Each configuration uses a nanowire/tube or nanoparticle (sphere). Hence, possible contact states between the gripper and the nano object being grasped can be classified as:

1. Contact between two cylinders (C-C), as seen in Fig. 3.13(a), where the contact is between a nanowire/tube and cylindrical fingers.
2. Contact between a cylinder and a sphere (C-S), as seen in Fig. 3.13(b), where a nanoparticle is grasped by a cylindrical fingers.
3. Contact between a flat surface and a cylinder (FS-C) contact. Fig. 3.13(a) and (c) show contacts between a nanowire/tube and the substrate or rectangular fingers.

4. Contact between a flat surface and a sphere (FS-S), as seen in Fig. 3.13(b) and (d) where the contact is between a nanoparticle and the substrate or rectangular grippers.

The contacting states depend on the surface profile of the gripper's fingers and the nano object with which they are in contact. Contact mechanics of these four states will be discussed in detail in the following parts.

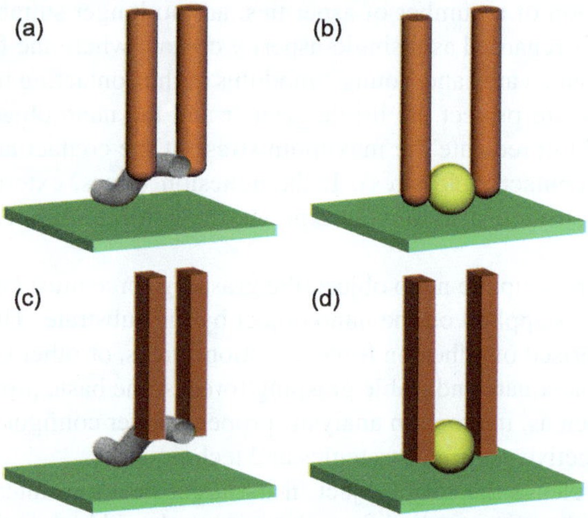

Fig. 3.13 Contact configurations of nanowire/tube and nanoparticle (sphere) grasping with two-finger grippers. In (a) and (c), the nanowire/tube is grasped by grippers with rectangular and cylindrical fingers, respectively. In (b) and (d), the nanoparticle is grasped by grippers with rectangular and cylindrical fingers, respectively.

3.3.2 *Contact Mechanics of Nanoscale Grasping*

3.3.2.1 Hertz Theory for Elastic Contact

More than one hundred years ago, Hertz investigated the contact between two smooth elastic microspheres (as seen in Fig. 3.14) of radius R_1 and R_2 and demonstrated that both the size and the shape of the contact area under a load P follow a classic law:

$$S = \pi \left(\frac{3PR}{4E^*} \right)^{2/3} \tag{3.34}$$

where S is the size of the circular contact area, E^* and R are the elastic constants and relative radius of the microspheres:

$$\frac{1}{E^*} = \frac{1-\nu_1^2}{E_1} + \frac{1-\nu_2^2}{E_2} \qquad R = \frac{R_1 R_2}{R_1 + R_2} \tag{3.35}$$

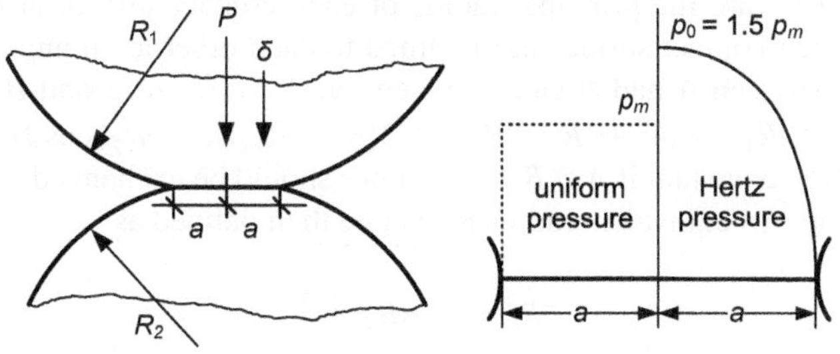

Fig. 3.14 The contact between two elastic microspheres of radius of R_1 and R_2 under a load P.

where ν and E are the Poisson's ratio and the Young modulus of the microspheres. Due to the contact pressure, the surfaces of microspheres penetrate each other along the pressure. The sum of the penetration δ can be given by:

$$\delta = \frac{S}{\pi R} = \left(\frac{9P^2}{16RE^{*2}}\right)^{1/3}. \tag{3.36}$$

The Hertz theory also provides maximum pressure applied on the contact area:

$$p_0 = \frac{3P}{2S} = \left(\frac{6PE^{*2}}{\pi^3 R^2}\right)^{1/3}. \tag{3.37}$$

Hence, the maximum pressure p_0 is 1.5 times the mean pressure p_m applied on the contact area.

3.3.2.2 Mechanics of the Contact with General Profiles

In the general case, i.e., when two surfaces are brought into contact, within the defined frame, the separation h between two close surfaces is given by:

$$h = Ax^2 + By^2 = \frac{1}{2R_1}x^2 + \frac{1}{2R_2}y^2 \tag{3.38}$$

where A and B are constants, R_1 and R_2 are defined as the principal relative radius of curvature. R_1 and R_2 can be expressed as:

$$R_1 = \frac{R_1' R_1''}{R_1' + R_1''} \qquad R_2 = \frac{R_2' R_2''}{R_2' + R_2''} \tag{3.39}$$

where R' and R'' are the principal radius of each contact surface at the origin. If the axes of each contact surface are inclined to each other at an angle ψ, then the relationship between A and B can be given by: $A = (s-t)/2$ and $B = (s+t)/2$, where $t = 1/2[(R_1' - R_1'')^2 + (R_2' - R_2'')^2 + 2(R_1' - R_1'')(R_2' - R_2'')\cos 2\psi]^{1/2}$ and $s = 1/2R_1 + 1/2R_2$. Note that if $A > B$, their values should be exchanged for the following calculation. An equivalent radius R_e can be then defined as:

$$R_e = \frac{1}{2}(AB)^{-1/2}. \tag{3.40}$$

We assume that the contact area is elliptical in shape and with semi-axis a and b. To satisfy the contact condition making use of Eq. (3.38) and (3.40), we obtain [50]:

$$A = (p_0/E^*)(b/e^2a^2)[\mathbf{K}(e) - \mathbf{E}(e)], \tag{3.41}$$

$$B = (p_0/E^*)(b/e^2a^2)[(a^2/b^2)\mathbf{E}(e) - \mathbf{K}(e)], \tag{3.42}$$

$$\delta = (p_0/E^*)b\mathbf{K}(e) \tag{3.43}$$

where **K**(e) and **E**(e) are complete elliptic integrals of argument $e = (1 - b^2/a^2)^{1/2}$, $b < a$. From Eq. (3.41) and (3.42), the size of the elliptical contact area S can be calculated from:

$$S = \pi ab = \pi\left(\frac{3PR_e}{4E*}\right)^{2/3}[F_1(e)]^2 \tag{3.44}$$

where

$$F_1(e) = \frac{4}{\pi e^2}(b/a)^{3/2}$$
$$\left\{\left[(a/b)^2\mathbf{E}(e) - \mathbf{K}(e)\right]\left[\mathbf{K}(e) - \mathrm{E}(e)\right]\right\}^{1/2}$$

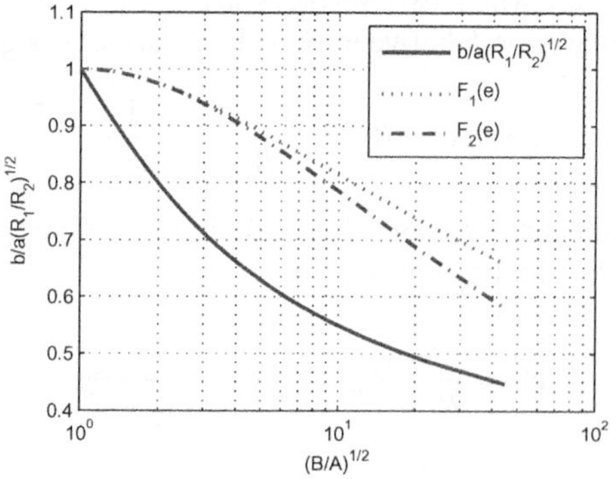

Fig. 3.15 The shape of the contact area of surfaces with general profiles.

and the maximum pressure p_0 is given by:

$$p_0 = \frac{3P}{2\pi ab} = \left(\frac{6PE^{*2}}{\pi^3 R_e^2}\right)^{1/3} [F_1(e)]^{-2}. \tag{3.45}$$

From Eq. (3.43) and (3.45), the penetration of the contact can be calculated by:

$$\delta = \left(\frac{9P^2}{16R_e E^{*2}}\right)^{1/3} \cdot F_2(e) \tag{3.46}$$

where $F_2(e) = 2/\pi(b/a)^{1/2}[F_1(e)]^{-1}\mathbf{K}(e)$.

From Eq. (3.41) and (3.42), the shape of the elliptical contact area is dependent on the ratio of the relative curvatures $(R1/R2)$. Fig. 3.15 plots the variation of $(b/a)(B/A)^{1/2}$ as a function of $(B/A)^{1/2}$. It shows that $(b/a)(B/A)^{1/2}$ decreases from 1.0 as the ratio of relative curvatures $(R^{'}/R^{''})$ increases, indicating that the contact ellipse becomes more slender.

3.3.2.3 Mechanics of C-C Contact

As two cylinders are brought into contact, e.g., a gripper with cylindrical fingers is used to grasp cylindrical nanowires, we may say $R_1^{'} = R_f$ and $R_1^{''} = \infty$ for the gripper $R_2^{'} = \infty$ and $R_2^{''} = R_c$ in Eq. (3.38), here, R_f and R_c are respectively the radii of the gripper finger and the nanowire at the contact locations. In the next step, A and B will be calculated by Eq. (3.41) and (3.42), which in turn give R_e for the contact area S, the maximum pressure p_0 and the penetration δ through Eq. (3.44)–(3.46).

Adhesion forces of C-C contact at micro/nano scale cannot be neglected as they contribute significantly to the contact deformation. Thus, in a contact mechanics simulation exercise, if the radii of the contacting cylinders are of the same order of magnitude, the adhesion forces at the C-C contact can be estimated as the "pull-off" force using [51]:

$$F_s = \frac{3\pi\sqrt{R_f R_c}\Delta\gamma}{2\sin\psi} \tag{3.47}$$

where $\Delta\gamma$ is the work of adhesion and ψ is the angle that the cylinder axes are inclined to each other.

3.3.2.4 Mechanics of C-S Contact

In this case, we may say $R_1^{'} = R_f$ and $R_1^{''} = \infty$, both $R_2^{'}$ and $R_2^{''}$ equal to the radius of the sphere R_s, thus, $R_1 = R_f$ and $R_2 = R_s/2$. Similarly, after A and B are calculated, the eccentricity R_e for contact area S, maximum pressure p_0 and δ using Eq. (3.44)–(3.46).

From Eq. (3.47), if we set $\psi = 90°$, an approximate adhesion force of C-S contact is given by:

$$F_s = \frac{3\sqrt{2R_s R_f}\pi\Delta\gamma}{4}. \tag{3.48}$$

3.3.2.5 Mechanics of FS-C Contact

This case can be considered the limit of the elliptical contact by rewriting (5):

$$h = Ax^2 = \frac{1}{2R_c}x^2 \tag{3.49}$$

where R_c is the radius of the cylinder. When a force q per unit length is applied on the cylinder, the contact is over a long strip of length l' and of width $2w'$ along the axis of the cylinder. The contact width w' and the maximum pressure stress p_0 can be concluded from:

$$w' = \left(\frac{4qR_c}{\pi E^*}\right)^{1/2}, \tag{3.50}$$

$$p_0 = \frac{4}{\pi}p_m = \left(\frac{qE^*}{\pi R_c}\right)^{1/2} \tag{3.51}$$

where p_m is the mean pressure stress on the contact area. Thus, the contact area is given by:

$$S = 2w'l' = 2l\left(\frac{4qR_c}{\pi E^*}\right)^{1/2}. \tag{3.52}$$

The adhesion force at the flat surface and the cylinder contact without external loads is given by [52]:

$$F_s = 3.16l'\left(\frac{4}{3}E^*\Delta\gamma^2 R_c\right)^{1/3}. \tag{3.53}$$

3.3.2.6 Mechanics of FS-S Contact with the JKR Model

When two contacting surfaces are in the presence of adhesion forces, the deformation is more complicated. The Johnson-Kendall-Roberts (JKR) theory is a modern theory specifically used for the adhesion mechanics between two contacting solid surfaces of microspheres [53]. Thus, for the FS-S contact modelling, we use the JKR model for more an accurate calculation involving adhesion forces.

In the JKR model, two contacting spheres of radius R_1 and R_2 are deformed due to the adhesion forces even without an external force P. The circular contact area S can be given as a modified Hertz equation:

$$S = \pi\left[\frac{R}{K}(P + 3\pi R\Delta\gamma + \sqrt{6\pi R\Delta\gamma P + (3\pi R\Delta\gamma)^2})\right]^{2/3} \tag{3.54}$$

where $K = (4/3)[(1-\nu_1^2)/E_1 + (1-\nu_2^2)/E_2]^{-1}$ is the reduced elastic modulus for the contact interface, in which ν_1 and ν_2 are the Poisson's ratios, and E_1 and E_2 are the Young's modulus of each sphere; $R = R_1R_2/(R_1+R_2)$ is the relative radius of curvature. $\Delta\gamma$ is the work of adhesion between contacting surfaces. From Eq. (3.54), if we set $P = 0$, another important result of the JKR model, the "pull-off" force, is given by:

$$F_s = \frac{3}{2}\pi R \Delta \gamma. \tag{3.55}$$

For a sphere (R_s in radius) on a flat surface, we can say, for example, $R_1 = R_s$ and $R_2 = \infty$, thus, in Eq. (3.54) R will be replaced by R_s for contact area estimation. The "pull-off" between the sphere and the flat surface is also given by Eq. (3.55).

3.3.2.7 Friction Forces at the Contact Interface

The well-known Amontons' law shows that the friction force F_f is proportional to a normal force applied on two contacting surfaces with many asperities. However, as the contact area is reduced to the nanoscale, the friction with a single asperity contact is more suitably described by [54]:

$$F_f = \tau S \tag{3.56}$$

where τ is a effective friction coefficient, and S is the contact area that is associated with the sum of the external force and the adhesion force applied on the contact area. τ is related to the effective shear stress of the contact interface: $\tau \approx G^*/29$ [55], where $G^* = [(2-\nu_1)/G_1 + (2-\nu_2)/G_2]^{-1}$, ν and G are the Poisson's ratio value and shear strength of each of contacting surfaces, respectively.

Therefore, it is possible to calculate the friction force of the nanoscale contact, as the contact area is estimated by the modelling addressed above. From Eq. (3.44) and (3.56), the friction forces of the C-C and C-S contacts are estimated as results of joint actions of the adhesion force and the external load:

$$F_f = \tau\pi \left[\frac{3(P+F_s)R_e}{4E^*}\right]^{2/3} [F_1(e)]^2. \tag{3.57}$$

Similarly, from Eq. (3.52) and (3.56), the friction force of the FS-C contact is obtained from:

$$F_f = 2\tau l \left[\frac{4(P+F_s)R_c}{\pi E^*}\right]^{1/2}. \tag{3.58}$$

In addition, the friction forces for the S-S and FS-S contacts can be derived from Eq. (3.56) with a contact area S_{JKR} that is calculated by Eq. (3.54):

$$F_f = \tau S_{\mathrm{JKR}}. \tag{3.59}$$

3.3.3 *Nanoscale Grasping with Different Grippers*

Grippers with one or two fingers are widely proposed in micro/nano applications. However, grippers with more than two fingers will cause difficulties in tip alignment with nanoscale accuracy and particularly for nano object release. Thus, single- and two-finger grippers are discussed.

3.3.3.1 Grasping with the Single-Finger Gripper

Figure 3.16 shows two grasping schemes with a single-finger gripper, namely "tip grasp" with fingertip contact and "side grasp" using the finger side.

As shown in Fig. 3.16(a), in order to grasp and release a nano object with a single finger, the following inequalities should hold:

$$\begin{cases} {}^tF_a^{\max} > {}^sF_a & \text{grasp} \\ {}^tF_a < {}^sF_a & \text{release} \end{cases} \tag{3.60}$$

where the friction forces ${}^tF_a$ and the adhesion force ${}^sF_a$ respectively come from the gripper and the substrate. At present, it is difficult to obtain these inequalities since the contact area of the tip-nano object contact is often less than that of the nano object-substrate contact. Although grasping at the atomic scale has been demonstrated using the scanning probe microscope (SPM) tip with external energy induced by electric field trapping [56], tunneling current induced heating or inelastic tunneling vibration [57], or with functionalization for molecular characterization [58, 59, 60], pick-and-place at the nanoscale is still not well resolved with the "tip grasp" method. Difficulties in this case are weak force outputs while controlling over the applied forces, as well as release accuracy since the single finger has no sufficient geometric limits on the grasping operation.

Figure 3.16(b) shows the "Side Grasp" method using the following inequalities:

$$\begin{cases} {}^tF_f^{\max} > {}^sF_a & \text{grasp} \\ {}^tF_f < {}^sF_a & \text{release} \end{cases} \tag{3.61}$$

where ${}^tF_f$ is the friction force derived from the adhesion force ${}^tF_a$ from the finger side, and ${}^sF_a$ is the adhesion force applied on the nano object from the substrate.

It is obvious that this method aims to increase ${}^tF_a$ through contact with the side rather than contact with the tip. However, from Eq. (3.57)–(3.59), ${}^tF_a$ should generally be much larger than ${}^sF_a$ to produce enough friction force to satisfy the inequalities. In this case, the release seems more difficult than with the "tip grasp." The "side grasp" scheme is usually used for nanowire/tube grasping in the SEM, which allows many grasp or release attempts using visual feedback. However, "side grasp" has similar difficulties to those for the "tip grasp" method, and nano object release is somehow harder.

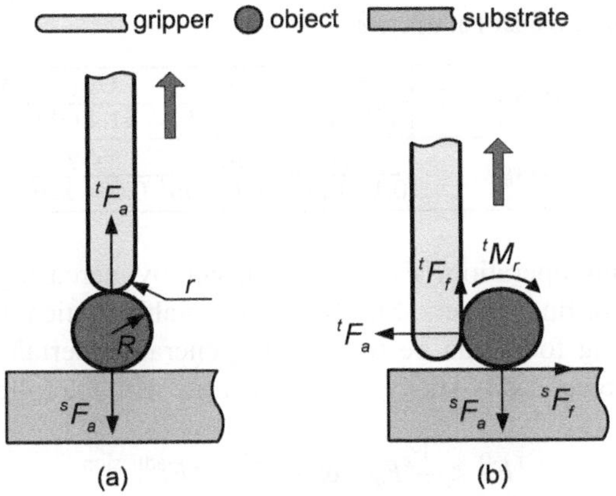

Fig. 3.16 Nanoscale grasping with the singer-finger gripper. (a) "Tip grasp" method. (b) "Side grasp" method.

3.3.3.2 Grasping with the Two-Finger Gripper

Figure 3.17 shows nanoscale grasping with two-finger grippers. Two configurations of the grippers with parallel- and nonparallel-finger are considered.

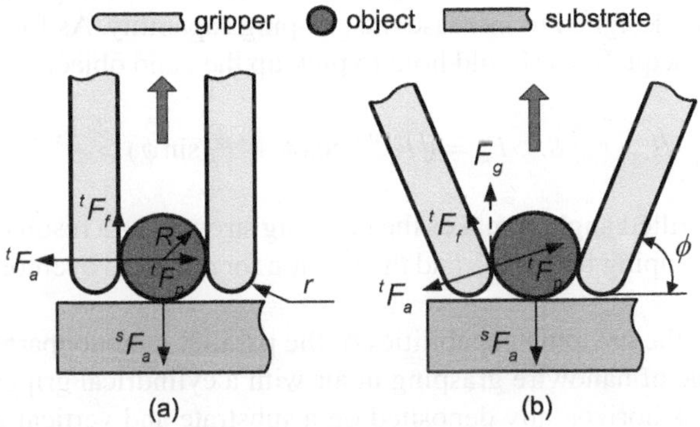

Fig. 3.17 Nanoscale grasping with two-finger grippers. (a) The gripper with parallel fingers. (b) The nonparallel gripper with a "V" configuration.

Figure 3.17(a) shows a gripper that has two parallel fingers with a tip radius of r. To vertically pick up the nano object with a radius of R, the following inequalities should hold:

$$R > r \quad \& \quad {}^tF_f^{\max} > \frac{1}{2}{}^sF_a. \tag{3.62}$$

Table 3.5 Interracial Contact Parameters

Gripper & Nanowire	E_{Si}	ν_{Si}	G_{SiO_2}	ν_{SiO_2}
	160 GPa	0.372	31.4 GPa	0.17
Surface energy	γ_{SiO_2}	γ_{H_2O}	$\Delta\gamma$	
	0.16 J/m^2	0.073 J/m^2	0.034 J/m^2	

During the grasping operation, ${}^tF_f$ can be adjusted by increasing or decreasing ${}^tF_p$ by driving single or dual fingers. Similarly, for a stable vertical release of the nano object, the grasping force can be reduced to generate a certain value of ${}^tF_f^0$ that makes the nano object easily sticky on the substrate with the following inequalities:

$$ {}^tF_f^0 < \frac{1}{2}{}^sF_a \quad \& \quad {}^tF_f^0 \geq {}^tF_f^{\text{adhesion}} \tag{3.63}$$

where ${}^tF_f^{\text{adhesion}}$ is the adhesive friction force derived from ${}^tF_a$ without any external clamping forces.

MEMS grippers have been used to demonstrate vertically aligned carbon nanotube pick-up in the SEM [41, 61]. However, this type of MEMS grippers might meet difficulties when grasping nanotubes attached horizontally to a substrate due to their coarse fingertips, as well as the small gripper-nanotube contact area, which cannot provide a sufficient grasping force (friction forces). This fact is explored in detail in the example of a nanotube nanotweezer evaluation given in the next section.

Fingers with a nonparallel alignment with "V" configuration shown in Fig. 3.17(b) are designed to increase the grasping capability. As for parallel fingers, the following inequalities should hold to pick up the nano object:

$$ R > r \quad \& \quad F_g = ({}^tF_f^{\max}\cos\phi + {}^tF_p\sin\phi) > \frac{1}{2}{}^sF_a. \tag{3.64}$$

Obviously, the tilted angle ϕ makes the grasping stronger as a result of the combined effect of the clamping forces ${}^tF_p$ and the friction forces ${}^tF_f$ to overcome the adhesion forces ${}^sF_a$.

To compare the grasping capabilities of the parallel- and nonparallel-finger gripper, an example of nanowire grasping in air with a cylindrical gripper is simulated. The nanowire is horizontally deposited on a substrate and vertical grasp makes an orthogonal contact between the gripper and the nanowire. Supposing the gripper and the nanowire are made of silicon with SiO_2 coating and with the same radius of 50 nm, ${}^tF_a$ and ${}^tF_f$ can be respectively estimated by Eq. (3.47) and (3.57) with parameters described in Table I.

The simulated grasping force F_g is plotted in Fig. 3.18 as a function of the angle ϕ and the clamping force ${}^tF_p$. The result shows that the fingers with a smaller tilted angle tends to produce a greater grasping force, e. g., with a same clamping force ${}^tF_p = 100$ nN, the nonparallel gripper with tilted angle $\phi = 70°$ produces 1.5 times more grasping force than the parallel gripper.

The release operation of the two-finger gripper is more complicated than the single-finger gripper. Particularly a two-finger gripper with rectangular fingers is

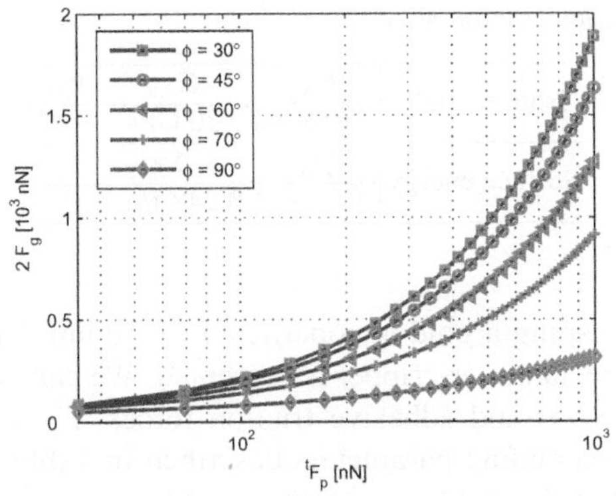

Fig. 3.18 Simulated grasping forces F_g with different clamping angles

used, which produces comparable adhesion forces ${}^tF_a$ and ${}^sF_a$. In contrast, cylindrical fingers may produce smaller adhesion forces, resulting in easier nano object release.

As shown in Fig. 3.19(a), in the first step of the release, the right finger moves to the right to separate from the nano object. In this case, the two opposing adhesion forces from the fingers are canceled each other out and the friction force ${}^sF_f$ is used to hold the nano object. When the right finger has been separated from the nano object, the gripper moves upward in Fig. 3.19(b) for the release operation with the following conditions: ${}^tF_f < {}^sF_a$. Note that in this last step, the nano object may roll with a small angle due to ${}^tM_r$ derived from ${}^tF_f$. However, apart from the release accuracy, this scheme may be effective for the two-finger gripper release operation. Moreover, this scheme can be extended to the nonparallel grippers applications.

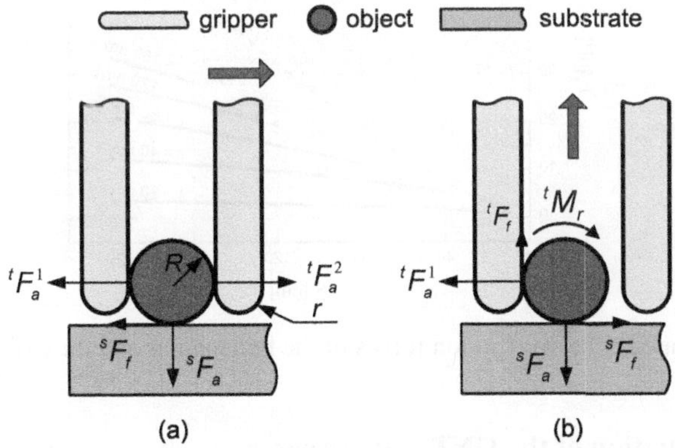

Fig. 3.19 Release operation of the two-finger gripper. (a) Horizontally open the gripper by driving the right finger. (b) Vertically release the nano object.

Table 3.6 Interfacial Contact Parameters

Nanoparticle	E_{Au}	G_{Au}	ν_{Au}
	79.5 GPa	30 GPa	0.42
Surface energy	γ_{Au}	$\Delta\gamma$	
	1.5 J/m^2	0.248 J/m^2	

An example of grasping a gold nanoparticle $R = 50$ nm deposited on a Si substrate with a silicon rectangular gripper is presented. We can calculate ${}^sF_a = {}^tF_a =$ 117 nN using Eq. (3.55) and adhesive friction forces ${}^sF_f = {}^tF_f = 63$ nN using Eq. (3.59) with corresponding parameters described in Table I and II. The calculated results show that the gold nanoparticle can be successively released by the rectangular gripper in air. We can infer from this that nanowire release using this scheme is easier than nanoparticle release because of the stronger adhesion force of the nanowire-substrate contact.

Theoretical calculation verifies that the rectangular gripper can release the nanoparticle. However, the comparable magnitudes of ${}^sF_a$ and ${}^tF_f$ introduces uncertainties into the release process, e.g., the nanoparticle skips from the substrate and sticks to the side of the gripper finger even with weak interferences. From Eq. (3.57), values simulated for adhesive friction forces ${}^tF_f$ with the rectangular and cylindrical silicon grippers are plotted in Fig. 3.20 as a function of the gold nanoparticle with different radii R. It shows that ${}^tF_f$ will be greatly reduced with a cylindrical finger (r in radius) under the same conditions. Moreover, ${}^tF_f$ can be further reduced by using thinner cylindrical grippers.

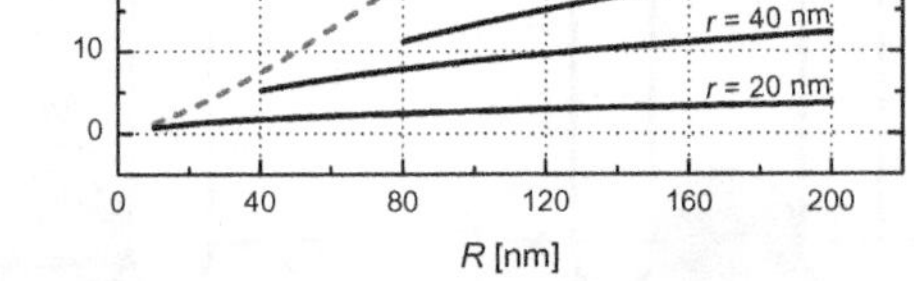

Fig. 3.20 Simulated adhesive friction forces of the rectangular and the cylindrical grippers.

3.3.3.3 Evaluation of the CNT Nanotweezer

For further understanding of the grasping capability of the parallel-finger gripper, the well-know CNT nanotweezer [45, 46] is taken as an example for grasping a gold nanoparticle deposited on a substrate under ambient environment conditions.

For the configuration shown in Fig. 3.21, the dimensions of the CNT nanotweezer are $r = 22.5$ nm and $L_{\mathrm{CNT}} = 5$ μm, and the parameters of the contact interface are shown in Table III. A natural distance $d_0 = 100$ nm between fingers, so a gold nanoparticle radius $R = d_0/2 = 50$ nm are adopted.

To process the hyperstatic electrode-finger-nanoparticle system, as shown in the top inset of Fig. 3.21, a force of constraint ${}^oF_p^1$ is applied on the upper finger that makes the displacement on the end of the CNT finger equal to zero. Assuming that the voltage V applied between the two CNT fingers produces a clamping force q per unit length, ${}^oF_p^1$ is calculated as:

$$ {}^oF_p^1 = \frac{3}{8} q L_{\mathrm{CNT}}. \tag{3.65} $$

Due to a complicated deflection expression, $\Delta d_{\max}$ is assumed to be obtained at $5L_{\mathrm{CNT}}/8$, which is given by:

$$ \Delta d_{\max} = \frac{615 q L_{\mathrm{CNT}}^4}{32768 E_{\mathrm{CNT}} I} \tag{3.66} $$

where E_{CNT} is Young's modulus of the CNT; I, the moment of inertia of the finger, is given by:

$$ I = \frac{\pi r^4}{4}\left[1 - \left(\frac{r_{in}}{r}\right)^4\right] = \pi r^4/4 \quad (r \gg r_{in}) \tag{3.67} $$

where r_{in} and r are respectively the internal and outer radius of the CNT finger.

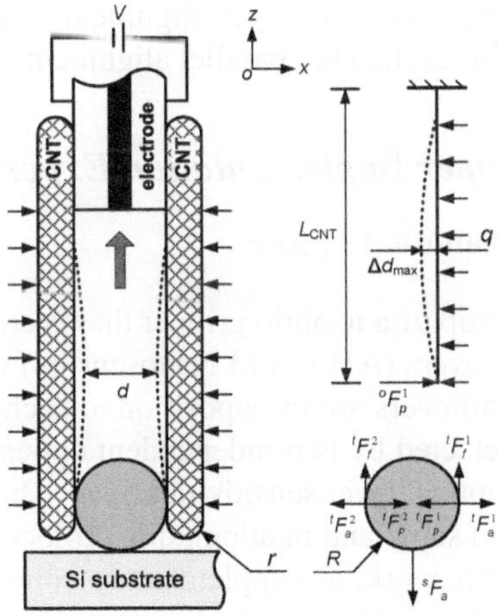

Fig. 3.21 Nanoscale grasping with the CNT nanotweezer.

In ambient conditions, ${}^sF_a = 58.4$ nN and adhesive friction ${}^tF_f = {}^tF_f^1 + {}^tF_f^2 = 16$ nN are computed respectively by Eq. (3.55) and Eq. (3.57). The result shows

Table 3.7 CNT Gripper Grasping Parameters

Gripper	E_{CNT}	G_{CNT}	ν_{CNT}
	1 TPa	430 GPa	0.165
Surface energy	γ_{CNT}	$\Delta\gamma_{(\text{CNT}-\text{Au})}$	$\Delta\gamma_{(\text{SiO}_2-\text{Au})}$
	0.24 J/m^2	0.419 J/m^2	0.248J/m^2

that the grasping inequality Eq. (3.62) is unsatisfied with just the adhesive friction forces. However, the clamping stiffness of the CNT nanotweezer is less than 0.01 N/m, which is too soft to produce sufficient grasping forces to pickup of the gold nanoparticle.

Assuming the CNT grippers are stiff enough to produce enough clamping forces, e.g., ${}^tF_p^1 = {}^tF_p^2 = 249$ nN for friction forces ${}^tF_f^1 = {}^tF_f^2 = 29.2$ nN to break ${}^sF_a$. In this case, the maximum contact stress $p_0 = 8.6$ GPa exceeds the yield stress of gold at the nanoscale (around 5 $\sim$ 6 GPa at 50 nm due to scale effects [62]) and damage the gold nanoparticle. Again, this example proves that main difficulty for nanoscale grasping is the fabrication of sharp end-effectors with enough grasping force output.

As discussed above, the following items are recommended to fabricate a practical two-finger nano gripper:

1. Cylindrical fingers (with a circular or elliptical section) with nanoscale radii are proposed for generating low adhesion forces between the grippers and the nano object for reliable release.
2. Enough clamping stiffness of the Gripper is required for producing sufficient clamping forces.
3. Nonparallel configuration of the finger alignment is recommended for producing more grasping forces than the parallel alignment.

3.3.4 Nanotip Gripper Implementation Experiments

3.3.4.1 Setup of the Nanotip Gripper

Figure 3.22 shows the setup of a nanotip gripper that is comprised of two individually actuated AFM cantilevers (ATEC-FM Nanosensors) with a protruding tip (see the inset). These two cantilevers are in opposition to each other and forces applied on each cantilever are detected by two independent optical levers. The cantilevers' normal stiffness k_b^n and optical lever sensitivities S_n are shown in Table IV. Detailed descriptions of the system setup and manipulation protocols can be seen in our previous research [63]. In this work, as supplementary contents, mechanics analyses, quantitative calculation of grasping limit improvement and interactive force estimation are detailed based on the contact mechanics research addressed in section 3.

Nanoscale grasping with the proposed nanotip gripper benefits from followings:

1. The AFM tip is very tiny (typically with an apex radius of 10 nm or less) compared with the normal size of the nano object to be manipulated, which leads to smaller adhesive forces that favor the release operation.

Table 3.8 Cantilevers Stiffness and optical Lever Sensitivities

Parameters	Cantilever I	Cantilever II
k_b^n	2.72 N/m	2.78 N/m
S_n	0.51 μm/V	0.49 μm/V

2. The tetrahedral (conical shape close to the very end) cantilever tip reduces the adhesion forces of the contacting interface compared with a rectangular gripper, while simultaneously enhancing gripper stiffness. In addition, the tip, tilted at an angle of 63°, can be used to make a nonparallel two-finger gripper that provides a larger grasping force compared with parallel fingers.
3. More importantly, the gripper can be used as a normal AFM to image the nano object, as well as locate the fingers for gripper alignment by scanning the end of one cantilever tip and the nano object with the other cantilever. This function makes it possible to perform nanoscale grasping without visual feedback (normally SEM or TEM vision) in ambient conditions.

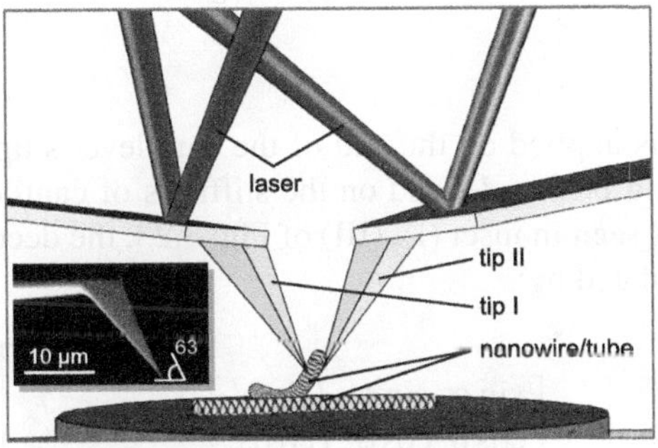

Fig. 3.22 Nonparallel configuration of the nanotip gripper.

3.3.4.2 Analysis of Grasping Mechanics

Figure 3.23 shows a schematic diagram of the analysis of the mechanics of a cantilever used as a gripper finger. In experiments, a cantilever's beam length L, beam width w and tip length l are measured under an optical microscope. The beam thickness t is determined using the forced oscillation method [64]. Thus, the normal stiffness k_b^n and lateral stiffness k_b^l of the cantilever's beam are calculated by:

$$k_b^n = \frac{Ewt^3}{4L^3}, \quad k_b^l = \frac{Gwt^3}{3L(l\sin\phi')^2} \tag{3.68}$$

where E and G are, respectively, the Young's and Shear modulus of the cantilever, $\phi' = 60°$ is the tilted angle through the rotation axis of the tip relative to the substrate (as shown in Fig. 3.24(a)).

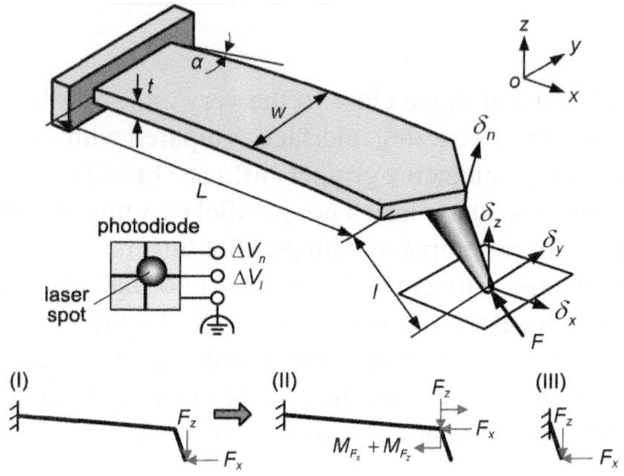

Fig. 3.23 Analysis of the mechanics of the gripper finger (AFM cantilever) during a grasping operation.

When a force F is applied on the end of the cantilever's tip, it moves with displacements δ_x, δ_y and δ_z that depend on the stiffness of cantilever on each axis in the defined fame. As seen in inset (I)–(III) of Fig. 3.23, the decoupled displacement on each axis is calculated by:

$$\begin{bmatrix} \delta_x \\ \delta_z \end{bmatrix} = \begin{bmatrix} \sin\alpha \ \sin\phi' \ l\sin\phi' \\ \cos\alpha \ \cos\phi' \ l\cos\phi' \end{bmatrix} \begin{bmatrix} d_b^{F_x} + d_b^{F_z} \\ d_t^{F_x} + d_t^{F_z} \\ \theta_b^{F_x} + \theta_b^{F_z} \end{bmatrix}, \tag{3.69}$$

$$\delta_y = F_y \left(\frac{1}{k_b^l} + \frac{1}{k_t} \right) \tag{3.70}$$

where α is the mounting angle of the cantilever, θ is the angular deflection of the beam and d is deflection of cantilever's beam and tip (respectively labeled by subscripts b and t, and with superscripts of force F_x and F_z), k_t is the stiffness of the cantilever's tip. The calculation of the deflections is detailed in Appendix B.

The tip close to its very end is assumed to be symmetrical in shape, thus, it has the same stiffness on each axis: $k_t \approx 20$ N/m simulated by the finite element method. From Eq. (3.70), the overall lateral stiffness of the cantilever is about 19 N/m, which makes the tip alignment more stable during the grasping operation.

3.3.4.3 Grasping Capabilities

Figure 3.24(a) shows a force simulation for a nano object (with a circular section, e.g., nanowires or nanoparticles) grasp operation using the proposed nanotip. Equations can be obtained for a static equilibrium:

$$\begin{cases} F_x = ({}^tF_p \sin\phi - {}^tF_f \cos\phi) \\ F_z = ({}^tF_p \cos\phi + {}^tF_f \sin\phi) = \frac{1}{2}{}^sF_a \end{cases} \tag{3.71}$$

where $\phi = 68°$. ${}^tF_f$ can be calculated by (24):

$${}^tF_f = \tau\pi \left[\frac{3({}^tF_p + {}^tF_a)R_e}{4E^*}\right]^{2/3} [F_1(e)]^2 \tag{3.72}$$

where ${}^tF_a = {}^oF_a$ is the adhesion force that is estimated by Eq. (3.47). ${}^tF_p$ is significant for estimating the maximum stress on the contacting area using Eq. (3.45) to avoid damage on both the gripper and the nano object.

The normal force F_z and lateral force F_l applied on the end of the tip can be calculated from voltage output ΔV_n and ΔV_l of optical levers by:

$$k_b^n \delta_n = k_b^n (d_b^{F_x} + d_b^{F_z}) = \Delta V_n S_n, \tag{3.73}$$

$$F_y = \Delta V_l S_l \tag{3.74}$$

where δ_n is the normal displacement on the end of the cantilever beam, S_n and S_l are, respectively, the normal and lateral sensitivities of the optical lever. When ΔV_n is detected and ${}^tF_a$ is calculated by Eq. (3.47), F_x and F_z can be then calculated by Eq. (3.71)–(3.73).

Assuming no preload is applied before grasp, as shown in Fig. 3.24(b), for a successful grasping, the angle η as well as the dig-in distance ξ should be positive. Their relation is given by:

$$\xi = R(1 - \cos\eta) = R\left[1 - \sqrt{1 - \frac{(R-r)^2}{(R+r)^2}}\right]. \tag{3.75}$$

However, during the pickup, when the grasping force is not enough to break the adhesion force, the gripper will slide upward along the surface of the nano object due to the beam and tip deflections, resulting in a smaller value of ξ'. The grasp is probably lost as ξ is reduced to zero and slide further resulting in complete grasp failure. The minimum radius of the nano object $R_{\min}$ can be estimated by assuming that $\delta_x^{\max} \leq \xi$ with known ${}^sF_a$, parameters of the gripper and the contacting interface, e.g., Young's modulus, shear modulus and surface energy. $R_{\min}$ can be estimated from the following procedure:

1. Calculate the adhesion force ${}^sF_a$ with a pre-estimated size limit R_0, then calculate corresponding ${}^tF_a$ and ${}^tF_f$ using Eq. (3.71) and (3.72).

Table 3.9 Nanotip Gripper Parameters

k_b^n	k_t	L	l	r	α	${}^sF_a$
4 N/m	20 N/m	250 μm	10 μm	8 nm	5°	100 nN

2. When F_x and F_z are calculated by Eq. (3.71), $R_{\min}$ can be then computed using Eq. (3.69) and (3.75). Then calculate a new estimation $R_1 = (R_0 + R_{\min})/2$ and repeat the steps with R_1 until the difference between the estimation R_1 and the calculated $R_{\min}$ in successive iterations is less than the predefined $\Delta R = 0.5$ nm.

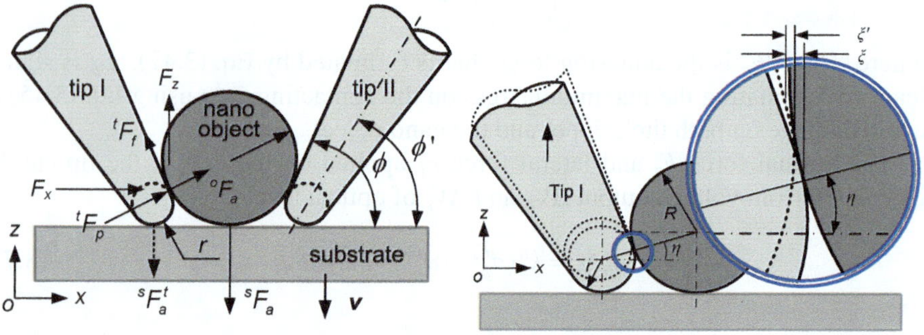

Fig. 3.24 (a) Force simulation using the nonparallel two-tip gripper (b) Simulation of grasping limit on the size of the nano object

For the case of silicon nanowire (SiNW) grasp with gripper parameters defined in Table V, the final value of $R_{\min} = 19.7$ takes no account of contributions of F_x and k_t is in accordance with the result estimated in [63]. Considering $k_t = 20$ N/m, $R_{\min}$ increases to 21.5 nm, and reaches 34.3 nm when adding the tip deflection deduced by F_x. The results show that F_x is one of the significant factors on determining the minimum size of the nano object that can be grasped.

To improve the grasping limit, cantilevers with larger k_b^n and k_t can be used to effectively reduce $R_{\min}$. In addition, if a preload of ${}^tF_p$ is applied before grasping, the gripper will hold the SiNW more tightly producing a stronger ${}^tF_f$, thereby significantly reducing $R_{\min}$. The simulation in Fig. 3.25(a) shows that the grasping limit is improved by increasing the cantilever stiffness. The simulated results indicate that a stiffer tip is more effective than a stiffer beam for decreasing $R_{\min}$. The former provides a lower limit of 21.5 nm and the later 28.9 nm.

Figure 3.25(b) shows the grasping limit plotted as a function of the preload ${}^tF_p$. The grasping limit can theoretically equal the radius of the tip apex with a proper preload. However, when the radius of the nano object decreases to less than 17.6 nm, the nano object becomes difficult to grasp because of the contact with the tip apex (sphere-sphere contact). Thus, in this case, $R_{\min}$ reaches the limit of 17.6 nm, while

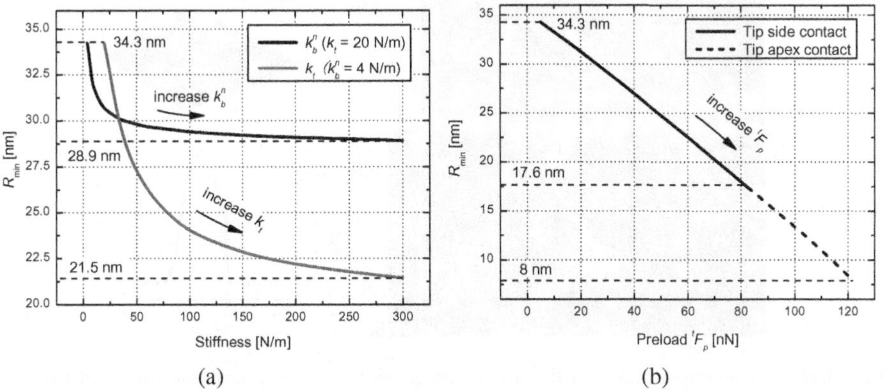

Fig. 3.25 Improving grasping limit by (a) increasing stiffness of the cantilever beam k_b^n or tip k_t and (b) ^tF_p preloading.

$^tF_p = 83$ nN. However, the preloading involves great risks in damaging the tips as well as the nano objects. In this case, the maximum stress p_0 on the contact area should be guaranteed less than the yield stress of the contact before the preloading.

In a word, selecting stiffer cantilevers to build a gripper and preloading a proper clamping force are two effective means for improving grasping capabilities.

3.3.4.4 Pick-and-Place Silicon Nanowires

In experiments, silicon nanowires (SiNWs) were deposited on a freshly cleaned silicon wafer coated with 300 nm silicon dioxide. A pre-scanned image (8 μm $\times$ 8 μm) is shown in Fig. 3.26, which includes the topographic image of SiNWs, and the local image of tip II (see the zoomed inset). A grasping location on the left SiNW is marked A–A, where the SiNW has a height of 153 nm. The left SiNW will be transported to the target position and released onto the right SiNW to build a nano crossbar.

Figure 3.27 shows an example of contact detection with tip II: Seen as icons in the graph, the tip starts to dig into the root of the SiNW as the tip makes contact with the SiNW. Further movement leads to pushing the SiNW without any obvious change in the voltage output. During retraction, after contact with the substrate breaks, the bending force sharply reaches a positive peak with a similar response to that of the approach. Eventually, the tip pulls off the SiNW and reaches zero.

When the tip digs into the SiNW, the cantilever produces a pre-grasping force $\Delta F = 27$ nN with a voltage difference of about 20 mV. The corresponding preload on the tip II is estimated as $^tF_p \approx 26$ nN using Eq. (3.71)–(3.73).

Figure 3.27 shows the curve of the peeling force spectroscopy on Tip II during the pick-and-place manipulation of the same SiNW. The curve starts from contact state between the nanotweezer, the SiNW and the substrate. As the gripper is moved up to

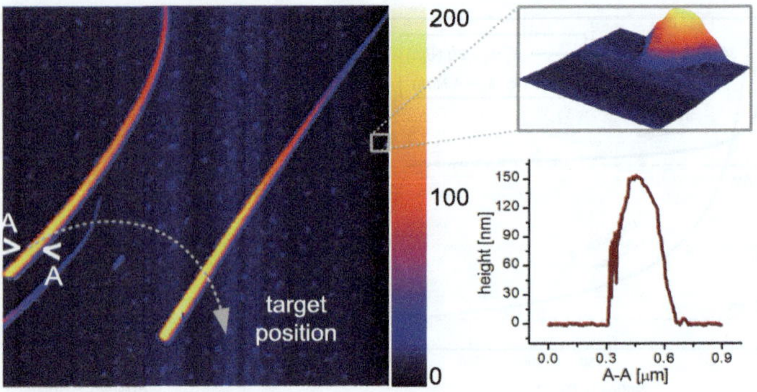

Fig. 3.26 Pre-scanned image of the SiNW. Insets show a 3D topographic image of the tip II and the nanowire height at location A–A.

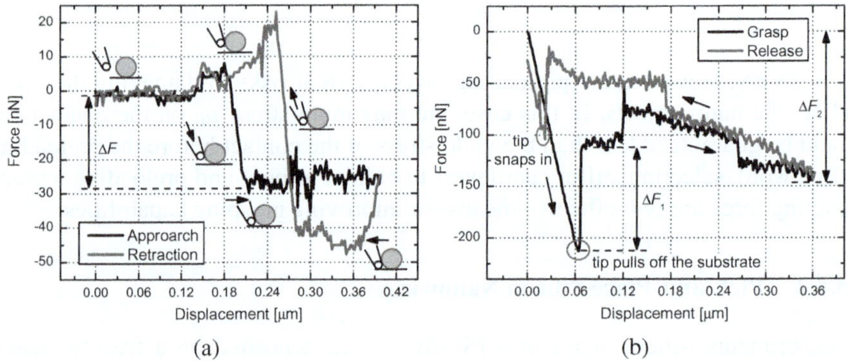

Fig. 3.27 (a) Contact detection by normal force sensing on tip II (b) Force detection on Tip II during the grasp and release operation

pick up the SiNW, the cantilever is bent downwards creating negative forces until the cantilever pulls off the substrate with a voltage difference of 75 mV indicating a pull-off force $\Delta F_1 = 103$ nN. As the gripper is moved up further, the force magnitude gradually keeps increasing with the SiNW peeling force responses. Retraction leads to a continuous decrease except for a weak fluctuation at 178 nm. Snap-in occurs at 25 nm after a mild force decrease. With even further retraction, the magnitude of the normal force of Tip II approaches the prior state before grasping.

During pickup, the maximum SiNW peeling force occurs at retraction start, where the voltage is about -105 mV indicating a grasping force of $\Delta F_2 = 144$ nN. At this point, from calculation, ${}^tF_f \approx 31$ nN that is much smaller than the SiNW-substrate adhesion force, and ${}^tF_p \approx 218$ nN that generates a maximum contact stress $p_0 = 7.1$ Gpa with $R \approx 19.5$ nm at the contact location of 55 nm from the tip end. Fortunately, this contact stress is still below the yield stress of the silicon at the nanoscale (around 12 $\sim$ 13 GPa at the nanoscale) due to size effects on the hardness [54].

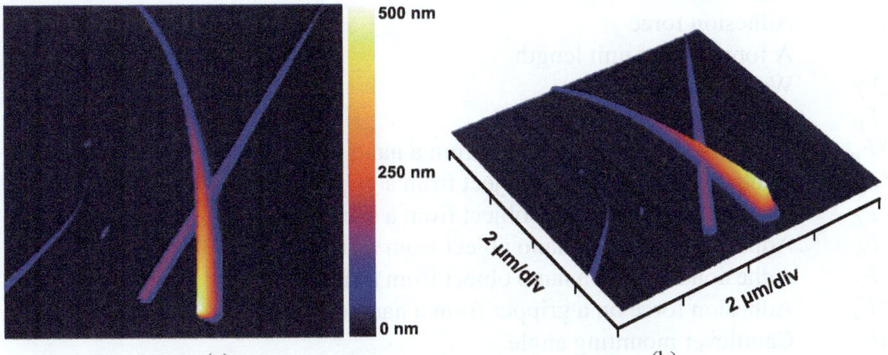

(a) (b)

Fig. 3.28 Pick-and-place manipulation results for the SiNWs. (a) A post-manipulation image verifies that the manipulated SiNW is piled on another SiNW. (b) 3D topographic image of the manipulation result.

The post-manipulation image in Fig. 3.28 verifies that the SiNW has been successfully transported and piled on another SiNW, building a nanocrossbar with a maximum height about 500 nm. During the pick-and-place manipulation, once the SiNW was reliably grasped, the gripper moved up 800 nm at a velocity of 80 nm/s, then the SiNW was transported a distance of 4.05 μm on the X-axis at a velocity of 150 nm/s and 1.95 μm on the Y-axis at a velocity of 72 nm/s.

As a result, the theoretical analysis and the experimental results validate the schemes for nanoscale grasping enabling more practical 3-D nanomanipulation and nanoassembly, such as constructing 3-D nano-structures, building 3-D nano-devices, and revealing nano-mechanic phenomena.

3.3.5 List of Selected Symbols

Symbol	*Description*
r	Radius of end of a gripper's finger
R	Radius of a nano object being manipulated
S	Contact area
p_0	Maximum pressure on a contact area
δ	Contact penetration
L	Cantilever beam length
w	Cantilever beam width
t	Cantilever beam thickness
l	Cantilever tip length
E	Young's modulus

G	Shear modulus
E^*	Combined elastic modulus
τ	Effective friction coefficient
P	External load
F_s	Adhesion force
q	A force on per unit length
$\Delta\gamma$	Work of adhesion
${}^tF_p$	Clamping force of a gripper
${}^oF_p$	Repulsive force on a gripper from a nano object
${}^tF_f$	Friction force on a nano object from a gripper
${}^sF_f$	Friction force on a nano object from a substrate
${}^tF_a$	Adhesion force on a nano object from a gripper
${}^sF_a$	Adhesion force on a nano object from a substrate
${}^oF_a$	Adhesion force on a gripper from a nano object
α	Cantilever mounting angle
ϕ	Tilted angle of a tip through its front edge
ϕ'	Tilted angle of a tip through its rotation axis
ψ	Inclined angle between axes of contact surfaces
η	Effective grasping angle relative to the substrate
ξ	Effective grasping distance
k_b^n	Normal stiffness of a cantilever's beam
k_t	Stiffness of the cantilever's tip
S_n	Normal sensitivity of a optical lever
S_l	Lateral sensitivity of a optical lever

3.3.6 Deflections on the Cantilever

Deflections on the cantilever's beam and tip can be calculated by:

$$d_b^{F_x} = \frac{F_x}{k_b^n}\left(\sin\alpha + \frac{3l\sin\phi'}{2L}\right) \tag{3.76}$$

$$d_b^{F_z} = \frac{F_z}{k_b^n}\left(\cos\alpha + \frac{3l\cos\phi'}{2L}\right) \tag{3.77}$$

$$d_t^{F_x} = \frac{F_x\sin\phi'}{k_t} \tag{3.78}$$

$$d_t^{F_z} = \frac{F_z\cos\phi'}{k_t} \tag{3.79}$$

$$\theta_b^{F_x} = \frac{3F_x}{k_b^n L}\left(\frac{\sin\alpha}{2} + \frac{l\sin\phi'}{L}\right) \tag{3.80}$$

$$\theta_b^{F_z} = \frac{3F_z}{k_b^n L}\left(\frac{\cos\alpha}{2} + \frac{l\cos\phi'}{L}\right) \tag{3.81}$$

3.4 Conclusion

In the first part of the chapter, several models have been summarized, from surface forces to contact forces. In the second part of this chapter, novel techniques for tribological characterization of nano-scale object-substrate interfaces are proposed. By pushing micro/nano-objects or attaching them on the nano-probe, frictional forces can be directly or indirectly estimated, and sliding, spinning, or rolling frictional parameters can be estimated in any environment. These methods could be directly applied for other nano-materials, e.g. carbon nano-tubes, nano-crystals, DNA, nanowires, etc., for their frictional characterization on various substrates with different pushing speeds and environmental conditions. However, modeling of the contact and pushing mechanics for these various geometries is needed to be improved. Besides of characterization applications, these measurements would open new controlled means of understanding nano-scale friction by observing the effects of lattice mismatch between the nano-scale object and substrate, surface roughness, temperature, speed, humidity, contact mechanics, chemical interactions, etc.

In order to understand the interactive phenomena between a gripper and a nano object, contact mechanics were modeled for different cylinder-cylinder, cylinder-sphere, flat surface-sphere and flat surface-cylinder contact surfaces. Contact modelling made it easy to estimate the interfacial adhesion forces, deduced contact friction forces and contact stress, thereby providing a theoretical analysis for the gripper design and task planning for successful nanoscale grasping. To further improve our understanding, grasping strategies with one-finger and two-finger grippers were discussed. The analysis shows that the gripper with a nonparallel configuration has better grasping capabilities than the parallel configuration. A homemade gripper constructed from two AFM cantilevers with a "V" configuration was introduced. The grasping capabilities of the proposed nanotip gripper were analyzed in detail and ways for improving the grasping limitation were presented. Finally, the nanotip gripper's capabilities were validated by a successful pick-and-place manipulation of silicon nanowires to build a nano crossbar.

References

1. Schafer, D., Reifenberger, R., Patil, A., Andres, R.: Fabrication of two-dimensional arrays of nanometer-size clusters with the Atomic Force Microscope. Appl. Physics Letters 66, 1012–1014 (1995)
2. Junno, T., Deppert, K., Montelius, L., Samuelson, L.: Controlled manipulation of nanoparticles with an Atomic Force Microscopy. Appl. Physics Letters 66, 3627–3629 (1995)
3. Stark, R.W., Thalhammer, S., Wienberg, J., Heckl, W.M.: The AFM as a tool for chromosomal dissection - the influence of physical parameters. Appl. Phys. A 66, 579–584 (1998)
4. Sitti, M., Hashimoto, H.: Controlled pushing of nanoparticles: Modeling and experiments. IEEE/ASME Trans. on Mechatronics 5(2), 199–211 (2000) 107
5. Sitti, M., Hashimoto, H.: Tele-nanorobotics using Atomic Force Microscope as a robot and sensor. Advanced Robotics 13(4), 417–436 (1999)

6. Sitti, M., Aruk, B., Shintani, K., Hashimoto, H.: Scaled teleoperation system for nanoscale interaction and manipulation. Advanced Robotics 17(3), 275–291 (2003)
7. Falvo, M., Taylor II, R.M., Helser, A., Chi, V., Brooks, F.P., Washburn, S., Superfine, R.: Nanometer-scale rolling and sliding of carbon nanotubes. Nature 397, 236–238 (1999)
8. Falvo, M., Steele, J., Taylor II, R.M., Superfine, R.: Gearlike rolling motion mediated by commensurate contact: Carbon nanotubes on HOPG. Phys. Review B 62, R10665–667 (2000)
9. Sitti, M., Hashimoto, H.: Pushing micro/nanoscale particles on substrates using Atomic Force Microscope probe towards tribological characterization of particle-substrate interfaces. In: STM 1999 Conference, Korea, pp. 385–386 (July 1999)
10. Meyer, E., Overney, R., Howald, L., Luthi, R., Frammer, J., Guntherodt, H.J.: Friction and wear of Langmuir-Blodgett films observed by Frictional Force Microscopy. Physical Review Letters 69, 1777–1780 (1992)
11. Dedkov, G.: Experimental and theoretical aspects of the modern nanotribology. Phys. Stat. Sol. 179(3), 3–35 (2000) 99
12. Falvo, M., Superfine, R.: Mechanics and friction at the nanometer scale. J. of Nanoparticle Research 2, 237–248 (2000) 99
13. Mailhot, B., Komvopoulos, K., Ward, B., Tian, Y., Somorjai, G.: Mechanical and friction properties of thermoplastic polyurethanes determined by Scanning Force Microscopy. J. of Applied Physics 89, 5712–5719 (2001) 99
14. Bhushan, B.: Handbook of micro/nano tribology, 2nd edn. CRC Press, Boca Raton (1999) 99, 101
15. Bouhacina, T., Desbat, B., Aime, J.: FTIR spectroscopy and nanotribological comparative studies: Influence of the absorbed water layers on the tribological behavior. Tribology Letters 9, 11–117 (2000) 101
16. Heim, L., Blum, J., Preuss, M., Butt, H.: Adhesion and friction forces between spherical micrometer-sized particles. Phys. Rev. Lett. 83, 3328–3331 (1999) 101
17. Brilliantov, N.V., Poschel, T.: Rolling as a continuing collision. Eur. Phys. J. B 12, 299–301 (1999) 101
18. Butt, H.: A technique for measuring the force between a colloidal particle in water and a bubble. Journal of Colloid and Interface Science 166, 109–117 (1994) 102
19. Sitti, M., Hashimoto, H.: Tele-touch feedback of surfaces at the nanoscale: Modeling and experiments. IEEE/ASME Trans. on Mechatronics 8(2) (June 2003) (to appear) 103
20. Israelachvili, J.: Intermolecular and Surface Forces, 2nd edn. Academic Press, London (1992) 104
21. Sitti, M., Hashimoto, H.: Two-dimensional fine particle positioning under optical microscope using a piezoresistive cantilever as a manipulator. Journal of Micromechatronics 1(1), 25–48 (2000) 101, 106
22. Dedkov, G.: Friction on the nanoscale: New physical mechanisms. Materials Letters 38, 360–366 (1999) 101
23. Meyer, E., Overney, R.M., Dransfeld, K., Gyalog, T.: Nanoscience: Friction and Rheology on the Nanometer Scale. World Scientific Pub., Singapore (1998) 101
24. Sitti, M., Bozma, I., Denker, A.: Visual tracking for moving multiple objects: Integration of vision and control. In: IEEE Int. Symp. on Industrial Electronics, pp. 535–540 (July 1995) 106
25. Sitti, M.: Teleoperated 2-D micro/nanomanipulation using Atomic Force Microscope. Ph.D. Thesis, Dept. of Electrical Engineering, University of Tokyo, Tokyo (September 1999) 107

26. Autumn, K., Sitti, M., Liang, Y.A., Peattie, A., Hansen, W., Sponberg, S., Kenny, T., Fearing, R., Israelachvili, J., Full, R.J.: Evidence for van der Waals adhesion in gecko setae. Proc. of the National Academy of Science 99(19), 12252–12256 (2002) 107
27. Mate, C.M.: Force Microscopy studies of the molecular origin of friction and lubrication. IBM J. Res. Develop. 39, 617–627 (1995) 108
28. Tanikawa, T., Arai, T.: Development of a micro-manipulation system having a two-fingered micro-hand. IEEE Trans. Robot. Autom. 15(1), 152–162 (1999)
29. Driesen, W., Varidel, T., Régnier, S., Breguet, J.M.: Micromanipulation by adhesion with two collaborating mobile micro robots. J. Micromech. Microeng. 15(10), S259–S267 (2005)
30. Xie, H., Rong, W.B., Sun, L.N.: A flexible experimental system for complex microassembly under microscale force and vision-based control. Int. J. Optomechatro. 1(1), 80–102 (2007)
31. Walle, B.L., Gauthier, M., Chaillet, N.: Principle of a submerged freeze gripper for microassembly. IEEE Trans. Robot. 24(4), 897–902 (2008)
32. Kim, K.Y., Liu, X.Y., Zhang, Y., Sun, Y.: Nanonewton force-controlled manipulation of biological cells using a monolithic MEMS microgripper with two-axis force feedback. J. Micromech. Microeng. 18(5), 055013 (2008)
33. Chen, T., Chen, L.G., Sun, L.N., Li, X.X.: Design and Fabrication of a Four-Arm-Structure MEMS Gripper. IEEE Trans. Ind. Electron. 56(4), 996–1004 (2008)
34. Xie, H., Régnier, S.: Three-dimensional automated micromanipulation using a nanotip gripper with multi-feedback. J. Micromech. Microeng. 19(7), 75009 (2009)
35. Menciassi, A., Eisinberg, A., Izzo, I., Dario, P.: From 'macro' to 'micro' manipulation: models and experiments. IEEE/ASME Trans. Mechatro. 9(2), 311–320 (2004)
36. Resch, R., Lewis, D., Meltzer, S., Montoya, N., Koel, B.E., Madhukar, A., Requicha, A.A.G., Will, P.: Manipulation of gold nanoparticles in liquid environments using scanning force microscopy. Ultramicroscopy 82(1-4), 135–139 (2000)
37. Sitti, M.: Atomic force microscope probe based controlled pushing for nanotribological characterization. IEEE/ASME Trans. Mechatron. 9(2), 343–349 (2004)
38. Chen, H., Xi, N., Li, G.Y.: CAD guided automated nanoassembly using automic force microscopy-based nanorobotics. IEEE Trans. Auto. Sci. Eng. 3(3), 208–217 (2006)
39. Bordag, M., Ribayrol, A., Conache, G., Fröberg, L.E., Gray, S., Samuelson, L., Montelius, L., Pettersson, H.: Shear stress measurements on InAs nanowires by AFM manipulation. Small 3(8), 1398–1401 (2007)
40. Fukuda, T., Arai, F., Dong, L.X.: Assembly of nanodevices with carbon nanotubes through nanorobotic manipulations. Proc. IEEE 91(11), 1803–1818 (2003)
41. Andersen, K.N., Petersen, D.H., Carlson, K., Mølhave, K., Sardan, O., Horsewell, A., Eichhorn, V., Fatikow, S., Bøggild, P.: Multimodal Electrothermal Silicon Microgrippers for Nanotube Manipulation. IEEE Trans. Nanotech. 8(1), 76–85 (2009) 120
42. Bosanac, L., Aabo, T., Bendix, P.M., Oddershede, L.B.: Efficient optical trapping and visualization of silver nanoparticles. Nano Letters 8(5), 1486–1491 (2008)
43. de Vries, A.H.B., Krenn, B.E., van Driel, R., Kanger, J.S.: Micro magnetic tweezers for nanomanipulation inside live cells. Biophysical J. 88(3), 2137–2144 (2005)
44. Sitti, M., Aruk, B., Shintani, H., Hashimoto, H.: Scaled teleoperation system for nanoscale interaction and manipulation. Advanced Robotics 17(3), 275–291 (2003)
45. Kim, P., Lieber, C.M.: Nanotube nanotweezers. Science 286(5447), 2148–2150 (1999) 122
46. Jang, J.E., Cha, S.N., Choi, Y., Kang, D.J., Hasko, D.G., Jung, J.E., Kim, J.M., Amaratunga, G.A.J.: A nanogripper employing aligned multiwall carbon nanotubes. IEEE Trans. Nanotech. 7(4), 389–393 (2008) 122

47. Hertz, H.: On the contact of elastic solids. J. Reine Angew. Math. 92, 156–171 (1881) 110
48. Riedo, E., Brune, H.: Young modulus dependence of nanoscopic friction coefficient in hard coatings. App. Phys. Lett. 83(10), 1986 (2003) 111
49. Ritter, C., Heyde, M., Stegemann, B., Rademann, K.: Contact-area dependence of frictional forces: Moving adsorbed antimony nanoparticles. Phys. Rev. B 71(8), 085405 (2005) 111
50. Johnson, K.L.: Contact Mechanics. Cambridge University Press, Cambridge (1985) 114
51. Israelachvili, J.N.: Intermolecular and surface forces. Academic, New York (1992) 115
52. Chaudhurya, M.K., Weaver, T., Hui, C.Y., Kramer, E.J.: Adhesive contact of cylindrical lens and a flat sheet. J. Appl. Phys. 80(1), 30–37 (1996) 116
53. Johnson, K.L., Kendall, K., Roberts, A.D.: Surface energy and the contact of elastic solids. Proc. R. Soc. Lond. A 324, 301–313 (1971) 116
54. Bhushan, B.: Springer handbook of nanotechnology. Springer, Heidelberg (2004) 117, 131
55. Dedkov, G.: Friction on the nanoscale: New physical mechanisms. Materials Letters 38(5), 360–366 (1999) 117
56. Whitman, L.J., Stroscio, J.A., Dragoset, R.A., Cellota, R.J.: Manipulation of adsorbed atoms and creation of new structures on room-temperature surfaces with a scanning tunneling microscope. Science 251(4998), 1206–1210 (1991) 118
57. Avouris, P.: Manipulation of matter at the atomic and molecular-levels. Accounts. Chem. Res. 28(3), 95–102 (1995) 118
58. Rief, M., Oesterhelt, F., Heymann, B., Gaub, H.E.: Single molecule force spectroscopy on polysaccharides by atomic force microscopy. Science 275(5304), 1295–1297 (1997) 118
59. Strunz, T., Oroszlan, K., Schäfer, R., Güntherodt, H.J.: Dynamic force spectroscopy of single DNA molecules. PNAS 96(20), 11277–11282 (1999) 118
60. Manohar, S., Mantz, A.R., Bancroft, K.E., Hui, C.Y., Jagota, A., Vezenov, D.V.: Peeling single-stranded DNA from graphite surface to deter-mine oligonucleotide binding energy by force spectroscopy. Nano Lett. 8(12), 4365–4372 (2008) 118
61. Molhave, K., Wich, T., Kortschack, A., Boggild, P.: Pick-and-place nanomanipulation using microfabricated grippers. Nanotechnology 17(10), 2434–2441 (2006) 120
62. Wu, B., Heidelberg, A., Boland, J.J.: Mechanical properties of ultrahigh-strength gold nanowires. Nature Materials 4(7), 525–529 (2005) 124
63. Xie, H., Haliyo, D.S., Régnier, S.: A versatile atomic force microscope for three-dimensional nanomanipulation and nanoassembly. Nanotechnology 20(21), 215301 (2009) 124, 128
64. Xie, H., Vitard, J., Haliyo, D.S., Régnier, S.: Enhanced accuracy of force application for AFM nanomanipulation using nonlinear calibration of optical levers. IEEE Sensor J. 8(8), 1478–1485 (2008) 125
65. Adamson, A.W., Gast, A.P.: Physical Chemistry of Surfaces, 6th edn. John Wiley and Sons, Chichester (1997) 89, 90, 91, 92
66. Arai, F., Ando, D., Fukuda, T., Nonoda, Y., Oota, T.: Micro manipulation based on micro physics. In: Proc. of IEEE/RSJ Conf. on Intelligent Robots and Systems, Pittsburgh, vol. 2, pp. 236–241 (1995) 90
67. Arai, F., Fukuda, T.: A new pick up and release method by heating for micromanipulation. In: IEEE Workshop on Micro Electro Mechanical Systems, Nagoya, Japan, pp. 383–388 (1997) 89
68. Bark, K.-B.: Adhsives Greifen von kleinen Teilen mittels niedrigviskoser Flssigkeiten. Springer, Heidelberg (1999)

69. Bastin, N., Chau, A., Delchambre, A., Lambert, P.: Effects of relative humidity on capillary force and applicability of these effects in micromanipulation. A Poster at the Proc. of the IEEE/RSJ 2008 International Conference on Intelligent RObots and Systems, Nice, September 22-26 (2008)
70. Belaidi, S., Girard, P., Leveque, G.: Electrostatic forces acting on the tip in atomic force microscopy: modelization and comparison with analytic expressions. J. Appl. Phys. 81(3), 1023–1029 (1997) 96, 97
71. Benmayor, B.: Dimensional Analysis and Similitude in Microsystem Design and Assembly. PhD thesis, Ecole Polytechnique Fdrale de Lausanne (2000) 89
72. Berge, B.: Electrocapillarit et mouillage de films isolants par l'eau. Acadmie des sciences 317(II), 157 (1993)
73. Berge, B., Peseux, J.: Variable focal lens controlled by an external voltage: An application of electrowetting. Eur. Phys. J. E. 3, 159–163 (2000)
74. Biganzoli, F., Fassi, I., Pagano, C.: Development of a gripping system based on capillary force. In: Proceedings of ISATP 2005, Montreal, Canada, July 19-21, pp. 36–40 (2005)
75. Borno, R.T., Steinmeyer, J.T., Maharbiz, M.M.: Transpiration actuation: the design, fabrication and characterization of biomimetic mciroactuators driven by the surface tension of water. J. Micromech. Microeng. 16, 2375–2383 (2006)
76. Allen Bowling, R.A.: A theoretical review of particle adhesion. In: Proc. of Symposium on particles on surfaces 1: Detection, Adhesion and Removal, San Francisco, pp. 129–142 (1986) 88, 89, 91, 95, 96, 97
77. Butt, H.-J., Cappella, B., Kappl, M.: Force measurements with atomic force microscope: Technique, interpretation and applications. Surf. Sci. Rep. 59, 1–152 (2005) 96, 97
78. Cappella, B., Dietler, G.: Force-distance curves by atomic force microscopy. Surf. Sci. Rep. 34, 1–104 (1999) 96, 97
79. Casimir, H.B.C., Polder, D.: The influence of retardation of the London - van der Waals forces. Phys. Rev. 73(4), 360 (1948) 90
80. Chau, A.: Theoretical and experimental study of capillary condensation and of its possible use in micro-assembly - Etude thorique et exprimentale des forces dues la condensation capillaire. PhD thesis, Universit libre de Bruxelles (2007)
81. Colchero, J., Gile, A., Bar, A.M.: Resolution enhancement and improved data interpretation in electrostatic force microscopy. Phys. Rev. B 64, 1–11 (2001)
82. Dafflon, M.: Prhenseurs, conditions et stratgies pour une micromanipulation de prcision. PhD thesis, EPFL (2008) 91
83. de Gennes, P.-G., Brochart-Wyard, F., Qur, D.: Gouttes, bulles, perles et ondes, Belin (2002)
84. Derjaguin, B.V., Muller, V.M., Toporov, Y.P.: Effect of contact deformations on the adhesion of particles. J. Colloid Interface Sci. 53(2), 314–326 (1975) 88, 97
85. Durand, E.: Electrostatique: Tome II Problmes gnraux. Conducteurs, Masson (1966) 95
86. Fearing, R.S.: Survey of sticking effects for micro parts handling. In: Proc. of IEEE/RSJ Conf. on Intelligent Robots and Systems, August 5-9, pp. 212–217 (1995) 88, 89, 95, 96, 97
87. Feddema, J.T., Xavier, P., Brown, R.: Micro-assembly planning with van der Waals force. In: Proc. of IEEE Int. Symposium on Assembly and Task Planning, Porto, pp. 32–38 (July 1999) 89, 91
88. Feddema, J.T., Xavier, P., Brown, R.: Micro-assembly planning with van der waals force. Journal of Micromechatronics 1(2), 139–153 (2001) 91
89. Feng, Y., Zhou, Z., Ye, X., Xiong, J.: Passive valves based on hydrophobic microfluidics. Sensors and Actuators A 108, 138–143 (2003)

90. Good, R.J.: Theory of adhesive forces accross interfaces: 1. the Lifshitz - van der Waals component of interaction and adhesion. In: Lee, L.-H. (ed.) Fundamentals of Adhesion. Plenum Press, New York (1991) 89
91. Grutzeck, H.: Fluidisches Greifen in der Mikrosystemtechnik. PhD thesis, TU Cottbus (1999)
92. Gutowsky, W.: Thermodynamics of adhesion. In: Lee, L.H. (ed.) Fundamentals of Adhesion. Plenum Press, New York (2001)
93. Hao, H.W., Baro, A.M., Saenz, J.J.: Electrostatic and contact forces in force microscopy. J. Vacuum Sci. Technol. B 9(2), 1323–1328 (1991) 96, 97, 98
94. Harper, W.: Contact and frictional electrification. Clarenton Press, Oxford (1967) 94
95. Van Hauwermeiren, R.: Cours d'lectricit gnrale. Cours de 2me Candidature, ULB (1993)
96. Hendriks, B.H.W., Kuiper, S., Van As, M.A.J., Renders, C.A., Tukker, T.W.: Electrowetting-based variable-focus lens for miniature systems. Optical Review 12(3), 255–259 (2005)
97. Herman, M.C., Papadopoulos, K.D.: [65]. J. Colloid Interface Sci. 136(285), 234 (1990)
98. Herman, M.C., Papadopoulos, K.D.: [65]. J. Colloid Interface Sci. 142(331), 234 (1991)
99. Hertz, H.: Uber die berührung fester elastischer körper. J. Reine Angew. Math. 92, 156–171 (1882) 97
100. Hoffman, R.L.: A study of the advancing interface i. interface shape in liquid-gas systems. J. Colloid Interface Sci. 50(2), 228–241 (1975)
101. Hudlet, S., Saint Jean, M., Berger, J.: Evaluation of the capacitive force between an atomic force microscopy tip and a metallic surface. Eur. Phys. J. B 2, 5–10 (1998) 96, 97
102. Israelachvili, J.N.: The nature of van der Waals forces. Contemporary Physics 15(2), 159–177 (1974) 89, 90, 91
103. Israelachvili, J.N.: Intermolecular and Surface Forces, 2nd edn. Academic Press, London (1992) 89, 90, 91, 93
104. Israelachvili, J.N., Gee, M.L.: Contact angles on chemically heterogeneous surfaces. Langmuir 5(1), 288 (1989)
105. Johnson, K.L., Kendall, K., Roberts, A.D.: Surface energy and the contact of elastic solids. Proc. R. Soc. London Ser. A 324, 301 (1971) 88, 97
106. Johnson, K.L.: Contact Mechanics. Cambridge University Press, London (1985)
107. Johnson, R.E., Dettre, R.H.: Contact angle, wettability and adhesion. Adv. Chem. Ser. 43(8) (1998)
108. Jonassen, N.: Electrostatics. Kluwer Academic, Dordrecht (2002) 94
109. Greenwood, J.A., Johnson, K.L.: An adhesion map for the contact of elastic spheres. J. Colloid Interface Sci. 192(2), 326–333 (1997) 97
110. Klimchitskaya, G.L., Mostepanenko, V.M.: Experiment and theory in the casimir effect. Contemporary Physics 47(3), 131–144 (2006) 89
111. Komvopoulos, K.: Surface engineering and microtribology for microelectromechanical systems. Wear 200(1), 305–327 (1996)
112. Kondo, T., Juodkazis, S., Misawa, H.: Reduction of capillary force for high-aspect ratio nanofabrication. Appl. Phys. A 81(8), 1583–1586 (2005)
113. Kralchevsky, P.A., Nagayama, K.: Particles at fluid interfaces and membranes. Studies in interface science, vol. 10. Elsevier, Amsterdam (2001) 93
114. Krupp, H.: Particle adhesion theory and experiment. Adv. Colloid Interface Sci. 1, 111–239 (1967) 94, 95

115. Lambert, P.: Capillary Forces in Microassembly: Modeling, Simulation, Experiments, and Case Study. In: Microtechnology and MEMS. Springer, Heidelberg (2007) 92, 93
116. Lambert, P., Chau, A., Delchambre, A., Régnier, S.: Comparison between two capillary forces models. Langmuir (2008)
117. Lambert, P., Delchambre, A.: Parameters ruling capillary forces at the submillimetric scale. Langmuir 21, 9537–9543 (2005)
118. Lambert, P., Régnier, S.: Surface and contact forces models within the framework of microassembly. Journal of Micromechatronics 3(2), 123–157 (2006) 88, 90, 91
119. Lambert, P., Seigneur, F., Koelemeijer, S., Jacot, J.: A case study of surface tension gripping: the watch bearing. J. Micromech. Microeng. 16(7), 1267–1276 (2006)
120. Law, B.M., Rieutord, F.: Electrostatic forces in atomic force microscopy. Phys. Rev. B 66(035402-1) (2002) 97
121. Lee, J., Kim, C.-J.: Surface-tension-driven microactuation based on continuous electrowetting. J. Microelectromechanical Syst. 9(2), 171–180 (2000)
122. Lee, L.-H.: The chemistry and physics of solid adhesion. In: Lee, L.-H. (ed.) Fundamentals of Adhesion. Plenum Press, New York (1991) 88, 94
123. Lenders, C., Valsamis, J.-B., Desaedeleer, M., Delchambre, A., Lambert, P.: Assembly of a micro ball-bearing using a capillary gripper and a microcomponent feeder. In: Micro-Assembly Technologies and Applications, International Precision Assembly Seminar (IPAS). Springer, Heidelberg (2008)
124. Lhernould, M.S.: Theoretical and experimental study of electrostatic forces applied to micromanipulation: influence of surface topography. PhD thesis, Université libre de Bruxelles (2008) 96
125. Lhernould, M.S., Delchambre, A., Régnier, S., Lambert, P.: Electrostatic forces and micromanipulator design: on the importance of surface topography parameters. In: Proceeding of the 2007 IEEE/ASME International Conference on Advanced Intelligent Mechatronics (2007) 96
126. Lhernould, M.S., Delchambre, A., Régnier, S., Lambert, P.: Electrostatic forces in micromanipulations: Review of analytical models and simulations including roughness. Appl. Surf. Sci. 253, 6203–6210 (2007) 96
127. Lifshitz, E.M.: Unknown title. Sov. Phys. - JETP 73(2) (1956) 90
128. London, F.: The London - van der Waals attraction between spherical particles. Physica 4(10), 1058–1072 (1937) 89
129. Lowell, J., Rose-Innes, A.: Contact electrification. Advances in Physics 29(6), 947–1023 (1980) 94
130. Marmur, A.: Tip-surface capillary interactions. Langmuir 9, 1922–1926 (1993) 92
131. Mastrangelo, C.H.: Mechanical stability and adhesion of microstructures under capillary forces - part i: Basic theory. J. Microelectromechanical Syst. 2(1), 33–43 (1993)
132. Mastrangelo, C.H.: Surface force induced failures in microelectromechanical systems. In: Bhushan, B. (ed.) Tribology Issues and Opportunities in MEM, pp. 367–395. Kluwer Academic Publishers, Netherlands (1998)
133. Mastrangelo, C.H.: Suppression of stiction in mems. In: Spring MRS Meeting, MRS, Boston, MA (1999)
134. Mathew Mate, C.: Tribology on the Small Scale. Oxford University Press, Oxford (2008)
135. Maugis, D.: Adhesion of spheres: the JKR-DMT transition using a Dugdale model. J. Colloid Interface Sci. 150(1), 243–269 (1992) 88, 97, 98
136. Moore, A.D.: Electrostatics and its applications. John Wiley, Chichester (1973) 94
137. Obata, K.J., Motokado, T., Saito, S., Takahashi, K.: A scheme for micro-manipulation based on capillary force. J. Fluid Mech. 498, 113–121 (2004)

138. Oh, K.W., Ahn, C.H.: A review of microvalves. J. Micromech. Microeng. 16(5), 13–39 (2006)
139. Ohlerich, U.: Tension superficielle, interfaciale et angle de contact. In: Workshop, Anvers, October 15 (2002)
140. Pagano, C., Zanoni, L., Fassi, I., Jovane, F.: Micro-assembly: design and analysis of a gripper based on capillary force. In: Proc. of the 1st CIRP - International Seminar on Assembly Systems, Stuttgart, Germany, November 15-17 pp. 165–170 (2006)
141. Papadopoulos, K.D., Kuo, C.-C.: [65]. Colloids Surf. 46(115), 234 (1990)
142. Patil, S., Dharmadhikari, C.V.: Investigation of the electrostatic forces in scanning probe microscopy at low bias voltage. Surf. Interf. Anal. 33, 155–158 (2002) 97
143. Patil, S., Kulkarni, A.V., Dharmadhikari, C.V.: Study of the electrostatic force between a conducting tip in proximity with a metallic surface: Theory and experiment. J. Appl. Phys. 88(11), 6940–6942 (2000) 97
144. Peirs, J.: Design of micromechatronic systems: scale laws, technologies, and medical applications. PhD thesis, KUL, Belgium (2001) 88
145. Rabinovich, Y.I., Esayanur, M.S., Mougdil, B.M.: Capillary forces between two spheres with a fixed volume liquid bridge: Theory and experiment. Langmuir 21, 10992–10997 (2005)
146. Russel, W.B., Saville, D.A., Schowalter, W.R.: Colloidal dispersions. Cambridge University Press, Cambridge (1989) 90
147. Sacha, G.M., Verdaguer, A., Martinez, J., Senz, J.J., Ogletree, D.F., Salmeron, M.: Effective radius in electrostatic force microscopy. Appl. Phys. Lett. 86, 123101 (2005) 96, 98
148. Schmid, D., Koelemeijer, S., Jacot, J., Lambert, P.: Microchip assembly with capillary gripper. In: Proc. of the 5th International Workshop on Microfactories, October 25-27, 4 pages (2006)
149. Segupta, A.K., Papadopoulos, K.D.: [65]. J. Colloid Interface Sci. 152(534), 234 (1992)
150. Sitti, M., Hashimoto, H.: Teleoperated touch feedback from the surfaces at the nanoscale: Modelling and experiments. IEEE-ASME Trans. Mechatron. 8(1), 1–12 (2003) 98
151. Smythe, W.R.: Static and dynamic electricity. Mc Graw-Hill, New York (1968) 97
152. van den Tempel, M.: [65]. Adv. Colloid Interface Sci. 3(137) (1972) 90
153. Vgeli, B., von Knel, H.: AFM-study of sticking effects for microparts handling. Wear 238(1), 20–24 (2000) 90
154. Vora, K.D., Peele, A.G., Shew, B.-Y., Harvey, E.C., Hayes, J.P.: Fabrication of support structures to prevent su-8 stiction in high aspect ratio structures. Microsyst. Technol. (2006)
155. Wu, D., Fang, N., Sun, C., Zhang, X.: Stiction problems in releasing of 3d microstructures and its solution. Sensors and Actuators A 128, 109–115 (2006)
156. Yang, B., Lin, Q.: A latchable microvalve using phase change of paraffin wax. Sensors and Actuators A 134, 194–200 (2007)
157. Yang, S., Zhang, H., Nosonovsky, M., Chung, K.-H.: Effects of contact geometry on pull-off force measurements with colloidal probe. Langmuir 24, 743–748 (2000)
158. Surface profile parameters, `http://www.predev.com/smg/parameters.htm`
159. Adamczyk, Z.: Particle adsorption and deposition: role of electrostatic interactions. Advances in Colloid and Interface Science 100–102, 267–347 (2003)
160. Arai, F., Ando, D., Fukuda, T., Nonoda, Y., Oota, T.: Micro manipulation based on micro physics. In: Proc. of IEEE/RSJ Conf. on Intelligent Robots and Systems, Pittsburgh, vol. 2, pp. 236–241 (1995)

161. Bell, G.M., Levine, S., McCartney, L.N.: Approximate methods of determining the double-layer free energy of interaction between two charged colloidal spheres. Journal of Colloid and Interface Science 33, 335–359 (1970)
162. Bhattacharjee, S., Elimelech, M., Borkovec, M.: DLVO interaction between colloidal particles: beyond Derjaguin's approximation. Croatica Chemica Acta 71(4), 883–903 (1998)
163. Bhushan, B.: Handbook of Micro/Nano Tribology, 2nd edn. CRC Press, Boca Raton (1998)
164. Bhushan, B.: Modern Tribology Handbook, Two Volume Set. CRC Press, Boca Raton (2000)
165. Bhushan, B.: Adhesion and stiction: mechanisms, measurement techniques, and methods for reduction. J. Vacuum Sci. Technol. B 21(6), 2262–2296 (2003)
166. Bobji, M.S., Biswas, S.K.: Estimation of hardness nanoindentation of rough surfaces. J. Mater. Res. 13(11), 3227 (1998)
167. Allen Bowling, R.: A theoretical review of particle adhesion. In: Proc. of Symposium on particles on surfaces 1: Detection, Adhesion and Removal, San Francisco, pp. 129–142 (1986)
168. Brant, J.A., Childress, A.E.: Assessing short-range membrane.colloid interactions using surface energetics. Journal of Membrane Science 203, 257–273 (2002)
169. Brant, J.A., Childress, A.E.: Colloidal adhesion to hydrophilic membrane surfaces. Journal of Membrane Science 241, 235–248 (2004)
170. Danuser, G., Pappas, I., Vgeli, B., Zesch, W., Dual, J.: Manipulation of microscopic objects with nanometer precision: potentials and limitations in nano robot design. Int. J. Rob. Res. (1998)
171. de Gennes, P.-G., Brochart-Wyard, F., Qur, D.: Gouttes, bulles, perles et ondes. Belin (2002)
172. Derjaguin, B.V.: Untersuchungen über die reibung und adhäsion, IV. Kolloid Zeitschrift 69(2), 155–164 (1934)
173. Fearing, R.S.: Survey of sticking effects for micro parts handling. In: Proc. of IEEE/RSJ Conf. on Intelligent Robots and Systems, August 5-9, pp. 212–217 (1995)
174. Fowkes, F.M.: Role of acid-base interfacial bonding in adhesion. J. Adh. Sci. Tech. 1(1), 7–27 (1987)
175. Gagnepain, J.J., Roques-Carmes, C.: Fractal approach to two-dimensional and three-dimensional surface roughness. Wear 109, 119–126 (1986)
176. Ganti, S., Bhushan, B.: Generalized fractal analysis and its applications to engineering surfaces. Wear 180, 17–34 (1994)
177. Good, R.J.: Theory of adhesive forces accross interfaces: 1. the Lifshitz - van der Waals component of interaction and adhesion. In: Lee, L.-H. (ed.) Fundamentals of Adhesion. Plenum Press, New-York (1991)
178. Good, R.J., Van Oss, C.J.: Modern Approaches to Wettability, Theories and Applications. Plenum Press, New York (1992)
179. Hasegawa, M., Liu, J., Okuda, K., Nunobiki, M.: Calculations of the fractal dimensions of machined surface profiles. Wear 192, 40–45 (1996)
180. He, L., Zhu, J.: The fractal character of processed metal surfaces. Wear 208, 17–24 (1997)
181. Herman, M.C., Papadopoulos, K.D.: Effects on the van der waals and electric double layers interactions of two parallel flay plates. J. Colloid Interface Sci. 136(2), 385 (1990)
182. Hoek, E.M.V., Agarwal, G.K.: Extended DLVO interactions between spherical particles and rough surfaces. Journal of Colloid and Interface Science 298, 50–58 (2006)

183. Hogg, R., Healy, T.W., Fuerstenau, D.W.: Mutual coagulation of colloidal dispersions. Transactions of the Faraday Society 62, 1638–1651 (1966)
184. Hunter, R.J.: Foundations of Colloid Science. Oxford University Press, Oxford (2001)
185. Israelachvili, J.N.: Intermolecular and Surface Forces, 2nd edn. Academic Press, London (1992)
186. Jahn, R., Truckenbrodt, H.: A simple fractal analysis method of the surface roughness. Journal of Materials Processing Technology 145, 40–45 (2004)
187. Komvopoulos, K.: Surface engineering and microtribology for microelectromechanical systems. Wear 200(1), 305–327 (1996)
188. Komvopoulos, K.: A fractal analysis of stiction in microelectromechanical systems. J. Tribology 119(3), 391–400 (1997)
189. Kostoglou, M., Karabelas, A.J.: Effect of roughness on energy of repulsion between colloidal surfaces. J. Colloid Interface Sci. 171, 187–199 (1995)
190. Krupp, H.: Particle adhesion theory and experiment. Adv. Colloid Interface Sci. 1, 111–239 (1967)
191. Lambert, P.: A Contribution to Microassembly: a Study of Capillary Forces as a gripping Principle. PhD thesis, Universit libre de Bruxelles, Belgium (2004)
192. Lhernould, M.S.: Theoretical and experimental study of electrostatic forces applied to micromanipulation: influence of surface topography. PhD thesis, Universit libre de Bruxelles (2008)
193. Lhernould, M.S., Delchambre, A., Régnier, S., Lambert, P.: Electrostatic forces in micromanipulations: Review of analytical models and simulations including roughness. Appl. Surf. Sci. 253, 6203–6210 (2007)
194. Ling, F.F.: Fractals, engineering surfaces and tribology. Wear 136, 141–156 (1990)
195. Majumdar, A., Bhushan, B.: Role of fractal geometry in roughness characterization and contact mechanics of surafces. J. Tribology 112, 205–216 (1990)
196. Majumdar, A., Tien, C.L.: Fractal characterization and simulation of rough surfaces. Wear 136, 313–327 (1990)
197. Mandelbrot, B.B.: The fractal geometry of nature. W.H. Freeman, New York (1982)
198. McCormack, D., Carnie, S.L., Chan, D.Y.C.: Calculations of electric double-layer forces and interaction free energy between dissimilar surfaces. Journal of Colloid and Interface Science 169, 177–196 (1995)
199. Montmitonnet, P., Delamare, F., Felder, E., Bucaille, J.-L.: Cours de tribologie de la mise en forme. Ecole des Mines de Paris, Centre de Mise en forme des matriaux (2003)
200. Morrow, C.A.: Adhesive rough surface contact. PhD thesis, University of Pittsburg (2003)
201. Ohshima, H.: Effective surface potential and double-layer interaction of colloidal particles. Journal of Colloid and Interface Science 174, 45–52 (1995)
202. Ouyang, Q., Ishida, K., Okada, K.: Investigation of micro-adhesion by atomic force microscopy. Applied Surface Science 169, 644–648 (2001)
203. Patrikar, R.M.: Modeling and simulation of surface roughness. Appl. Surf. Sci. 228, 213–220 (2004)
204. Peirs, J.: Design of micromechatronic systems: scale laws, technologies, and medical applications. PhD thesis, KUL, Belgium (2001)
205. Provder, T., Kunz, B.: Application of profilometry and fractal analysis to the characterization of coatings surface roughness. Prog. Org. Coat. 27, 219–226 (1996)
206. Rabinovich, Y.I., Adler, J.J., Ata, A., Singh, R.K., Moudgil, B.M.: Adhesion between nanoscale rough surfaces I role of asperity geometry. J. Colloid Interface Sci. 232(1), 10–16 (2000)

207. Gelb, A., Eichenlaub, S., Beaudouin, S.: Roughness models for particle adhesion. J. Colloid Interface Sci. 280, 289–298 (2004)
208. Sader, J.E.: Accurate analytic formulae for the far field effective potential and surface charge density of a uniformly charged sphere. Journal of Colloid and Interface Science 188, 508–510 (1997)
209. Sader, J.E., Carnie, S.L., Chan, D.Y.C.: Accurate analytic formulae for the double layer interaction between spheres. Journal of Colloid and Interface Science 171, 46–54 (1995)
210. Shulepov, S.Y., Frens, G.: Surface roughness and the particle size effect of slow perikinetic coagulation. J. Colloid Interface Sci. 170, 44–49 (1994)
211. Sitti, M.: Atomic force microcope probe based controlled pushing for nanotribological characterization. IEEE/ASME Transactions on Mechatronics 9(2), 343–349 (2004)
212. Sitti, M., Hashimoto, H.: Teleoperated touch feedback from the surfaces at the nanoscale: Modelling and experiments. IEEE-ASME Trans. Mechatron. 8(1), 1–12 (2003)
213. Suresh, L., Walz, J.Y.: Effect of surface roughness on the interaction energy between a colloidal sphere and a flat plate. J. Colloid Interface Sci. 183, 199–213 (1996)
214. Tabor, D.: Surface forces and surface interactions. J. Colloid Interface Sci. 58(1), 2–13 (1977)
215. van der Wal, E.: Bioactivity and surface reactivity of RF-sputtered calcium phosphate thin films. PhD thesis, Utrecht University (2003)
216. van Oss Carel, J.: Interfacial Forces in Aqueous Media. Taylor & Francis, Abington (2006)
217. Vervey, E.J.W., Overbeek, J.T.G.: Theory of the Stability of Lyophobic Colloids. Elsevier, Amsterdam (1948)
218. Vgeli, B., von Knel, H.: AFM-study of sticking effects for microparts handling. Wear 238(1), 20–24 (2000)
219. Weisenhorn, A.L., Hansma, P.K., Albrecht, T.R., Quate, C.F.: Forces in atomic force microscopy in air and water. Appl. Phys. Lett. 54(26), 2651 (1989)
220. Whitehouse, D.J.: Fractal or fiction. Wear 249(5-6), 345–353 (2001)
221. Whitehouse, D.J.: Handbook of surface metrology. Institute of Physics, London (1994)
222. Wu, J.-J.: Structure function and spectral density of fractal profiles. Chaos Solitons Fractals 12, 2481–2492 (2001)
223. Wu, W., Giese, R.F., van Oss, C.J.: Stability versus flocculation of particle suspensions in water — correlation with the extended DLVO approach for aqueous systems, compared with classical DLVO theory. Colloids and Surfaces B: Biointerfaces 14, 47–55 (1999)
224. Zhou, G., Leu, M., Blackmore, D.: Fractal geometry modeling with applications in surface characterisation and wear prediction. Int. J. Mach. Tools Manufac. 35(2), 203–209 (1995)
225. Zhu, H., Ge, S., Huang, X., Zhang, D., Liu, J.: Experimental study on the characterization of worn surface topography with characteristic roughness parameter. Wear 255, 309–314 (2003)

Chapter 4
Teleoperation Based AFM Manipulation Control

4.1 Introduction

Teleoperation with haptic feedback is a promising solution for nanoscale operations, for both educational purposes and high precision manipulation tasks. It is well understood that interactions and forces among nanoscale objects are very different from objects at larger scales. The reason for these differences stem from the fact that forces do not scale linearly with the characteristic length of an entity. Inertial (volumetric) forces, such as momentum and weight that are significant at macro-scales, become negligible against areal (adhesive) and peripheral (capillary) forces, particularly at micro/nanoscale [1, 2]. Therefore, experiencing the nanoscale environment by means of a bilateral teleoperation system is, first and foremost, an educational step that can be employed for training on nanoscale phenomena, especially nanotribology, for researchers.

Moreover, some applications include fragile and sensitive samples, which have a lot of variation, such as biological specimens. Automated manipulation has been proposed as a solution, and give interesting results when a task must be performed repeatedly. However, exploratory experiments cannot be easily solved using automated systems, and complex applications require a user's high-level control capabilities. For such experiments, teleoperation would be imperative and force feedback simply enhances the experience of the operator for better success rates.

Bilateral teleoperation implies that forces and positions are scaled and transferred back and forth between the master and slave sides of the teleoperation system, where the master side includes the direct interface between the operator and the teleoperation system (*i.e.* a haptic device), and the slave side includes the manipulation mechanism (*i.e.* a nanomanipulator) [3]. Bilateral teleoperation enables the operator to interact with the environment on the slave side, and feel reaction forces as if the operator is actually feeling the slave forces. Important issues in bilateral control include stability, transparency, and quality of teleoperation that must take into account ease of manipulation and limits of comfortable human perception. Unfortunately, these issues are in conflict with each other and hence, there exists a trade-off between them [4, 5, 6].

H. Xie et al.: Atomic Force Microscopy Based Nanorobotics, STAR 71, pp. 145–235.
springerlink.com

Works in teleoperation began in the 1990s, where Hollis *et al.* used a scanning tunneling microscope as a nanomanipulator [7]. To improve performance, many studies have been conducted to ensure transparency and stability. On transparency and reflected impedance issues, Yokokohji and Yoshikawa defined the ideal response of bilateral control systems in [4]. However in practice, the system becomes unstable if this ideal response is attempted. This is an important discovery, which states that it is impossible to have simultaneously a stable and transparent system; in essence, stability and transparency are incompatible with each other. The trade-off between stability and transparency means that it is impossible to reflect the remote environment exactly to the master side, in a stable manner, which alters the operator's feeling of the slave side. Cavusoglu *et al.* [5] defined a novel fidelity measure based on the sensitivity of transmitted impedance changes for soft environments. Also, Hogan's work [6] on controlling impedance was especially important, considering the limits of the master system or the operator.

One promising controller to ensure stability is to use a passivity observer and an adaptive energy dissipation force [8, 9], which forces the system to remain passive and effectively cancels instability. Recent literature on passivity based control schemes generally investigate bilateral control under variable time delay and packet losses in the communication link, such as teleoperation over the internet [10, 11, 12]. The study of Boukhnifer and Ferreira [13] applies wave variables to achieve a passive controller for a micro-teleoperation system, which is also designed for transparency.

Recent teleoperation systems dealing with micro and nanoscale employ an atomic force microscope (AFM) as a nanomanipulator, which enables direct force measurements between the end-effector and the sample, and the ability to mechanically interact with a wide range of materials [14]. A review of teleoperated nanomanipulation approaches can be found in [15]. A problem arrising in all AFM based teleoperation system is the determination of haptic force based on available measurements. Since AFM gives only two measurements, and can be used in either static or dynamic mode, a strategy must be defined to compute haptic feedback. It includes force reconstruction from physical models, and conversion from amplitude measurements to force information in the case of dynamic mode.

In this chapter, particularities of AFM based teleoperation are reviewed, and classical control schemes are analysed and compared in terms of transparency, stability, and ease of manipulation as well as limits of human force sensing. Strategies are proposed to derive haptic feedback from AFM measurements for both 1D, 2D and 3D teleoperations. Both understanding of nanoscale properties and assembly tasks are considered, and experimental results are presented to validate the analyses.

This chapter is organized as follows. In Section 4.2, theoretical issues of AFM based bilateral teleoperation systems are considered. Section 4.3 presents experimental platforms on which experiments of Section 4.4, 4.5, 4.6 and 4.7 are conducted. These experiments deal respectively with 1D, 2D, and 3D haptic feedback rendering teleoperations. Section 4.8 concludes this work. Notations used in this Chapter can be found in Section 4.9.

4.2 Bilateral Scaled Teleoperation Control Using an AFM

4.2.1 *Architecture of a Teleoperation System*

The architecture of a teleoperation system is depicted in Fig. 4.1. It is composed of three main blocks, the haptic interface, the coupling scheme, and the environment.

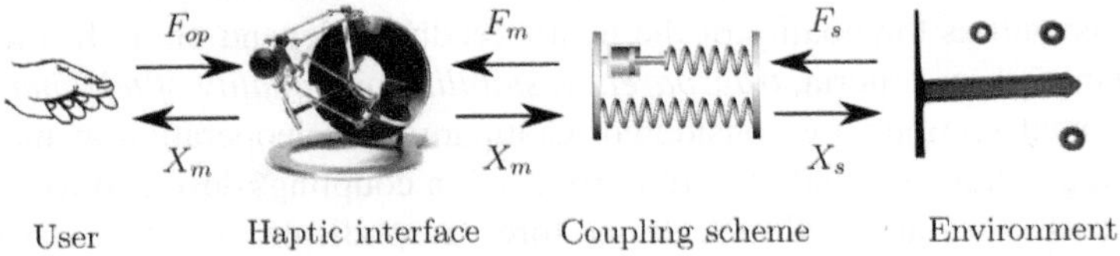

Fig. 4.1 Architecture of a teleporation system and exchanged variables

Several types of haptic interfaces have been developped. They present complementary characteristics in terms of maximum force sent to operators, available workspace, number of degrees of freedom, or maximum stiffness of the haptic feedback. This field is still open for research developments, since ease of manipulation and quality of force rendering greatly depend on the haptic interface used, as it will be seen through this chapter. Haptic interfaces specially designed for nanoscale applications are currently investigated [16]. However, to ensure a wide variety of applications, only commercially available devices are considered here. These interfaces are modeled by an inertia M_h, and a damping B_h [17]. Their transfer functions $H(s)$ whose input, resp. output, is the force applied on the haptic handle, resp. its velocity are written in the Laplace domain:

$$H(s) = [(B_h + M_h s)]^{-1} \tag{4.1}$$

where s is the Laplace variable. Numerical values of M_h and B_h depend on the haptic interface characteristics. The values of the principal ones can be found in [17]. In particular, the Virtuose characteristics used for stability analysis in the following sections are:

$$M_h = 0.4\text{kg} \; ; \; B_h = 0.1\text{N}\cdot\text{s}\cdot\text{m}^{-1} \tag{4.2}$$

The environment is composed of the AFM platforms described in previous chapters. The end-effectors used in this chapter are tip or tipless cantilevers. The displacements with respect to the substrate are made by micro and nanotranslators, or slave robots. At each time step, the position of the cantilever is known from the nanotranslators' sensors. The closed loop transfer function of the nanotranslator on the

z axis is a second order function, with two time constants τ_1 and τ_2 Numerical values of the parameters have been identified as:

$$N(s) = [(1+\tau_1 s)(1+\tau_2 s)]^{-1} \tag{4.3}$$

where $\tau_1 = 1.35 \cdot 10^{-3}$ s, $\tau_2 = 0.57 \cdot 10^{-3}$ s and s is the Laplace variable.

The coupling must connect the haptic interface, or master device, to the AFM platform to control the translators. The desired position of the nanotranslators is based on the position of the haptic handle. While manipulating the master device, the user feels a force feedback based on measured interaction forces applied on the tool.

Coupling is a critical issue in teleoperation since ease and precision of manipulation, as well as the quality of the haptic feedback depend on it. To measure its performance, three criteria, *transparency*, *stability*, and *quality of teleoperation* detailed in next sections are considered. Compared to teleoperation at macroscale, two scaling factors, α_F and α_p, are introduced in couplings designed for nanoscale experiments. They are used to scale the force, resp. displacement variables. These coefficients may lead to instability as it will be detailed in following sections.

4.2.2 Performance Measurement

To compare and evaluate control scheme's performances, three criteria are considered: *transparency*, *stability* and *quality of the teleoperation*.

A transparent control scheme ensures that what the user feels while manipulating the haptic handle is the same as what he or she would have felt by directly manipulating the tool. In particular, when there is no interaction between the tool and its environment, the user should feel a null force. On the contrary, while touching a stiff object, the feeling should be as stiff as possible. More details can be found in Section 4.2.2.1.

In addition to being transparent, a coupling must be stable to ensure safety and usability. Different solutions to ensure stability are discussed in Section 4.2.2.2.

The two previous criteria are the most commonly used to analyze a control scheme. However, they do not take into account ease of manipulation and limits of comfort for haptic feedback. Section 4.2.2.3 will thus detail several issues that must be adressed to design a teleoperation system that can efficiently peform a wide variety of tasks.

4.2.2.1 Transparency

Definition

Transparency is the ability to transmit to operators the phenomena which occur in the environment as faithfully as possible. It is usually measured in terms of impedance. It is defined in [18] and [3] as a comparison between user impedance

$Z_{op} = \frac{F_{op}}{V_m}$ and environment impedance $Z_e = \frac{F_s}{V_s}$ where V is the velocity variable. Ideal transparency is achieved if:

$$Z_{op} = Z_e \tag{4.4}$$

However, for submicron scales, it is necessary to consider the scaling factors α_F and α_p such that the impedances on master and slave sides can be compared. In our context, perfect transparency will be achieved if:

$$\frac{F_{op}}{V_m} = \frac{-\alpha_F F_s}{\alpha_p V_s} \Leftrightarrow Z_{op} = \frac{\alpha_F}{\alpha_p} Z_e \tag{4.5}$$

Transparency is an important issue for teleoperation systems. In particular, tasks involving small force variations, or complex force fields, require high transparency. However, a trade-off between stability and transparency might be necessary and must be adressed considering the application.

It has to be noted that the force sensor used modifies the profile of the measured forces, and therefore the operator's feeling. This issue is presented in the next paragraph.

Influence of the sensor

Humans perceive the hardness of a surface by observing the equivalent stiffness of it with their hands or skin. Therefore, it is necessary not to add any other non-rigid object between the hand of the operator and the surface that is being felt. With an AFM as the slave nanomanipulator, a cantilever exists that touches the surface; this is essentially a non-rigid object that effectively lies between the hand of the operator and the surface that will be felt. For force transparency, the effect of the cantilever must be removed. Using a quasi-static assumption, the measured forces can be modeled by an equivalent system consisting of two springs in series, shown in Fig. 4.2.

If the vertical base position of the cantilever is $z_p \in \Re$ with reference $z_p = 0$ when the tip touches the sample surface, deflection of the cantilever is $\zeta \in \Re$, and indentation of the surface is $\delta \in \Re$, then:

$$z_p = \zeta + \delta. \tag{4.6}$$

The force measured on the slave side ($F_s \in \Re$) would be given as:

$$F_s = K_p \zeta = K_s \delta = K_{eq} z_p, \tag{4.7}$$

where $K_p \in \Re^+$, $K_s \in \Re^+$, and $K_{eq} \in \Re^+$ are the probe, surface, and equivalent stiffness, respectively, from two springs in series:

$$K_{eq} = \frac{K_p K_s}{K_p + K_s} \tag{4.8}$$

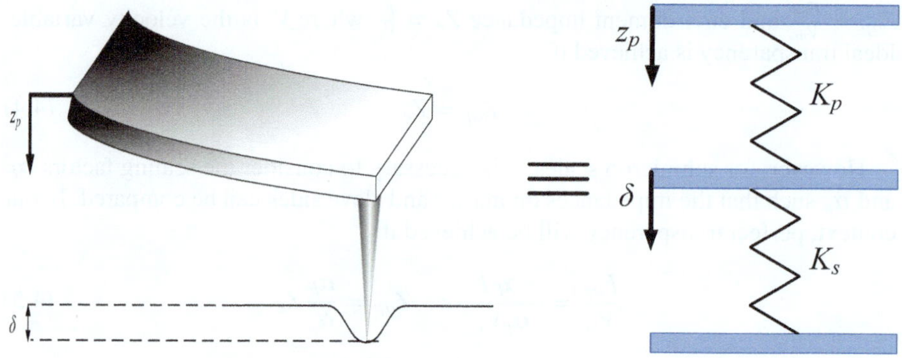

Fig. 4.2 An AFM probe in contact with a surface and its static model; K_p and K_s are the stiffness of the probe and surface, respectively.

Hence, the probe stiffness should be as large as possible to achieve an equivalent stiffness close to the surface stiffness for a large range of materials, while being as small as possible to achieve high sensitivity force measurements.

If the probe stiffness was infinite, the equivalent stiffness would be equal to the surface stiffness, and the impedance felt by the operator would be the surface stiffness. A simple solution to having a finite stiffness cantilever (a requirement for an AFM) is to mathematically transform the measured force to remove the effect of the cantilever. In this case, the force reflected would feel as though the cantilever is a rigid structure and the entire motion of the base is used to indent the substrate; mathematically, this becomes:

$$F_t = K_s z_p = \frac{K_s}{K_{eq}} F_s = \frac{K_p}{K_p - K_{eq}} F_s, \tag{4.9}$$

where $F_t \in \Re$ is the transparent force. If the substrate is comprised of a single material, an initial force-distance curve would be sufficient to calculate the equivalent stiffness. Assuming the probe stiffness is known, Equation (4.9) provides a means to achieve transparent forces from the AFM. Note that this mathematical transformation assumes a constant K_{eq}, meaning a linear force-distance relationship. While this is not always true, using a nominal value of K_{eq} for the force range the AFM can measure, would still improve transparency.

If the surface topography is known (a possibility when using the AFM to image the surface), another possibility to get an online reading of the transparent force is by using (4.6) and (4.7):

$$F_t = \frac{z_p}{\delta} F_s = \frac{z_p}{z_p - \zeta} F_s. \tag{4.10}$$

where δ is determined from the surface topography and z_p.

Therefore, with these methods, a correction factor can be implemented that converts the force measurements from the AFM into transparent forces, improving the feeling of small-scale surfaces to the operator through the tele-nanomanipulation system.

4.2.2.2 Stability

Stability is a critical issue for the control scheme usability, as an unstable system might cause damage on tools and manipulated objects. Two approaches are considered: stability and passivity. They are detailed in the following paragraphs.

Stability criteria

To analyze the control scheme's stability, classical tools such as the Routh-Hurwitz or the Jury criteria are used. Studies are made in continuous or discrete time depending on the time delays.

To analyze stability, the coupling transfer function, whose input, resp. output, is the position of the haptic handle, resp. the haptic force applied to the user is considered. To compute this transfer function, the environment must be considered. To model it, Equation (4.8) is used. The worst case for the issue of stability is when the equivalent stiffness is the highest, which corresponds to $K_{eq} = K_p$. In following sections this approximation will be considered.

Even if this approximation is restrictive, it is sufficient to point out the inherent problems of the proposed control scheme in the nano-world.

Passivity

In addition to stability, passivity can be considered. Passivity ensures the stability of a system, as long as the control scheme is connected to passive environments. The operator is commonly considered as passive, or at least that he or she will not destabilize the system [6], [9]. The environment is also considered as passive, even if this condition is restrictive at nanoscales.

Passivity can be analyzed using either frequency or energy based definitions. The last one enables to design observers and controllers to monitor the energy of the system and add damping if necessary to avoid instability [19], [20]. It will therefore be detailed here.

From an energetic point of view, a system is passive if and only if it does not create energy. Based on signals in time domain (either continuous or discrete), passivity is given by [19]:

Definition 4.1. An M-port system with an initial energy $E(0)$ is passive if and only if:

$$\int_0^t f_1(\tau)v_1(\tau) + ... + f_M(\tau)v_M(\tau)d\tau + E(0) \geq 0 \; \forall t \geq 0 \quad (4.11)$$

for all admissible forces and velocities.

The most important drawback of this approach is conservativity. Passivity is restrictive as it ensures stability for any environment and operator.

Less restrictive pasivity based stability conditions

As stated above, passivity is restrictive (only sufficient stability conditions are derived, but they might be not necessary). Less restrictive criteria based on passivity analysis have been developped.

The first approach detailed in [21] is to consider the particular case of a control scheme connected to one port systems (such as operators or environments), instead of the general case of a two port system. The analysis is carried out based on unconditional stability. A necessary and sufficient condition is given by Llewelyn criterion [22]:

Theorem 4.1. *A system with an immittance matrix* **P** *is unconditionally stable if and only if the following conditions are verified:*

$$c_1 : Re(p_{11}) \geq 0 \quad (4.12)$$
$$c_2 : Re(p_{22}) \geq 0 \quad (4.13)$$
$$c_3 : 2Re(p_{11})Re(p_{22}) - |p_{12}p_{21}| - Re(p_{12}p_{21}) \geq 0 \quad (4.14)$$

where:

$$\mathbf{P} = \begin{bmatrix} p_{11} & p_{12} \\ p_{21} & p_{22} \end{bmatrix} \quad (4.15)$$

This approach, less conservative than passivity remains stricter than stability as no hypothesis is made on operators and environments apart from the fact that they are passive [23].

The second approach consists in restricting conditions on operators and environments, so that not all passive systems are considered, but only effective situations [24]. In particular, a minimum damping is introduced on the operator side.

Conclusion

Two main approaches might be considered to analyze a teleoperation system's stability. The first one, stability, requires to model the environment. Necessary and sufficient stability conditions are derived. The second approach is applied without any knowledge about the environment, apart from a passivity hypothesis. It is thus more restrictive as it ensures stability for any environment. Only sufficient stability conditions are derived.

4.2.2.3 High Quality Teleoperation

To ensure high quality teleoperation, two main issues must be addressed, dealing respectively with ease of manipulation and quality of force feedback. These two points are detailed in this section.

Ease of manipulation

When an operator manipulates the haptic device, coordinates representing the handle position are scaled linearly (by a scaling factor, $\alpha_p \in \Re^+$) and fed forward to the nanopositioning stage on the slave side, which moves the sample. Selection of α_p is made using the ratio of motion ranges between the master and slave subsystems. The choice of this coefficient also depends on the required precision. As it will be seen in following sections, stability conditions should also be taken into account, and limit the nanostage displacements.

Because of the limited workspace of the haptic interface, it might be impossible to reach the entire nanopositionning stage workspace. A clutching function, which enables to move the haptic handle without moving the nanostage, is used in the following experiments.

The choice of α_p should thus be made considering the required precision, and the duration of experiments. Tasks which require efforts for manipulating an object because of excessively small displacements do not enable users to concentrate on the experiment.

Quality of haptic feedback

Haptic feedback must be easily perceived by untrained users to provide useful information. However, it is limited by haptic interfaces characteristics, which are usually limited to a few newtons, and by human capabilities. To adapt haptic feedback to forces measured on the slave side, impedance transform is proposed here.

Defining $\alpha_p \in \Re^+$ and $\alpha_F \in \Re^+$ as the position (or velocity) and force scaling factors, respectively:

$$z_s = \alpha_p z_m \tag{4.16}$$

$$F_m = \alpha_F F_t \tag{4.17}$$

where $z_m \in \Re$ and $z_s \in \Re$ are the master and slave vertical-axis positions, respectively, $F_m \in \Re$ is the force on the master, and F_t is the transparent force, found in Section 4.2.2.1.

The master and slave stiffnesses are defined as:

$$K_m = \frac{dF_m}{dz_m} = \frac{\alpha_F dF_t}{dz_m} \quad (4.18)$$

$$K_s = \frac{dF_t}{dz_s} = \frac{dF_t}{\alpha_p dz_m} \quad (4.19)$$

for a quasi-static case, or explicitly in terms of each other:

$$K_m = \alpha_p \alpha_F K_s = \alpha_K K_s \quad (4.20)$$

where $\alpha_K \in \Re^+$ is the impedance scaling factor. As a result, for the slave stiffness to be reflected to the master side unchanged, $\alpha_K = 1$, implying:

$$\alpha_F = 1/\alpha_p. \quad (4.21)$$

Therefore, if the slave forces are corrected for transparency, as explained in Section 4.2.2.1, and the scaling factors are adjusted as mentioned above, the operator should feel the stiffness of the surface with very high fidelity.

With respect to the human perception of stiffness, some materials encountered at the nanoscale can have a very small stiffness, which can prevent an operator on the master side from distinguishing the surface from free motion (especially considering that there is an inherent friction on the master device). On the other hand, a very hard surface can have a reflected stiffness that is uncomfortably high.

One solution to this problem is to linearly transform the slave stiffness to a comfortable range for the human operator. Taking the largest (K_{II}) and smallest (K_I) stiffness values encountered on the slave side, and using two values as maximum and minimum stiffness (K_{max} and K_{min}) that are comfortable to work with, a line can be drawn that passes through these two points in the master/slave stiffness plane. The slope of this line is:

$$m = \frac{K_{max} - K_{min}}{K_{II} - K_I} \quad (4.22)$$

yielding a linear transformation as:

$$K_m = K_{min} + m(K_s - K_I) \quad (4.23)$$

from slave stiffness to master stiffness (as indicated by the indices). If a surface with a stiffness outside this range is encountered, either the K_I and K_{II} values can be adjusted or the transformation can be saturated to the K_{max} and K_{min} values.

With this transformation, Equation (4.9) becomes:

$$F_{tr} = K_m z_p = \frac{K_m}{K_s} F_t = \frac{K_m}{K_{eq}} F_s, \quad (4.24)$$

where F_{tr} is the force after impedance transformation, which is scaled with α_F to achieve the uncompensated force $F_{un} = \alpha_F F_{tr}$ on the master side.

4.2.2.4 Conclusion

The three criteria detailed above, transparency, stability, and quality of teleoperation, will be used in next sections to analyze and compare control schemes.

4.2.3 *Direct Force Feedback*

In this section the first control scheme, namely Direct Force Feedback (*DFF*) is introduced and analysed [25].

4.2.3.1 Control Scheme Structure

This control scheme, depicted in Fig. 4.3, is the most intuitive formulation to provide amplified forces to operators [26]. Basically, the user operates a haptic device in the macro-world to impose the displacements of the slave device in the nano-world. The controller scales down the motions provided by the user by a coefficient α_p, and magnifies environmental forces by a factor α_F to provide haptic feedback.

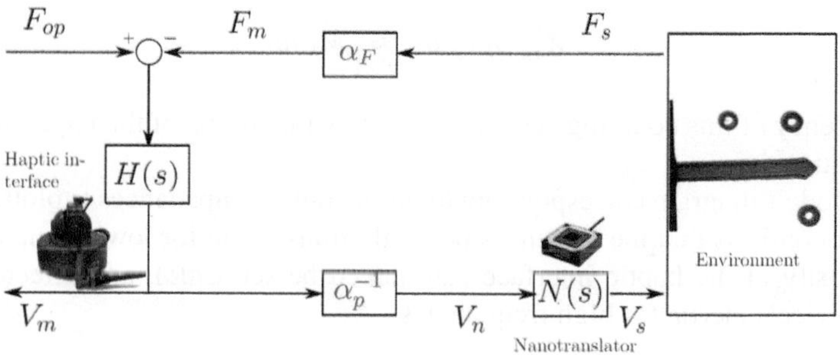

Fig. 4.3 Direct Force Feedback control scheme

4.2.3.2 Transparency

Using the control scheme depicted in Fig. 4.3, the impedance on the master side is derived:

$$Z_{op}(s) = \frac{F_{op}(s)}{V_m(s)} = \frac{\alpha_F}{\alpha_p} Z_e(s) N(s) + \frac{1}{H(s)} \quad (4.25)$$

Contact

While in contact, the impedance on the master side will be that of Equation (4.25). According to Equation (4.5), this corresponds to the impedance he or she should ideally feel ($\frac{\alpha_F}{\alpha_p} Z_e(s)$) modulated by the nanotranslator dynamic. It is also influenced by the haptic device characteristics.

In the frequency domain ($s = j\omega$), Equation (4.25) can be rewritten as:

$$Z_{op}(j\omega) = \frac{\alpha_F}{\alpha_p} Z_e(j\omega) \frac{1}{\tau_1 \tau_2 \omega^2 + (\tau_1 + \tau_2) j\omega + 1} + (M_h j\omega + B_h) \tag{4.26}$$

For **low frequencies**, the impedance $Z_{op,LF}^{DFF}$ can be approximated by:

$$Z_{op,LF}^{DFF}(j\omega) \underset{\omega \ll 1}{\approx} \frac{\alpha_F}{\alpha_p} Z_e(j\omega) + B_h \tag{4.27}$$

Impedance on the environment side, as well as the viscosity of the haptic interface are sent to the user. However, B_h can be set aside compared to $\frac{\alpha_F}{\alpha_p} Z_e$, so this coupling tends to the ideal transparency for low frequencies.

For **high frequencies**, the impedance $Z_{op,HF}^{DFF}(j\omega)$ is:

$$Z_{op,HF}^{DFF}(j\omega) \underset{\omega \gg 1}{\approx} M_h j\omega \tag{4.28}$$

Transparency of this coupling is only affected by the inertia of the haptic device for high frequencies.

The Bode's diagram corresponding to these contact impedances is plotted in Fig. 4.4(a). It confirms that the system is perfectly transparent for low frequencies (and the viscosity of the haptic interface can indeed be set aside), and affected by the haptic device's inertia for high frequencies.

Non contact

When no force is applied to the cantilever, impedance on the environment side is $Z_e(j\omega) = 0$. According to Equation (4.26), it leads to:

$$Z_{op}(j\omega) = M_h j\omega + B_h \tag{4.29}$$

At **low frequencies**, the operator will mainly feel the viscosity of the haptic device. At **high frequencies**, the inertia of the master arm will be predominant. This can indeed be verified in the Bode's diagram depicted in Fig. 4.4(b).

Using the *DFF* control scheme, the haptic rendering only depends on the haptic device characteristics, for both contact and non contact modes.

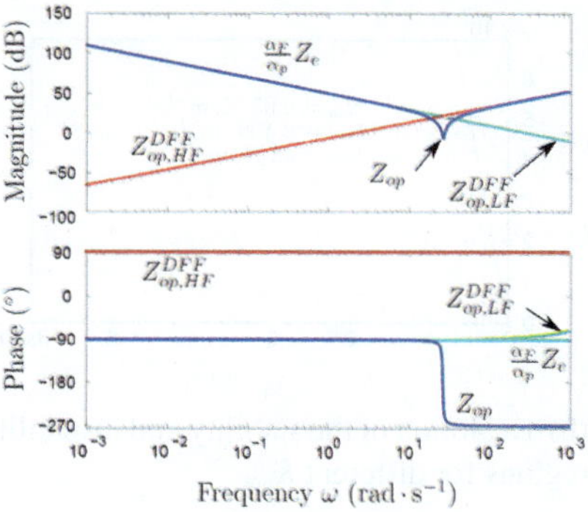

(a) **Contact** mode

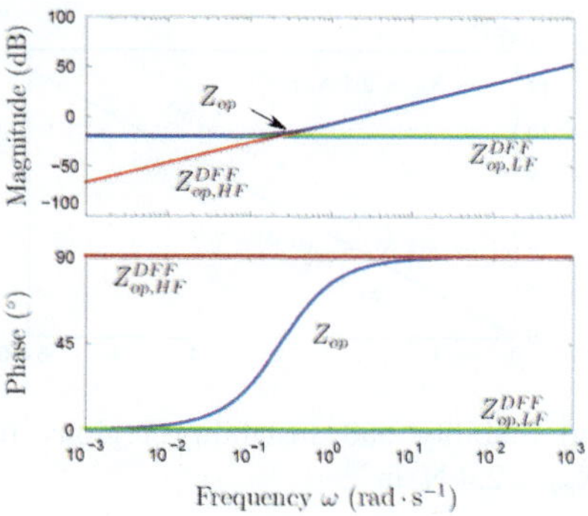

(b) **Non contact** mode

Fig. 4.4 Bode's diagram for the *DFF* control scheme. Numerical values used for these plots are: $K_{eq} = 2.4\ \mathrm{N \cdot m^{-1}}$, $\alpha_F = 6 \cdot 10^6$ et $\alpha_p = 0.05 \cdot 10^6$. These values are used to perform experiments in following sections.

4.2.3.3 Stability

To determine necessary and sufficient conditions of stability, the Routh-Hurwitz criterion is applied to the closed loop transfer function, since the system is considered to be linear, time invariant. It is directly deduced from the impedance on the user side computed for transparency analysis:

$$\frac{V_m(s)}{F_{op}(s)} = \frac{1}{Z_{op}(s)} = \frac{H(s)}{1 + H(s)N(s)\frac{\alpha_F}{\alpha_p}\frac{K_{eq}}{s}} \tag{4.30}$$

The corresponding characteristic equation is:

$$M_h \tau_1 \tau_2 s^4 + [M_h(\tau_1 + \tau_2) + B_h \tau_1 \tau_2]s^3 + [M_h + B_h(\tau_1 + \tau_2)]s^2 + B_h s + \frac{\alpha_F}{\alpha_p} K_{eq} = 0 \tag{4.31}$$

The Routh-Hurwitz criterion gives the following necessary and sufficient stability condition:

$$R = \frac{\alpha_F}{\alpha_p} < \frac{\gamma_{DFF}}{K_{eq}} = R_{max} \tag{4.32}$$

where $\gamma_{DFF} = \frac{B_h(\tau_1+\tau_2)\left[M_h^2+M_h B_h(\tau_1+\tau_2)+B_h^2\tau_1\tau_2\right]}{\left[M_h(\tau_1+\tau_2)+B_h\tau_1\tau_2\right]^2}$. The value of γ_{DFF} only depends on system parameters (haptic interface and nanotranslator), for a given environment. Thus, system's stability only depends on the ratio $\frac{\alpha_F}{\alpha_p}$.

Figure 4.5(a) represents this stability condition: for a given value K_{eq}, force amplification must be lower than a given value, and displacements must be scaled down to ensure stability. This condition is stricter for stiffer cantilevers (Fig. 4.5(b)).

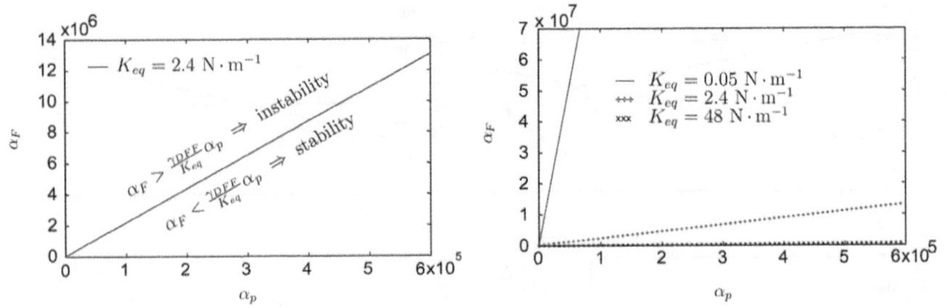

(a) Stability and instability regions for $K_{eq} = 2.4\ \mathrm{N \cdot m^{-1}}$

(b) Evolution of the stability and instability regions for different K_{eq}

Fig. 4.5 *DFF* control scheme's stability with respect to α_F and α_p

4.2.3.4 Determination of the Control Scheme Parameters

As seen in the previous stability and transparency analyses, control scheme parameters determine the performance of the system. Based on the previous paragraph conclusions, this section will highlight the influence of the gains on the coupling. For this control scheme, only two parameters must be tuned: α_F and α_p. The choice of these coefficients must take into account the following conditions, given in descending order of priority:

1. stability: the chosen ratio must verify Equation (4.32),
2. scaling requirements: as they ensure scaling of the variables, α_F and α_p must take into account scaling constraint. Their values depend on the master and slave devices used, and the task to be realized,
3. transparency: according to Section 4.2.3.2, the system transparency only depends on master and slave device characteristics. For this coupling the choice of controller gains do not depend on transparency considerations.

4.2.3.5 Conclusion

The *DFF* coupling scheme is the simplest structure to connect a haptic device to a manipulator. This coupling is highly transparent, but suffers from stability issues.

4.2.4 Force-Position Control

To avoid stability concerns pointed out for the *DFF* control scheme, a second coupling, the Force-Position (*FP*) control scheme, is considered. Compared to the previous one, damping is added.

4.2.4.1 Control Scheme Structure

As for the *DFF*, the inputs of the Force-Position control scheme are the velocity of the haptic device handle and the force applied by the environment on the cantilever (Fig. 4.6). The outputs are the velocity used as the desired reference for the nanotranslator and the force that will be fed back to the user by the haptic device. As previously, α_F and α_p are respectively the force and velocity scaling factors.

Two controllers are added in the *FP* coupling. G_n and $C(s)$ are respectively a proportional (*P*) and a proportional-integral (*PI*) controller:

$$C(s) = \frac{F_m}{V_m - \alpha_p^{-1} V_n} = B_{fp} + \frac{K_{fp}}{s} \tag{4.33}$$

where the integral gain K_{fp} and the proportional gain B_{fp} modify respectively the damping and the stiffness of the haptic force. The second controler, G_n, computes the slave device desired velocity.

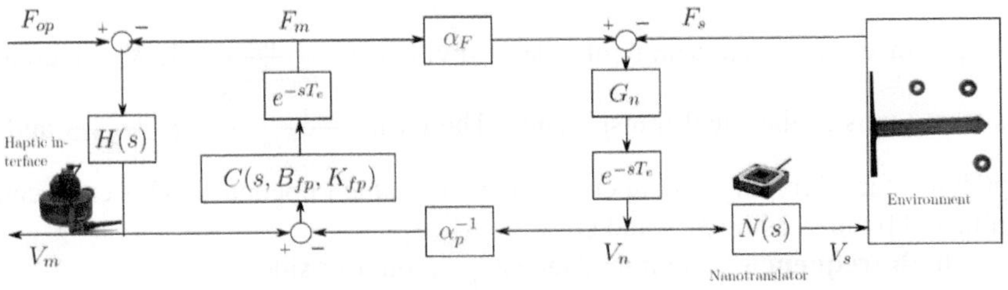

Fig. 4.6 Force-Position control scheme (*FP*)

Since the sampling period is small (a few milliseconds), the coupling is represented in the continuous time domain. Delays modeled by e^{-sT_e} are introduced in Fig. 4.6 to avoid algebraic loops when implementing the control scheme on a computer in the discrete time domain. Variables f_m and $\dot{x}_n$[1] are saved and used for computations made at the next time step.

4.2.4.2 Transparency

To analyze transparency, the impedance on the master side $Z_{op}(s) = F_{op}(s)/V_m(s)$ is computed using the control scheme depicted in Fig. 4.6:

$$Z_{op}(s) = \frac{n_{z_e}(s)Z_e(s) + \alpha_F C(s)H(s)e^{-sT_e} + \alpha_F + \alpha_p G_n C(s)e^{-2sT_e}}{d_{z_e}(s)Z_e(s) + \alpha_F H(s) + \alpha_p G_n C(s)H(s)e^{-2sT_e}} \tag{4.34}$$

[1] Lower case, resp. upper case, letters refer to time domain, resp. Laplace domain, variables.

where:

$$n_{z_e}(s) = \alpha_F G_n C(s) H(s) N(s) e^{-2sT_e} + \alpha_F G_n N(s) e^{-sT_e} \tag{4.35}$$

$$d_{z_e}(s) = \alpha_F G_n H(s) N(s) e^{-sT_e} \tag{4.36}$$

As T_e is small (several millisenconds), the Taylor series to the first order of $x \mapsto e^x$ around 0 ($T_e \to 0$), $e^{-sT_e} \approx 1 - sT_e$ is considered in the following sections.

Contact

When contact is established, $Z^{FP}_{op,LF}$ can be computed for **low frequencies**:

$$Z^{FP}_{op,LF}(j\omega) \underset{\omega \ll 1}{\approx} \frac{\alpha_F}{\alpha_p + \frac{\alpha_F K_{eq}}{K_{fp}}} Z_e(j\omega) + \frac{1}{1 + \frac{\alpha_p K_{fp}}{\alpha_F K_{eq}}} + \frac{1}{\frac{G_n K_{eq}}{K_{fp}} + \frac{\alpha_p G_n}{\alpha_F}} + B_h \tag{4.37}$$

As K_{fp} increases, the impedance felt by the operator tends to the ideal impedance $\frac{\alpha_F}{\alpha_p} Z_e$.

Impedance on the environment side is divided by $\frac{\alpha_F}{\alpha_p + \frac{\alpha_F K_{eq}}{K_{fp}}}$. If K_{fp} increases, this term tends to the ideal transparency. The term $\frac{1}{1+\frac{\alpha_p K_{fp}}{\alpha_F K_{eq}}} + \frac{1}{\frac{G_n K_{eq}}{K_{fp}} + \frac{\alpha_p G_n}{\alpha_F}}$ and the haptic interface damping add an error on the user side impedance. This error can be minimized by increasing K_{fp} and G_n.

For **high frequencies**, the impedance $Z^{FP}_{op,HF}$ on user side is:

$$Z^{FP}_{op,HF}(j\omega) \underset{\omega \gg 1}{\approx} M_h j\omega \tag{4.38}$$

For the same reasons as the ones exposed in section 4.2.3.2, the user mainly feels the inertia of the haptic device for $\omega \gg 1$.

Bode's diagrams represented in Fig. 4.7(a) illustrate analytical results, and enable to compare them to those obtained for the *DFF* control scheme.

For low frequencies, the variations in the Bode's magnitude are the same for the simulated contact impedance and for the environment. Impedances on user and environment sides only differ by a static gain which can be reduced by increasing K_{fp} and G_n. Consequently, when the user reaches the contact point, he or she is able to detect the variations of forces involved during the process.

Non contact

When no force is acting on the cantilever, impedance on the operator side is:

$$Z_{op}(j\omega) = \frac{\alpha_F \left(K_{fp} + B_{fp} j\omega\right) e^{-j\omega T_e}}{\alpha_F j\omega + \alpha_p G_n \left(K_{fp} + B_{fp} j\omega\right) e^{-2j\omega T_e}} + M_h j\omega + B_h \tag{4.39}$$

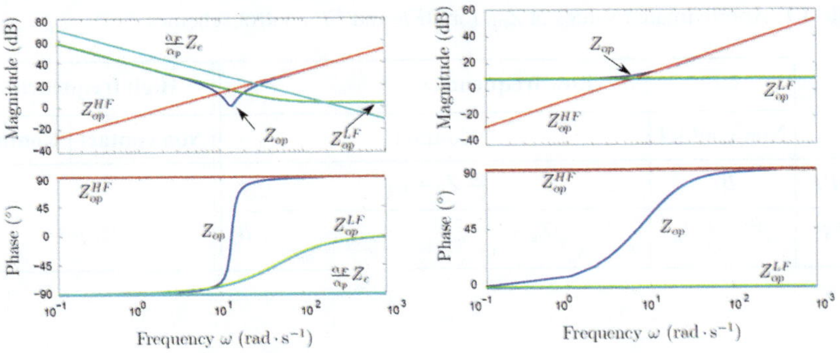

(a) **Contact** mode (b) **Non contact** mode

Fig. 4.7 Bode's diagram for the *FP* control scheme. Numerical values used for these plots are: $K_{eq} = 2.4\ \mathrm{N \cdot m^{-1}}$, $\alpha_F = 6 \cdot 10^6$, $\alpha_p = 0.05 \cdot 10^6$ (same values as the ones used in Fig. 4.4), $K_{fp} = 100\ \mathrm{N \cdot m^{-1}}$, $B_{fp} = 2\ \mathrm{N \cdot s \cdot m^{-1}}$, $G_n = 48.0\mathrm{m} \cdot^{-1} \cdot \mathrm{s}^{-1}$ et $T_e = 5$ ms. These values are used to perform experiments in following sections.

For **low frequencies**, the impedance $Z^{FP}_{op,LF}$ can be approximated from:

$$Z^{FP}_{op,LF}(j\omega) \underset{\omega \ll 1}{\approx} \frac{\alpha_F}{\alpha_p G_n} + B_h \tag{4.40}$$

To minimize the impedance on the operator side, without affecting the scaling factors, G_n is the only parameter that can be tuned. The higher it is set, the better the transparency is. As for the *DFF*, the user also feels the viscosity of the haptic interface.

At **high frequencies**, the impedance is: $Z^{FP}_{op,HF}(j\omega) \underset{\omega \gg 1}{\approx} M_h j\omega$. As for the *DFF* coupling, the operator will mainly feel the inertia of the haptic device.

The Bode's diagram (Fig. 4.7(b)) confirms the validity of the approximations made in previous equations. When compared to the *DFF* control scheme Bode's diagram (Fig. 4.4(b)) for non-contact mode, it highlights the lack of transparency of the *FP* control scheme for low frequencies. However, this difference can be reduced by increasing G_n.

Conclusion

As highlighted in Table 4.1 which summarizes approximated impedances for the *DFF* and *FP* control schemes, the *FP* coupling is less transparent than the *DFF*. To obtain the same transparency K_{fp} has to be high enough to ensure a stiff contact, while G_n will influence the contact and non-contact behavior of the coupling. The higher it is, the less viscous the feeling will be.

Table 4.1 Approximated values of Z_{op} for *DFF* and *FP* control schemes

	Low frequencies		High frequencies	
	Non contact	**Contact**	**Non contact**	**Contact**
DFF	B_h	$\frac{\alpha_F}{\alpha_p} Z_e + B_h$	$M_h j\omega$	
FP	$\frac{\alpha_F}{\alpha_p G_n} + B_h$	$\frac{\alpha_F}{\alpha_p + \frac{\alpha_F K_{eq}}{K_{fp}}} Z_e + \frac{1}{1 + \frac{\alpha_p K_{fp}}{\alpha_F K_{eq}}} + \frac{1}{\frac{G_n K_{eq}}{K_{fp}} + \frac{\alpha_p G_n}{\alpha_F}} + B_h$		

4.2.4.3 Stability

Applying the Routh-Hurwitz criterion to verify the stability of the *FP* coupling does not provide easily interpretable analytical conditions due to the complexity of this control scheme. Thus, the methodology applied to analyse the *DFF* coupling is not adapted here. A different approach is proposed. First, necessary stability conditions between controller gains are derived using simplified control schemes, corresponding to specific situations. Numerical values are then selected within the determined bounds, and according to transparency requirements given in Section 4.2.4.2. Stability of this coupling is then verified numerically using Llewelyn criterion.

4.2.4.4 Determination of the Control Scheme Parameters

Scaling factors

As for the first control scheme, α_F and α_p are chosen according to amplification constraints, due to the master and slave device characteristics, and the task to be performed. Precise positioning, and/or important force feedback might be needed.

Proportional integral controller B_{fp} and K_{fp}

To highlight the influence of the gains B_{fp} and K_{fp}, a simplification of the control scheme in the case of an infinitely stiff contact is proposed in [27]. The velocity V_n is null, and the simplified closed loop transfer function is then:

$$\frac{V_m(s)}{F_{op}(s)} = \frac{H(s)}{1 + H(s)C(s)e^{-sT_e}} \tag{4.41}$$

Using the first order Taylor development of $x \mapsto \exp(x)$ in the region of 0, $e^{-sT_e} \approx 1 - sT_e$. The control scheme characteristic equation is then:

$$(M_h - B_{fp}T_e)s^2 + (B_{fp} + B_h - K_{fp}T_e)s + K_{fp} = 0 \tag{4.42}$$

The system is stable if the Routh-Hurwitz criterion is achieved, i.e. if and only if all the coefficients of this equation have the same sign. This implies:

$$M_h > B_{fp}T_e \tag{4.43}$$

$$B_{fp} + B_h > K_{fp}T_e \tag{4.44}$$

These relations highlight the importance of the sampling period which must be kept as low as possible. The stiffness of the coupling is bounded by the inherent damping of the haptic interface and that added by the coupling. The maximum damping (and therefore, the maximum stiffness of the coupling), admissible is limited by the inertia of the master arm. These conclusions are consistent with [27].

Proportional controller G_n

Since the sampling period is low, stability analysis is conducted in the continuous time domain. However, while implementing the control scheme on a computer, discrete time computations are performed. The force f_m at time step $k+1$ is computed using position and velocity at time step k (see Fig. 4.6):

$$f_m(k+1) = B_{fp}\Delta\dot{x}(k) + K_{fp}\Delta x(k) \tag{4.45}$$

where:

$$\begin{aligned} \Delta\dot{x}(k) &= \alpha_p\dot{x}_n(k) - \dot{x}_m(k) \\ \Delta x(k) &= \alpha_p x_n(k) - x_m(k) \end{aligned} \tag{4.46}$$

For similar reasons, $\dot{x}_n$ is computed using:

$$\dot{x}_n(k+1) = G_n\left[f_s(k) - \frac{1}{\alpha_F}f_m(k)\right] \tag{4.47}$$

Considering (4.45) and (4.47) and the fact that the position x_n is computed using Tustin's discretization, when the probe is well above the substrate (no force applied on it, i.e. $f_s = 0$), f_m is given by:

$$f_m(k+1) = \lambda_1 f_m(k-1) + \lambda_2\dot{x}_m(k) + (\lambda_3 + \lambda_4)K_{fp} \tag{4.48}$$

where:

$$\begin{aligned} \lambda_1 &= -G_n\left[\frac{B_{fp}\alpha_p + K_{fp}\alpha_p\frac{T_e}{2}}{\alpha_F}\right] & \lambda_2 &= -\left[B_{fp} + K_{fp}\tfrac{T_e}{2}\right] \\ \lambda_3 &= \alpha_p\left[x_n(k-1) + \tfrac{\dot{x}_n(k-1)T_e}{2}\right] & \lambda_4 &= -\left[x_m(k-1) + \tfrac{\dot{x}_m(k-1)T_e}{2}\right] \end{aligned}$$

So that numerical computations do not diverge, λ_1 is limited. G_n must be lower than a maximum value (necessary stability condition):

$$|\lambda_1| < 1 \Leftrightarrow G_n < \frac{\alpha_F}{\alpha_p K_{fp} \frac{T_e}{2} + \alpha_p B_{fp}} = G_{n_{lim}} \tag{4.49}$$

This condition depends on how the control scheme is implemented. Solutions to solve the algebraic loop might be considered to increase this maximum value if necessary.

Summary

To determine the controller gains, both transparency considerations given in Section 4.2.4.2 and stability conditions given in previous sections must be taken into account. Table 4.2 summarizes these conditions, and gives minimal and maximal values admissible for each gain.

Table 4.2 Valid range of values for *FP* controller gains.

	Minimum value	Maximum value
K_{fp}	high for a stiff contact feeling (Sec. 4.2.4.2)	Eq. (4.43), Eq. (4.44) $\rightarrow \frac{M_h}{T_e^2} + \frac{B_h}{T_e}$
B_{fp}	Eq. (4.43)$\rightarrow K_{fp}T_e - B_h$	low for non contact transparency (Sec. 4.2.4.2) and lower than $\frac{M_h}{T_e}$ for stability
G_n	high for both contact and non contact transparency (Sec. 4.2.4.2)	Eq. (4.49) $\rightarrow \frac{\alpha_F}{\alpha_p K_{fp} \frac{T_e}{2} + \alpha_p B_{fp}}$

As seen in paragraph 4.2.4.2, G_n must be high to allow transparent contact and non contact behaviors. According to Equation (4.49), for stability reasons, G_n must be smaller than $\frac{\alpha_F}{\alpha_p K_{fp} \frac{T_e}{2} + \alpha_p B_{fp}}$. Therefore, smaller values for B_{fp} and K_{fp} allow for a higher $G_{n_{lim}}$, and increase transparency.

Concerning K_{fp}, a maximum value to verify the Routh-Hurwitz criterion is derived. However, as explained above, for transparency reasons K_{fp} should be low enough to limit the viscosity when there is no contact.

The factors α_F and α_p are chosen according to scaling requirements, and must be consistent with Table 4.2.

4.2.4.5 Numerical Stability Analysis with the Llewelyn Criterion

To derive stability conditions for the *FP* control scheme, the Llewelyn criterion is used. The admittance matrix $\mathbf{P}$ of the coupling is defined as:

$$\begin{aligned} \begin{bmatrix} V_m \\ V_s \end{bmatrix} &= \mathbf{P}(z) \cdot \begin{bmatrix} F_{op} \\ F_s \end{bmatrix} \\ &= \begin{bmatrix} p_{11}(z) & p_{12}(z) \\ p_{21}(z) & p_{22}(z) \end{bmatrix} \cdot \begin{bmatrix} F_{op} \\ F_s \end{bmatrix} \end{aligned} \tag{4.50}$$

where:

$$\begin{aligned}
p_{11}(z) &= V(z)\left[\alpha_F + \alpha_p G_n C(z)\right]/D(z) \\
p_{12}(z) &= \left[\alpha_F\, \alpha_p G_n C(z) V(z)\right]/D(z) \\
p_{21}(z) &= \left[G_n V(z) C(z)\right]/D(z) \\
p_{22}(z) &= \alpha_F\, G_n\left[1 + C(z) V(z)\right]/D(z) \\
D(z) &= \alpha_F + \alpha_p G_n C(z) + \alpha_F V(z) C(z)
\end{aligned}$$

Inequalities that these coefficients must verify are detailed in Section 4.2.2.1. They will be numerically verified for the chosen gains in experimental sections below.

4.2.4.6 Conclusion

Since damping is considered in this control scheme, it is more stable than the *DFF* coupling. However, transparency is decreased. To address these issues, a third control scheme is proposed in the next section.

4.2.5 Passivity Based Bilateral Control

In the previous section, a virtual linear damper with a constant damping value is added. With this virtual element, the stability is substantially improved. However, such a damper decreases the transparency, and causes an artificial viscous feeling during operation, even for free motion. For better transparency, such a damper should be added only when the system gets unstable.

As seen in Section 4.2.2.2, passivity control is an elegant way to include a virtual and adaptive dissipative term that forces the system to remain passive (and stable) at all times. Taking a mechanical plant on the master side with an initial energy E_0 into consideration, its energy at the kth step ($k \in Z^+$) in sampled-time is given as [8]:

$$E_k = E_{k-1} + F_m(t_{k-1})\Delta z_m(t_{k-1}), \tag{4.51}$$

where $\Delta z_m(t_{k-1}) = z_m(t_k) - z_m(t_{k-1})$. For a system to be passive, E_k should never be negative.

For teleoperation, it is sufficient to use an energy estimator on just the master side (haptic device), and add a dissipative force to remove any energy that could be generated. In other words, an energy estimator in the form:

$$\widehat{E}_{k+1} = E_k + F_{un}(t_k)\Delta z_m(t_k) \tag{4.52}$$

can be used to estimate the energy of the master side before any possible compensation to the force. According to the energy estimate, the output of the passivity controller is defined as the force:

$$F_c(t_k) = \begin{cases} -\frac{\widehat{E}_{k+1}}{\Delta z_m(t_k)}, & \text{if } \widehat{E}_{k+1} < 0 \\ 0, & \text{if } \widehat{E}_{k+1} \geq 0 \end{cases} \tag{4.53}$$

and

$$F_m(t_k) = F_{un}(t_k) + F_c(t_k). \tag{4.54}$$

Therefore, the energy of the system at the next time step becomes:

$$E_{k+1} = E_k + F_m(t_k)\Delta z_m(t_k) \tag{4.55}$$

$$E_{k+1} = \widehat{E}_{k+1} + F_c(t_k)\Delta z_m(t_k). \tag{4.56}$$

These relations are directly applicable for simulated or experimental teleoperation at macro-scales. However for teleoperated nanomanipulation, nanoscale physics must be considered. Consider a nanoprobe that is moving towards a nanosurface; as the tip approaches the surface, van der Waals forces will start to attract the tip and cause the cantilever to bend, resulting in a negative force reading by the AFM [14]. This phenomenon is significant at the nanoscale and is negligible at larger scales. Since the van der Waals force occurs in the same direction as the velocity of the probe, it is essentially active behavior, which the passivity controller will dissipate.

Here, the trade-off between stability and transparency for teleoperation is most apparent. To account for the active behavior while not compromising stability, there are three approaches. The first approach is to distinguish the active and desired behaviors from the undesired ones. This is a very complex task, as the van der Waals forces can be responsible for instability, for example after an indentation task. The attractive forces would increasingly pull the tip as it approaches the substrate, and will be pushed away by a stronger repulsive force after indentation, which can cause a self-sustaining loop and result in increasing oscillations. This is particularly true for very stiff surfaces, combined with the inherent time delay in teleoperation. A second approach is to give a positive initial energy to the system, which allows for some energy generation. This approach allows the active behavior to be observed initially until the initial energy is depleted. However, this is not a complete solution. The third approach, utilized in this work, is to relax the energy constraints by supplying a maximum damping coefficient, $c_{max} \in \Re^+$, which will never be exceeded. Hence, it provides a saturation to the adaptive dissipative force: $F_c = -cv_z$, where v_z is the vertical-axis velocity and $0 \leq c \leq c_{max}$ is the effective damping constant of the dissipation force. This approach enables the system to become active for limited amounts of time. Thus, generated energy will not be dissipated instantaneously, but over a time interval. In consequence, active forces will be reduced but not removed (depending on c_{max}), and the van der Waals interactions can still be felt by the operator. Selection of c_{max} is done experimentally for the hardest surface used in the experiments (a glass substrate). c_{max} is the tuned coefficient of a constant virtual damper determined from a previous experiment.

A passivity based controller is proposed in Fig. 4.8, which includes transparency adjustment detailed in Section 4.2.2.1 and impedance transform detailed in Section 4.2.2.3.

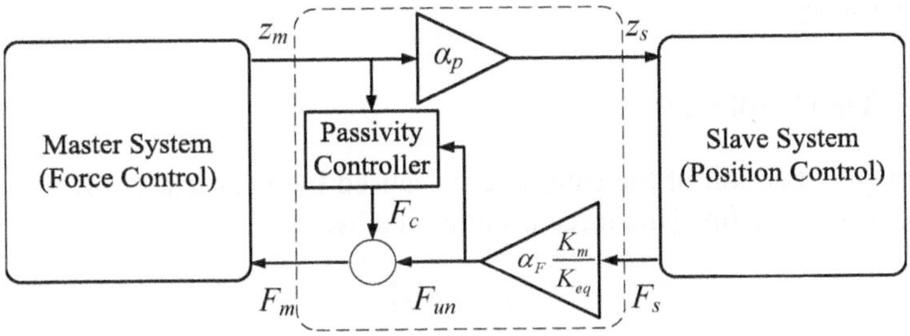

Fig. 4.8 The proposed teleoperated nanomanipulation scheme with transparency adjustment, impedance transform, and passivity control.

4.2.6 Conclusion

Three control schemes are proposed in this section. They present different performances in terms of transparency and stability. Before experimentally comparing them, teleoperation platforms are described.

4.3 Experimental Platforms

The experiments discussed in this chapter are performed on two custom-built main platforms in both our laboratories at ISIR and Carnegie Mellon. Both of these platforms use AFM force measurement to compute haptic feedback. A brief overview of this method, and issues for haptic coupling is presented. Then, the material used in each platform is detailed below.

4.3.1 AFM Force Measurement

As seen in previous chapters, in AFM force measurement deformations of the manipulation tool are measured using a laser beam, reflected onto a photodiode. Two voltage outputs are available, V_{A-B} and V_{LFM}, which correspond respectively to a displacement of the laser spot on the vertical (resp. horizontal) axis of the photodiode. After calibration (see Chapter 2), two deformations of the manipulation tool are measured from these outputs: its vertical deflection, ζ, and its angle of rotation ϕ. Interaction forces between the tool and its environment are computed from these measurements, and a model of the probe. In particular, available forces depend if the cantilever is a tip or a tipless one. A brief summary of the main results are given in the following.

4.3.1.1 Tip Cantilever

The vertical deflection of the cantilever is caused by a vertical effort f_s^z applied at the extremity of the tip. This force is computed by:

$$f_s^z = K_p \zeta \tag{4.57}$$

where K_p is the vertical stiffness of the cantilever.

The rotation of the cantilever is caused by a horizontal force f_s^y. This force is applied with a lever arm h corresponding to the tip height. If the angle of rotation is small, the force can be approximated by:

$$f_s^y = \frac{k_{t_\phi}}{h} \phi \tag{4.58}$$

where k_{t_ϕ} is the torsional stiffness of the cantilever.

Using the two voltage signals, assuming small beam deflections, two force readings can be measured. These two force readings are a combination of the 3-D forces applied to the AFM tip. For 3-D teleoperation experiments, a general decoupling procedure is applied to achieve the full force vector acting on the tip, as detailed in Section 4.6.1.

4.3.1.2 Tipless Cantilever

Compared to a tip cantilever, two main differences must be noted:

- the point on which the force is applied is not fixed, but can be situated at any point P (x_p, y_p) on the cantilever (see Fig. 4.9),
- no torsion can be applied on the cantilever since the lever arm due to the tip does not exist for a tipless cantilever.

Due to these differences, interaction forces that can be measured are modified.

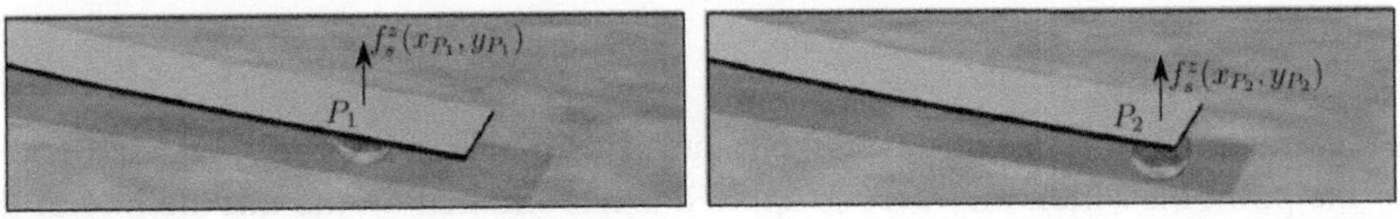

Fig. 4.9 Point on which the force is applied for a tipless cantilever: this position depends on the object position.

Deflection ζ

Similarly to the tip cantilever case, the deflection of a tipless cantilever is caused by a vertical interaction force. The point P on which this force is applied varies depending on the position of the object:

$$f_s^z(x_p, y_p) = k_z(x_p)\zeta(x_p, y_p) \tag{4.59}$$

where $K_p(x_p)$ is the normal stiffness of the cantilever at point x_p:

$$K_p(x_p) = \frac{Ewt^3}{4x_p^3} = \left(\frac{L}{x_p}\right)^3 K_p(L) \tag{4.60}$$

where $K_p(L)$ is the normal stiffness of the cantilever at its extremity, and corresponds to the stiffness used for a tip cantilever.

For a fixed-free cantilever, the deflection $\zeta(x, y_p)$ is:

$$\zeta(x, y_p) = \frac{f_s^z(x_p, y_p)}{EI_y}\left(x\frac{x_p^2}{2} - \frac{x_p^3}{6}\right) \quad \text{for } x_p \leq x \leq L \tag{4.61}$$

According to this equation:

$$\zeta(x_p, y_p) = \frac{2x_p}{3L - x_p}\zeta(L, y_p) \tag{4.62}$$

The interaction force is computed using Equations (4.60) and (4.62):

$$f_s^z(x_p, y_p) = \frac{2L^3}{x_p^2(3L - x_p)}K_p(L)\zeta(L, y_p) \tag{4.63}$$

For technical reasons, the laser spot is situated at the extremity on the cantilever, on its middle line (point $E(L, 0)$). The only available deflection is $\zeta(L, 0)$. Thus only $f_s^z(x_p, 0)$ can be known precisely.

Rotation ϕ

As the probe is a tipless cantilever, the rotation is not due to a force f_s^y, but to the fact that the point on which the force is applied is not situated on the middle line of the cantilever (i.e. $y_p \neq 0$). The computation of the cantilever's deformation is made by considering a cantilever along the y axis (free at its extremity). The hyptothesis of a monodimensional problem is made. The force $f_s^z(x_p, y_p)$ applied along the vertical axis causes a rotation $\phi(x_p, y_p)$ of the cantilever:

$$\phi(x_p, y_p) = \frac{f_s^z(x_p, y_p)}{2EI_x} y_p^2 + \phi(x_p, 0) \tag{4.64}$$

The momemtum $M_t(x_p, y_p)$ that would correspond to this angle is:

$$M_t(x_p, y_p) = k_{t_\phi}(x_p)\phi(x_p, y_p) \tag{4.65}$$

where $k_{t_\phi}(x_p) = \frac{Gwt^3}{3x_p}$ it the torsional stiffness. Since the rotation angle is measured at the extremity of the cantilever, and on its middle line, the only available measure is $M_t(L, 0)$.

As in the case of a tip cantilver, only some interaction forces applied on tipless cantilevers can be determined using laser measurement.

4.3.1.3 From AFM Force Measurement to Haptic Forces

All the measured forces on the AFM cantilever are scaled with the same factor and fed into one of the bilateral controllers designed in Section 4.2.3, 4.2.4, and 4.2.5; and output scaled forces are written to the haptic device. As a result, the operator feels and reacts to the scaled force from the slave side, closing the control loop. The haptic device, however, does not have a force sensor; therefore the force that is actually exerted on the operator is an estimate. The force is estimated by subtracting the inertial forces from the force reading of the haptic device (which solves the forward kinematics using joint torques) as: $\widehat{F_m} = F_h - m_h a_e$, where $\widehat{F_m}$ is the estimated force on the master side, F_h is the haptic device force reading, m_h is the mass of the haptic device's end-effector, and a_e is the estimated acceleration of the end-effector. a_e is estimated from [28]: $a_e = K_1(z - z_e) - K_2 v_e$, where K_1 and K_2 are proportional gains, z is the current end-effector position, z_e is the estimated end-effector position, and v_e is the estimated end-effector velocity.

4.3.2 *Material for Experiments*

4.3.2.1 ISIR Experimental Platform

The ISIR teleoperation platform is depicted on Fig. 4.10. The vision system is composed of an optical feedback (from an Olympus BX50WI microscope) with a 20× magnification.

To position the cantilever accurately with respect to the scene, three nanotranslators are available. The range (resp. resolution) provided by this system is equal to 50μm (resp. 0.1nm) for the x and y directions, 12μm (resp. 1.83nm) for the z direction.

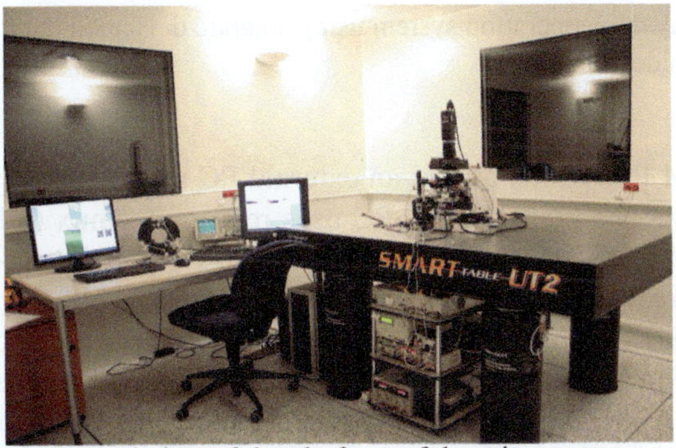

(a) General view of the platform of the micro-nano manipulation platform

(b) Virtuose haptic interface

Fig. 4.10 ISIR experimental setup for haptic feedback teleoperation

Available haptic devices are a Virtuose, depicted on Fig. 4.10(b) (Haption[2]), and a Omega (Force Dimension[3]). Both devices allow 3 degrees of freedom of force feedback. To avoid damages on the master device, forces sent back to the user are saturated to 6N and 3N respectively.

4.3.2.2 Carnegie Mellon Experimental Platform

Figure 4.11 displays the overall layout of the bilateral teleoperation system of the Carnegie Mellon platform. An Atomic Force Microscope (Veeco[4], Autoprobe M5) with an AFM cantilever (Pointprobe, NanoSensors[5], Neuchatel, Switzerland, K_p = 1.69 N/m, k_l = 2.46 N/m) is used as the slave nanomanipulator, which is accessed by a Windows 95-based PC (AFM PC) through the Veeco SPMAPI. A client-server

[2] http://www.haption.com
[3] http://www.forcedimension.com
[4] http://www.veeco.com
[5] http://www.nanosensors.com

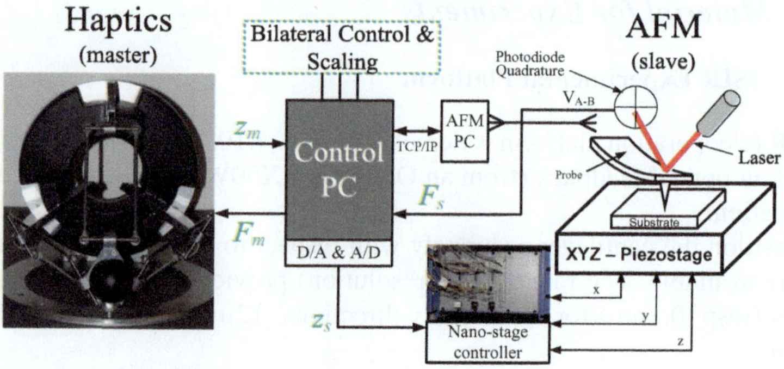

Fig. 4.11 System layout of the scaled teleoperation system using a haptic device on the master side and an AFM on the slave side.

program is created on the AFM PC, which allows an external PC to connect to the AFM through TCP/IP Ethernet. The main control PC uses real-time Linux (RTAI 3.3, Ubuntu Linux 2.6.15) and an interface program written in C++. It interfaces with the AFM PC through a direct Ethernet connection, and a 3-DOF haptic device (Force Dimension Inc[6], Omega) through USB. A 3-DOF piezoelectric nanopositioning stage (Physik Instruments[7] P-753, 12 μm range, <1 nm precision) is used as the slave positioner in the AFM, which allows for a faster control bandwidth compared to actuating the AFM's nanopositioner through TCP/IP. The three axes of the piezo stage are controlled through its dedicated controller (Physik Instruments E-612) by the main control PC using three D/A outputs (Adlink[8] PCI-6208). Positions are read from the amplifier using an A/D converter (National Instruments[9] PCI-6024E). In addition, the AFM's normal (A-B voltage signal) and lateral (LFM voltage signal) deflection signals are read directly from the AFM with the A/D converter. The total control bandwidth of the system is 1 kHz, which is limited by the data acquisition system.

For operation, a sample is placed on the custom nanopositioner and the AFM probe is automatically brought into contact with the sample, where contact-mode AFM is being used. By default, the AFM servos the z-position of the probe to maintain a constant force with the substrate. For our bilateral control experiments, this z-servo controller is disabled. Before beginning the experiments, we wait for any drift in the system to dissipate. Additionally, a cross-talk calibration routine, previously designed in [29], is utilized to remove this detrimental effect on the measured voltage signals.

---

6 http://www.forcedimension.com
7 http://www.physikinstrumente.com
8 http://www.adlinktech.com
9 http://www.ni.com

When a force is applied onto the AFM probe due to the sample motion, the deflection signals V_{A-B} and V_{LFM} reflect this. V_{A-B} and V_{LFM} are filtered by a second order 60 Hz analog low-pass-filter circuit. This value is chosen due to experiments on human compliance discrimination thresholds mentioned in [5].

4.4 1-D Teleoperated Touch Feedback Using AFM

In this section, bilateral couplings presented in Section 4.2 are compared, the influence of the controller gains is investigated, and experimental performance data are gathered on different substrates. To illustrate these points, one-dimensional manipulation experiments are performed on the platforms described in the previous section. 1-D teleoperation is the simplest case of bilateral touch feedback in the nanoscale. It consists of approaching the substrate with the probe, applying a force and finally retracting from the substrate. The force applied on the cantilever as a function of the distance from the substrate is depicted in Fig. 4.12(a) (arrows indicate the chronological path). The corresponding plot of the force applied on the cantilever as a function of time is given in Fig. 4.12(b). That is what the operator should ideally feel for the basic approach-retract experiments.

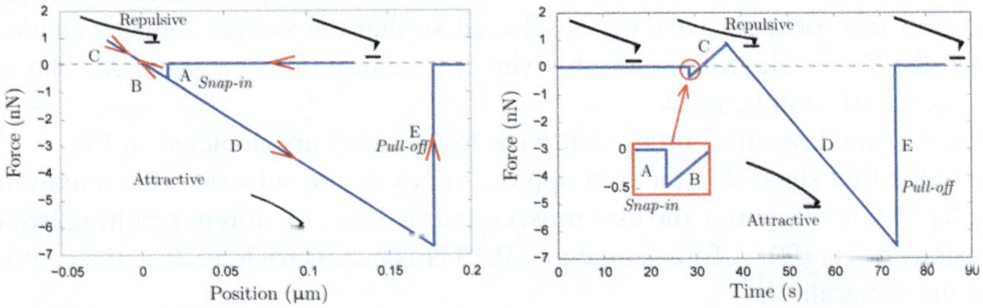

(a) Force as a function of position with respect to the substrate

(b) Force as a function of time

Fig. 4.12 Force applied on a cantilever while performing an approach-retract experiment

The bilateral coupling must render to the user the following phenomena present at the nanoscale:

- Two discontinuous phenomena (referenced as A and E on the plots), are respectively representative of the instants when the cantilever is attracted (pull-in) and released (pull-off) from the substrate.
- Two types of interacting forces are present: the very first (B) and the last (D) are attractive (after the pull-in and before the pull-off), while those in between are repulsive (C).
- Variations of the force magnitude are high between pull-in and pull-off phenomena (factor of 10 to 100).
- The pull-in force magnitude can be in the order of tens of nanonewtons (nN).

This experiment is performed with the three control schemes detailed in Section 4.2.1.

4.4.1 Approach Retract Experiment Using Direct Force Feedback Control

The approach-retract experiment is first realised using the *DFF* control scheme. Both contact, snap-in and pull-off forces detailed above should be transmitted through the haptic feedback.

4.4.1.1 Contact and Pull-Off Forces

The first experiment is made using the *DFF* control scheme. A cantilever with a spring constant $K_p = 2.4\text{N}\cdot\text{m}^{-1}$ is used. According to Equation (4.32) and Fig. 4.5(a), the maximum value allowed for the ratio R to assure a stable system is $R_{max} = 21.7$.

The scaling factor α_p is chosen according to the master and slave motion ranges. To get a good compromise between easiness and accuracy of manipulation, the value $\alpha_p = 0.05 \cdot 10^6$ is selected (the master motion range corresponds then to a displacement of 5 μm for the slave). α_F is selected so that the system remains stable, i.e. so that the Routh-Hurwitz criterion given in Equation (4.32) is satisfied. The value $\alpha_F = 0.12 \cdot 10^6$ will be used.

The obtained results (forces sent back to the user) are depicted in Fig. 4.13(a). During the first stage (C), the user applies forces on the substrate (the maximum is 0.30 N). While retracting, the user must counterbalance the forces resulting from the adhesion effects (D). A force equal to -0.13 N is necessary to release the cantilever from the substrate (E).

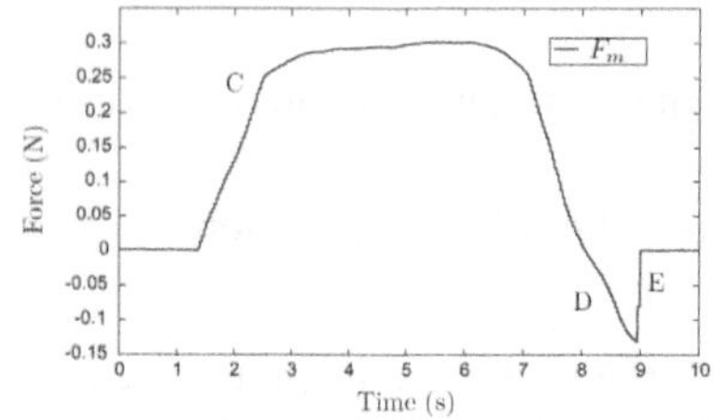

(a) $\alpha_p = 0.05 \cdot 10^6$, $\alpha_F = 0.12 \cdot 10^6$. $R < R_{max}$

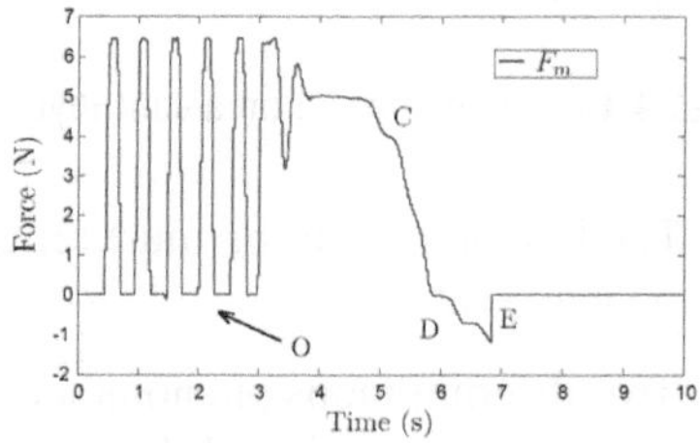

(b) $\alpha_p = 0.05 \cdot 10^6$, $\alpha_F = 6 \cdot 10^6$. $R > R_{max}$

Fig. 4.13 Approach-retract experiments for a cantilever of stiffness $K_p = 2.4\ \text{N}\cdot\text{m}^{-1}$. *DFF* control scheme

The system remains stable and the forces sent back to the user are equal to those measured by the photodiode scaled by α_F ($F_m = \alpha_F F_s$ according to the control scheme's design). Therefore, the operator indeed feels phenomena happening in the environment. However, using our haptic interface, forces fed back remain too low to increase the accuracy of the performed task. The force scaling factor has to be increased in order to provide better force feedback.

During this second experiment, the value of α_p remains the same, but α_F is increased: $\alpha_F = 6.0 \cdot 10^6$. With these values: $R = 120 > R_{max}$. The Routh-Hurwitz criterion is no longer satisfied. Consequently, the system is predicted to be unstable.

Experimental results are plotted in Fig. 4.13(b). Forces have been amplified compared to the ones in Fig. 4.13(a). It is then easier for the user to detect repulsive and attractive forces (C, D) and the high variation of the forces due to the pull-off (E). However, system's stability is affected. When the cantilever establishes contact with the substrate, it creates high amplitude oscillations in the system (O). This is very disturbing for the user who has to act like a damper to absorb the excessive energy responsible for the instability.

For a given velocity-scaling factor, the *DFF* control scheme suffers from a trade-off between stability and force amplification. However, based on its good transparency performances, experiments involving weak phenomena might be considered.

4.4.1.2 Pull-In, Contact and Pull-Off Forces

Snap-in forces are very weak (10 to 100 times smaller than pull-off ones), but are inherent to nanoscale. It is thus important to transmit them to users through haptic feedback. Because of their small amplitude (several dozens of nanonewtons), it is difficult to transmit these forces through haptic feedback. A transparent coupling must be used, and scaling factors must be adapted. Force amplification factor is thus set to $\alpha_F = 6700 \cdot 10^6$. To ensure system's stability, Equation (4.32) ($\frac{\alpha_F K_{eq}}{\gamma_{DFF}} < \alpha_p$) must be verified. Thus:

- to get less restrictive conditions, a cantilever of small stiffness is chosen ($K_p = 0.05\ \mathrm{N \cdot m^{-1}}$),
- the maximum value of the scaling factor ratio is: $R_{max} = 1040$. Compared to previous experiments, the displacement scaling factor must be increased to ensure system's stability ($\alpha_p = 50 \cdot 10^6$).

Results are depicted in Fig. 4.14. The amplified interaction force, as well as the force sent to the user (i.e. after saturation) are plotted. Snap-in forces are clearly felt by users since the corresponding haptic force is around 2 N. However, two points should be noted:

- due to the scale difference, it is not possible to transmit both snap-in forces and non staturated pull-off forces,

- to ensure stability, the position scaling factor α_p is increased. Displacements are very slow, and this basic approach-retract experiment lasted more than one minute and a half.

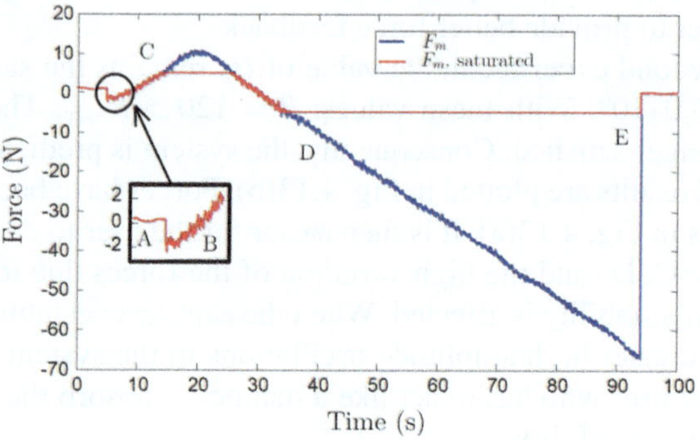

Fig. 4.14 Approach-retract experiment for a cantilever of stiffness $K_p = 0.05\ \mathrm{N \cdot m^{-1}}$, $\alpha_p = 50 \cdot 10^6$, $\alpha_F = 6700 \cdot 10^6$. $R < R_{max}$. *DFF* control scheme

Based on its high transparency, the *DFF* control scheme is adapted for experiments involving very weak forces. Snap-in forces (0.3 nN) have been transmitted to users. However, stability conditions are restrictive, and hardly compatible with nanoscale interaction force profile.

4.4.2 Approach Retract Experiment Using Force-Position Control

The same approach-retract experiment is now carried out using the *FP* control scheme. To compare the results with the ones obtained with the *DFF* control scheme, the same cantilever as in Section 4.4.1.1 ($K_p = 2.4\ \mathrm{N \cdot m^{-1}}$), and the same scaling factors ($\alpha_p = 0.05 \cdot 10^6$, $\alpha_F = 6 \cdot 10^6$) are used. Other parameters are chosen according to transparency and stability analyses (Table 4.2), and are given in Table 4.3.

Table 4.3 Numerical values of the *FP* coupling controller gains for approach-retract experiments

Raideur ($\mathrm{N \cdot m^{-1}}$)	**0.05**	**2.4**	**48.0**
α_F	$200 \cdot 10^6$	$6 \cdot 10^6$	$0.2 \cdot 10^6$
α_p	$0.05 \cdot 10^6$	$0.05 \cdot 10^6$	$0.05 \cdot 10^6$
B_{fp} ($\mathrm{N \cdot s \cdot m^{-1}}$)	1.2	2	1.5
K_{fp} ($\mathrm{N \cdot m^{-1}}$)	100	100	100
G_n ($\mathrm{m \cdot N^{-1} \cdot s^{-1}}$)	2329	48	1.98

To check if the system will be stable for the chosen gains, Llewelyn criterion is applied. Inequalities c_1, c_2 and c_3 (from Equations (4.12), (4.13) and (4.14)) are represented in the frequency domain. Results are given in Fig. 4.15 for different sampling periods. Llewelyn criterion is verified for low frequencies. However, for high frequencies, some of the inqualities are not verified anymore. The limit for which the system is stable is marked by a vertical line. The higher is the sampling period, the lower is this limit. If the sampling period is small enough (a few milliseconds), the system remains usable: due to its limited bandwidth high frequencies are attenuated.

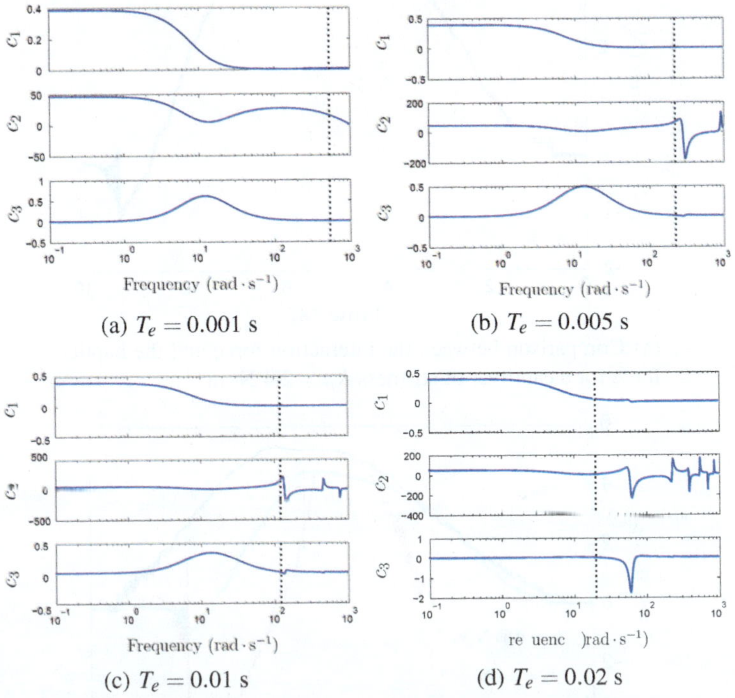

Fig. 4.15 Llewelyn criterion for a cantilever of stiffness $K_p = 2.4$ N·m^{-1}. Numerical values used for the plots are given in Table 4.3. Stability conditions are: $c_1 : Re(y_{11}) \geq 0$, $c_2 : Re(y_{22}) \geq 0$ and $c_3 : 2Re(y_{11})Re(y_{22}) - |y_{12}y_{21}| - Re(y_{12}y_{21}) \geq 0$.

To experimentally validate these conclusions and to compare this control scheme to the *DFF* one, the same approach-retract operation is performed. The obtained results (forces sent to users and amplified interaction forces) are plotted in Fig. 4.16(a). They must be compared to those in Fig. 4.13(b). The system remains stable during this experiment, contrary to the one performed with the *DFF* control scheme. Moreover, even if the force sent back to users is computed through the control scheme (and not directly fed back), the feeling that operators get reflects what happened in the remote environment since F_m and $\alpha_F F_m$ plots are similar. Oscillations that can be seen in F_m plot are induced by the virtual coupling. Since the haptic device bandwidth is limited, they are not felt by users. Therefore, although this control scheme is less transparent than the *DFF* one, the feeling is good enough to allow operators to feel the pull-off phenomenon with a force of 1 N. The analysis performed allows for an efficient tuning of the controller gains.

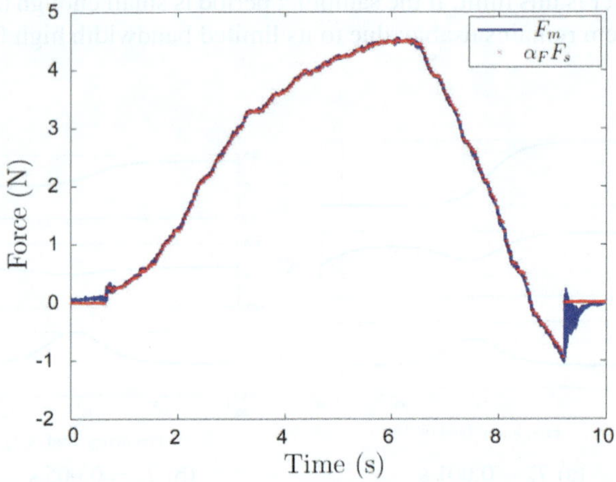

(a) Comparison between the interaction force and the haptic force for a cantilver of stiffness $K_p = 2.4\ \mathrm{N \cdot m^{-1}}$.

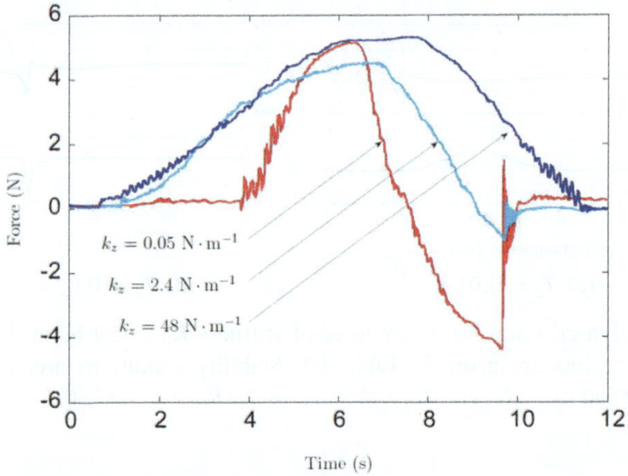

(b) Evaluation of the *Force Position* control scheme robustness and the gain tuning methodology.

Fig. 4.16 Approach-retract experiments, *FP* control scheme

To prove the robustness of our approach with respect to the environment stiffness, the same experiment is performed using cantilevers of different stiffnesses (from 0.05 to 48 $N \cdot m^{-1}$). The same displacement scaling factor is used for all the experiments. Force amplification is chosen so that users can clearly feel the contact. Other gains are chosen using Table 4.2, and numerically verified using the Llewelyn criterion. Chosen parameters are given in Table 4.3. Results are presented in Fig. 4.16(b). It should be noted that these experiments are performed in a non-controlled environment. Conditions of humidity and temperature may have changed between experiments. However, pull-off forces are indeed greater for cantilevers of low stiffness.

Transparency and stability analyses give satisfying rules for the gain tuning. The *FP* control scheme is suitable for nanomanipulations since the system remains stable for a wide variety of cantilevers (from $K_p = 0.05\ N \cdot m^{-1}$ up to $48\ N \cdot m^{-1}$). However, transparency is deteriorated.

4.4.3 Approach Retract Experiment Using Passivity Control

The last controller is the passivity controller designed in Section 4.2.5. As mentioned before, this controller has the advantage of adding an adaptive virtual damping to the system. The adaptive nature of the damping allows the controller to remove instabilities whenever they occur, while achieving better transparency than a constant virtual damping element [43]. For these experiments, an augmented reality interface is developed in the following section to compliment the experimental platform in Carnegie Mellon. This interface provides the user with side-view visual feedback of the motion by combining experimental force information with theoretical calculations.

4.4.3.1 Augmented Reality Interface

The designed bilateral controller enables a robust force feedback interface to the nanoscale world for one-dimensional touching, indenting experiments. However, the user still may not be able to visualize what really happens during these experiments. In this section, design of an augmented reality interface is explained to further enhance the user experience with a model-assisted OpenGL based visual feedback. The designed visual display is given in Fig. 4.17. A separate Linux PC is dedicated to this task. It communicates with the Haptic PC over an ethernet connection to receive the forces, material properties and animation parameters given by the operator, in a separate thread.

Assuming a spherical AFM probe tip is contacting a flat substrate slowly, measured normal forces on the slave side (F_s) can be related to the indentation (δ) and the contact radius (a) using the Maugis-Dugdale (MD) elastic contact mechanics model. MD model is superior to other contact mechanics models such as Johnson-Kendall-Roberts (JKR) and Derjaguin-Muller-Toporov (DMT) due to the fact that it works for all material combinations and sizes, using a non-dimensional parameter [30]

$$\lambda = 2\sigma_o \left(\frac{R}{\pi \Delta \gamma K^2} \right)^{1/3}. \tag{4.66}$$

Here σ_o is the constant adhesive stress defined in Dugdale model that acts over a larger circular area with radius $c \geq a$. $R = \left(\frac{1}{R_t} + \frac{1}{R_s} \right)^{-1}$ is the equivalent radius of contact between two elastic half spheres, calculated from the radii of the tip (R_t) and the substrate (R_s), respectively. For a flat substrate (*i.e.* $R_s \to \infty$), $R = R_t$. $\Delta\gamma \approx 2\sqrt{\gamma_t \gamma_s}$ is the relative surface energy of the interface. $K = 4/3 \left[(1 - \nu_t^2)/E_t + (1 - \nu_s^2)/E_s \right]^{-1}$ is the reduced elastic modulus of the interface, where ν_i is the Poisson's ratio and E_i is the elastic modulus for $i = t, s$.

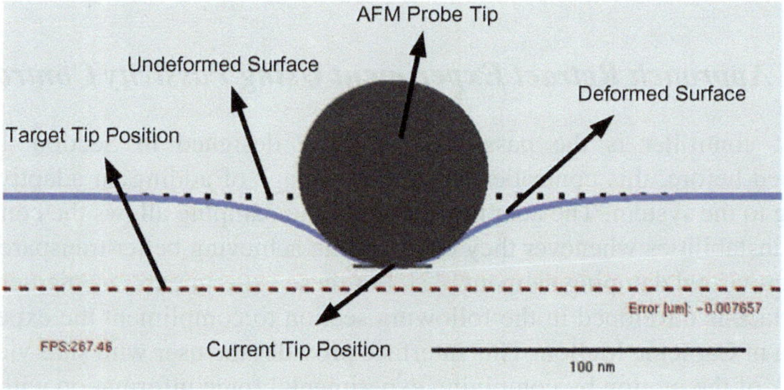

Fig. 4.17 A snapshot of the visual display during indentation of a silicone rubber substrate.

MD model approaches the DMT model for $\lambda < 0.1$ and JKR model for $\lambda > 5$, which means that the latter two models are designed for special cases of material combinations and sizes. The general equations of the MD model are three-fold [30]:

$$\begin{aligned} 1 = &\frac{\lambda \bar{a}^2}{2} \left[(m^2 - 2) \tan^{-1} \sqrt{m^2 - 1} + \sqrt{m^2 - 1} \right] \\ &+ \frac{4\lambda^2 \bar{a}}{3} \left[\sqrt{m^2 - 1} \tan^{-1} \sqrt{m^2 - 1} - m + 1 \right] \end{aligned} \tag{4.67}$$

$$\bar{F} = \bar{a}^3 - \lambda \bar{a}^2 \left[\sqrt{m^2 - 1} + m^2 \tan^{-1} \sqrt{m^2 - 1} \right] \tag{4.68}$$

$$\bar{\delta} = \bar{a}^2 - \frac{4\lambda \bar{a}}{3} \sqrt{m^2 - 1}, \tag{4.69}$$

where $m = c/a$ and

$$\overline{F} = \frac{F}{\pi R \Delta\gamma}, \tag{4.70}$$

$$\overline{\delta} = \delta \left(\frac{K^2}{\pi^2 R \Delta\gamma^2} \right)^{1/3}, \tag{4.71}$$

$$\overline{a} = a \left(\frac{K}{\pi R^2 \Delta\gamma} \right)^{1/3} \tag{4.72}$$

are in nondimensional form.

For visualization of one-dimensional experiments, the contact radius (a) and the tip-sample separation (δ) should be accurately displayed on screen. In an AFM system, with a calibrated probe, the most reliable signal is the force reading. Since all the experiments are held in ambient conditions, there should be sufficient filtering of this signal to reduce noise down to an acceptable range (a few nN in our case). Another reading is the base position of the probe (z_p), which has an arbitrary starting position. Using (4.6), substrate indentation (δ_{meas}) can be measured with an offset, which is due to this arbitrary starting position.

Using the force reading with (4.67) and (4.68), a and δ_{model} can be calculated from the model using a root finder. The indentation calculation is then compared to the measured value to account for the offset. This offset removal is done continuously when the probe is in contact with the substrate to remove possible drift effects as much as possible.

When contact is lost (*i.e.* $a = 0$), δ_{model} gives wrong results and the measured values (after offset correction) become the only reliable reading. Therefore, after pull-off, $\delta = \delta_{meas}$. Pull-off and snap-into-contact (pull-in) can be detected by sudden changes in the force reading.

The visual interface has also a scalebar for a sense of scale, and manual or automatic zooming capabilities. Automatic zooming comes handy for especially very soft surfaces that can be pulled up hundreds of nanometers before contact loss. In this case, the probe might come out of screen if not for an automated zoom-out feature based on the current δ value. After contact-loss, if the probe is brought back close to the surface, the system automatically and continuously zooms in back to the manual zoom setting by the operator.

For a dependable visual experience, the augmented reality interface must have an update rate of above 15 frames-per-second (fps), which means that other tasks (root finder, screen rendering) should be as fast as possible. For root finding, a Newton-Rhapson algorithm is utilized in this work due to its fast convergence capabilities. For rendering the surface deformation, splines are used to reduce the number of points used, and still have a smooth surface. The resultant update rate of the interface can go up to 400 fps and more than 60 fps at all times.

4.4.3.2 Force Feedback Experiments

This section displays the results of the proposed passivity-based bilateral teleoperation system for one-dimensional (vertical) touch experiments on two different substrates: a borosilicate glass surface, and a silicone rubber (SR) surface. These substrates are chosen to demonstrate teleoperation on a very hard substrate (glass) and a very soft substrate (silicone rubber).

As mentioned before, there are three modes of interest during tele-nanomanipulation. The first mode is the contact mode, where the flat substrate is indented and forces felt by the operator are positive (repulsive). The second mode is the attractive mode, where the weak inter-atomic (van der Waals) and possibly electrostatic forces between the probe tip and the substrate pulls the tip when it is sufficiently close to the substrate. Consequently, the cantilever bends downwards and the measured forces are negative (attractive). The third mode is the pulling mode, where adhesion between the tip and substrate causes upwards deformation of the substrate from its neutral position after indentation (the first mode). These forces are also negative, and can be much higher than the attractive (the second mode) forces, depending on the material properties of the substrate (surface energy, elastic modulus).

Table 4.4 displays the parameters used in experiments. The position scaling factor (α_p) was chosen according to the scaling of the vertical-axis range between the master and slave workspaces.

Table 4.4 Parameters used for teleoperation experiments

Parameter	Value
α_p	1.5×10^{-5}
E_0	0 J
c_{max}	1 Ns/m
K_{min}	40 N/m
K_{max}	200 N/m

Figures 4.18 and 4.19 show the vertical positions and forces for a teleoperation experiment on a glass substrate, respectively. Surface position is depicted as a horizontal dashed line in Fig. 4.18; since the nanopositioner is under the AFM probe, contact is achieved by moving the stage up, corresponding to moving the haptic device end-effector downwards. Only small indentations were made on the glass surface to maintain linear attractive and pulling regions on the force curve. In Fig. 4.19, a very large change of force is displayed upon contact. If the transparency adjustment was not made, these changes would not be as steep; thus, what would have been felt is the reflected probe stiffness, and not the larger surface stiffness. Similarly, if the impedance transformation was not made, force changes would be even steeper, and very uncomfortable to work with.

Figures 4.20 and 4.21 show the vertical positions and forces for a teleoperation experiment on a silicone rubber substrate, respectively. In these experiments, the

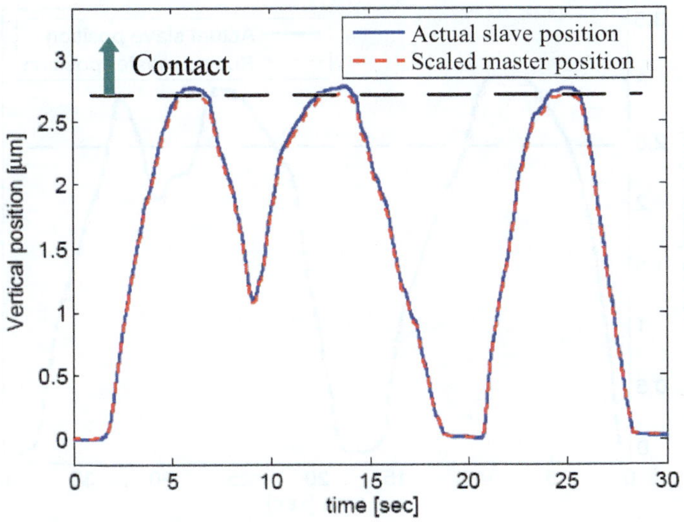

Fig. 4.18 Actual slave and scaled master vertical positions for a 3 µm range teleoperation experiment on a glass substrate.

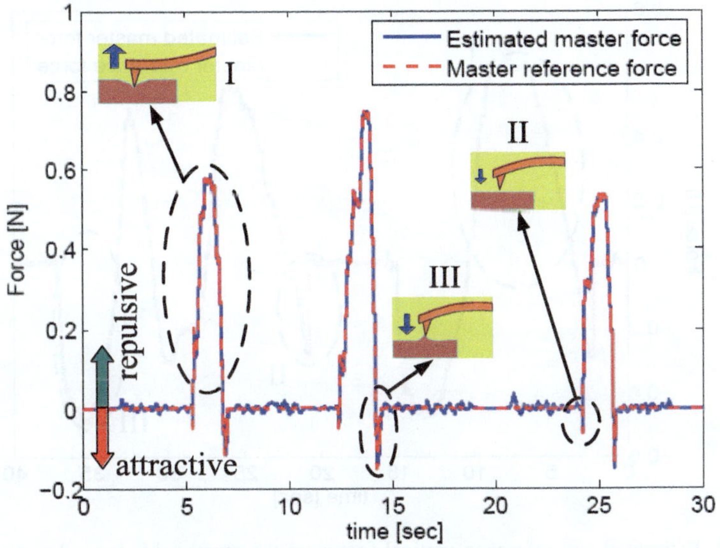

Fig. 4.19 Estimated and reference vertical forces on the master side for a 3 µm range teleoperation experiment on a glass substrate. I, II, and III indicate the three modes of interest during tele-nanomanipulation, with a graphic description.

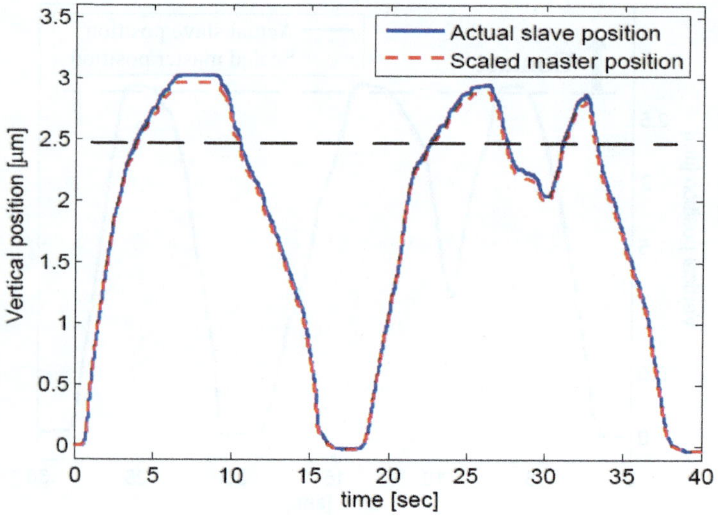

Fig. 4.20 Actual slave and scaled master vertical positions for a 3 μm range teleoperation experiment on a silicone rubber substrate. Dashed line represents the surface vertical position.

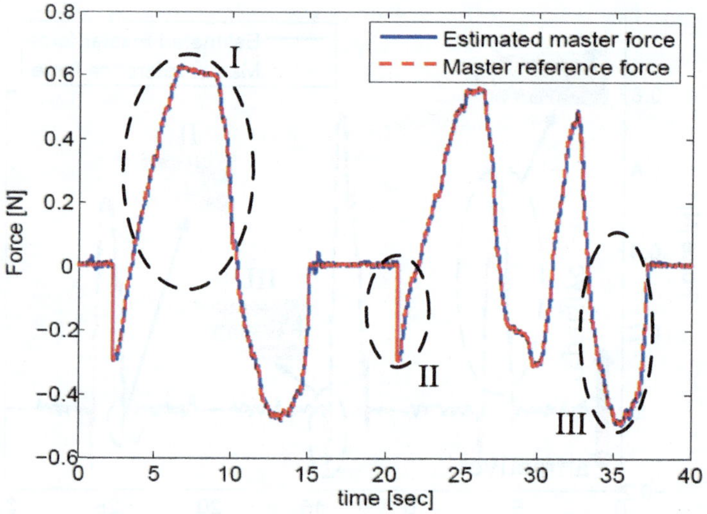

Fig. 4.21 Estimated and reference vertical forces on the master side for a 3 μm range teleoperation experiment on a silicone rubber substrate. I, II, and III indicate the three modes of interest during tele-nanomanipulation.

silicone rubber surface is indented more than a glass surface to achieve a comparable force. Since silicone rubber has stronger adhesive properties, the pulling mode is more significant than in glass. Force changes are not as rapid as in the glass case due to the low stiffness of silicone rubber. However, since silicone rubber is a very soft rubber, its untransformed force would have been actually much less than what the operator experiences in the experiment. The impedance transformation boosts the stiffness of silicone rubber such that the operator can distinguish the surface position, and hence, the amount of indentation.

4.4.3.3 Positioning Task Performance Experiments

To demonstrate the effectiveness of the proposed passivity-based control interface, three tip positioning tasks are defined as (Fig. 4.22):

Task A: Tip target is close to but above the surface, without contacting. This task is most useful for possible particle pushing experiments where tip-sample separation needs to be maintained [31].

Task B: Tip target is on the surface, in contact. This task represents topography measurements without indenting the surface.

Task C: Tip target is below the surface, in contact. This task represents normal topography measurements similar to contact mode AFM. It can also be used for the characterization of substrate material properties.

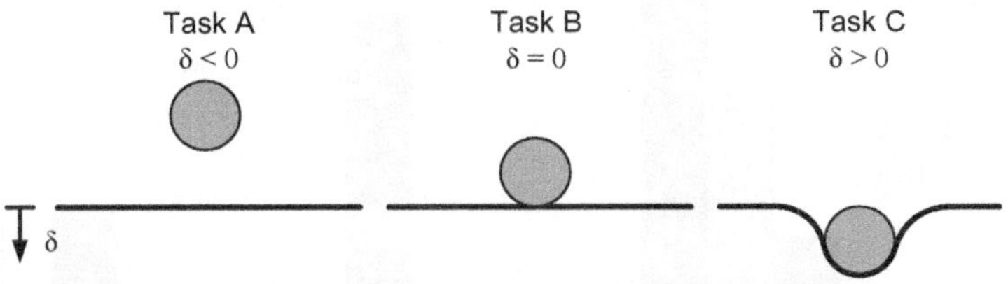

Fig. 4.22 A simple illustration of tasks with corresponding target tip positions.

Experiments are done by ten different users for at least ten target positions (of the three tasks) each, with a random order of tasks. Task A and C have three different target tip positions of $-(30, 40, 50)$ nm and $(10, 20, 30)$ nm, respectively, where positive values represent indentation. Note that, Task C is used for only soft substrates, since hard substrates may not be indented more than a few nanometers.

A norm of positioning error and completion time are the generally applied metrics for such teleoperation experiments [32]. Accordingly, two main performance metrics are chosen for the given tip positioning tasks. These metrics are namely, root-mean-squared (RMS) error between rise time and reaching time, and the average reaching speed to the target position. As each session of user experiments include a random order of tasks, the resulting profile resembles a staircase pattern with a number of steps. Consequently, rise time is defined as the time it takes for the initial error to diminish to its ten percent value. Reaching time is defined as the time after which the tip position remains in the 5 nm vicinity of the target position.

These metrics are chosen to quantify the effect of different substrates on task-based performance (*i.e.* a soft substrate is expected to give larger RMS errors and smaller reaching speeds in Task A, since it can be pulled up a lot by the tip after contact).

Table 4.5 Material properties used for tip positioning tasks [33]

	Glass	PS	SR
$E[GPa]$	72	3	0.0006
ν	0.17	0.33	0.5
$\gamma[mN/m]$	160	38	19.8

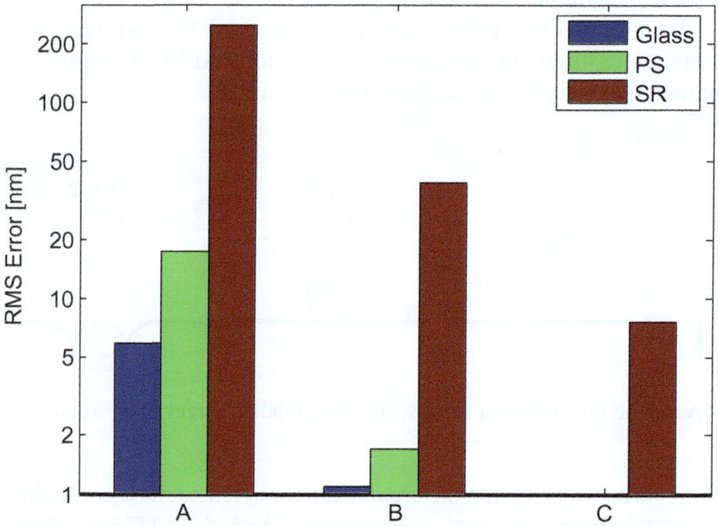

Fig. 4.23 Mean RMS error values of the vertical tip positioning experiments on three different substrates for the three different tasks defined in Fig. 4.22.

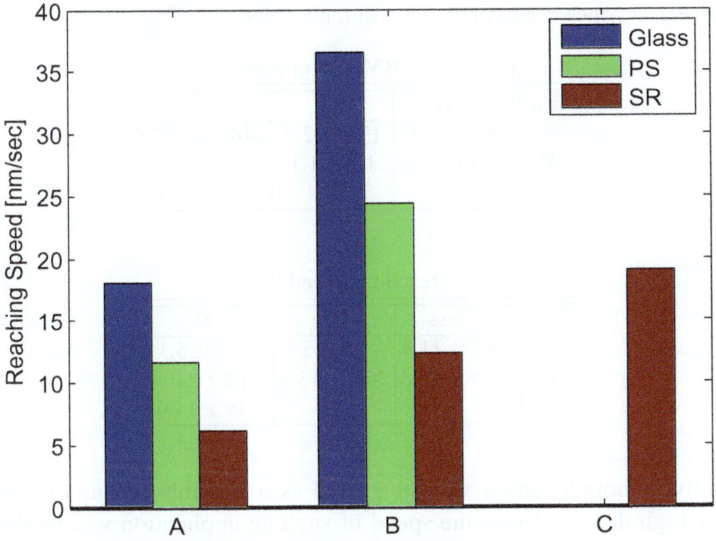

Fig. 4.24 Mean reaching speed values of the vertical tip positioning experiments on three different substrates for the three different tasks defined in Fig. 4.22.

Performance results of the experiments on three different substrates; namely glass, polystyrene (PS) and silicone rubber are given in Figs. 4.23 and 4.24. The results are also given in Table 4.6 with standard deviations included, for ease of comparison. As a general trend, RMS errors decrease and the average reaching speeds increase as the tip target gets larger, in every material. This is due to the fact that pull-off causes an overshoot in Task A. Also, as the tip is pressed more into the substrate, forces increase, which helps reduce vertical vibrations in the tip position.

Moreover, as the substrate gets softer (*i.e.* λ gets larger), positioning errors increase and reaching speeds decrease. This is also expected, since softer materials can be pulled up higher causing a larger overshoot and they provide less force, which allows more vibrations in the tip position. The previous work on similar tip positioning tasks on simulated surfaces did not yield such clear trends [33]. A possible reason for this may be that a side-view visual interface helped the users more than a 3-D view that needs to be rotated to get a clear view. Also, the performance metrics used in the two works were different, which might have a direct effect on the results.

4.4.3.4 Nanoindentation Experiments

An application of Task C for large force and penetration depth values is nanoindentation. If the substrate has plastic properties, it will permanently deform and a residual impression will remain, after the AFM tip is pressed onto it with a large enough force. This section demonstrates the possibility of teleoperated nanoindentation as a possible application of the designed teleoperation interface.

Table 4.6 Performance results of the tip positioning tasks

	RMS Error $[nm]$		
Task	Glass	PS	SR
A	5.9 ± 6.9	17.2 ± 2.8	248.9 ± 272.6
B	1.1 ± 1.3	1.7 ± 1.1	38.8 ± 15.7
C	-	-	7.6 ± 9.5

	Reaching Speed $[nm/sec]$		
Task	Glass	PS	SR
A	18.1 ± 11.4	11.6 ± 6.6	6.1 ± 5.1
B	36.5 ± 19.3	24.4 ± 13.4	12.3 ± 9.7
C	-	-	19 ± 14.4

Recently, nanoindentation was suggested as a possible means of data storage with very high density. Since the speed of such an application will be the limiting factor, current research focuses on many AFM tips working in parallel [34, 35]. A data storage application, which would certainly not be teleoperated, is beyond the scope of this work. However, the fact remains that a teleoperation interface is most useful to gain insight to nanoindentation, with a hands-on approach. Secondly, if how a surface is indented and how much force is required to achieve some amount of indentation is characterized, a more complicated automated nanomanipulation scheme can be built upon these findings to sculpt 3-D structures in the nanoscale.

In nanoindentation experiments a polymethyl methacrylate (PMMA) substrate, spin-coated on a silicon wafer for a thickness of about 1 μm is used. Spin-coating provides a flat film of polymer with RMS roughness on the order of a few nanometers. Fig. 4.25 displays a sequence of AFM images before and after teleoperated nanoindentation experiments at different locations. The operator presses the tip onto the surface for different amounts, with different speeds, different waiting times at maximum penetration, etc. to play with some parameters of and gain intuition for the mechanism of mechanical nanoindentation. The corresponding profiles of line scans over the indentations are shown as insets in the AFM images, above each corresponding indentation mark. As seen in these profiles, different amounts of plastic indentation can be achieved by different process parameters. Representative vertical position and force curves of a representative teleoperated nanoindentation experiment are given in Figs. 4.26 and 4.27, respectively.

A critical load value that separates elastic and plastic behavior (*i.e.* surface deforms elastically below the critical load and plastically above it) can be defined for a spherical indenter as [36]

$$F_c = \left(\frac{R_t}{K}\right)^2 (\pi H)^3, \tag{4.73}$$

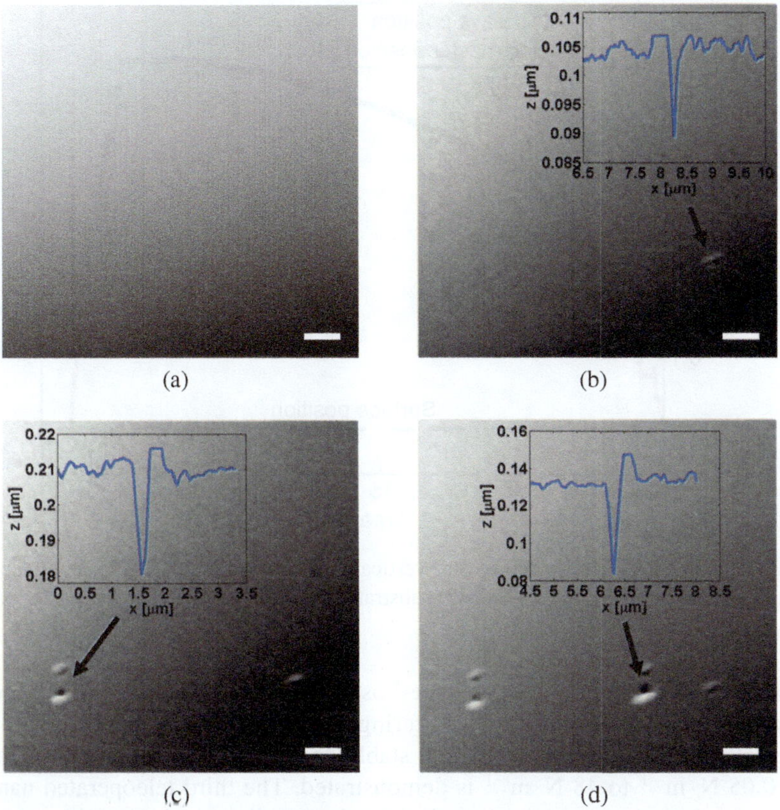

Fig. 4.25 10 μm × 10 μm AFM contact mode images of (a) before and (b), (c), (d) after a series of teleoperated nanoindentation experiments at the same area on a flat PMMA substrate. Scale bars indicate 1 μm length and insets provide the depth information of each indentation mark.

where H is the sample hardness. Using a tip radius of 20 nm, and PMMA hardness range of 1.58 - 3.28 GPa [37], the critical load range for indentation is found to be about 2.76 - 24.71 μN. The maximum load used in the indentation experiments was calculated to be about 4 μN, which is inside this critical load range.

4.4.4 Conclusion

Three different control schemes are analysed in this section, in the particular context of 1-D vertical nanorobotic applications. The Direct Force Feedback control scheme suffers from a trade-off between stability and force amplification if time consuming manipulations have to be avoided. However, transparency is high as proven analytically, and underscored by the fact that both pull-in (0.5 nN) and pull-off forces

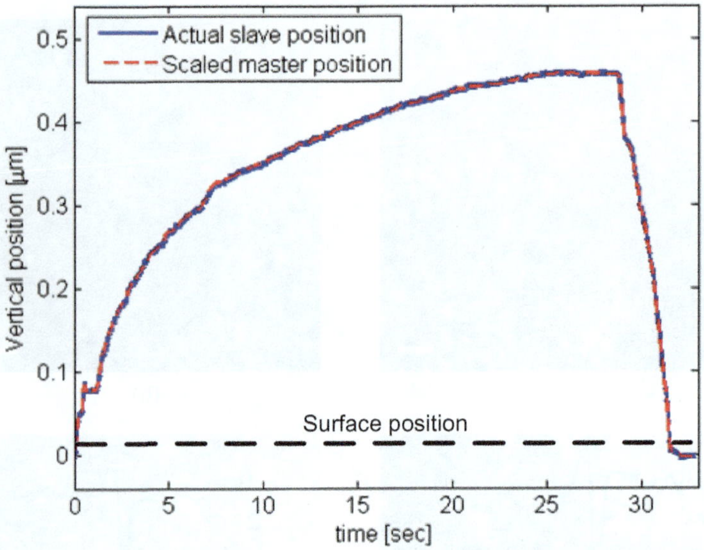

Fig. 4.26 Actual slave and scaled master vertical positions for a 0.5 μm range teleoperated nanoindentation experiment on a PMMA substrate.

are haptically transmitted. Using Force-Position control scheme, higher force amplification can be achieved without suffering from the duration of the manipulation. Forces of 10 nN are felt by users and stable contact for cantilevers of stiffnesses from 0.05 $N \cdot m^{-1}$ to 48 $N \cdot m^{-1}$ is demonstrated. The third teleoperated nanomanipulation system is a passivity-based bilateral controller which explicitly addresses stability, transparency, and comfortable human perception ranges issues to adjust forces sent to users. A constant position scaling factor is determined based on the master and slave workspace ranges. On the other hand, force scaling factor is adjusted by means of control theory for stability and mathematical transformations for transparency and impedance transformation. Visual feedback is generated by augmenting a general elastic contact mechanics model with the AFM measurements and displayed using OpenGL.

Vertical touch feedback experiments are performed with the three controllers and their capabilities are discussed. In addition, a statistical study is done on tip positioning performances for the third controller, on a variety of surfaces. Finally, a sample nanoindentation experiment is shown for the same controller as a demonstration of nanomanipulation.

Based on these results, 2D and 3D teleoperation systems can be developed. They are detailed in following sections.

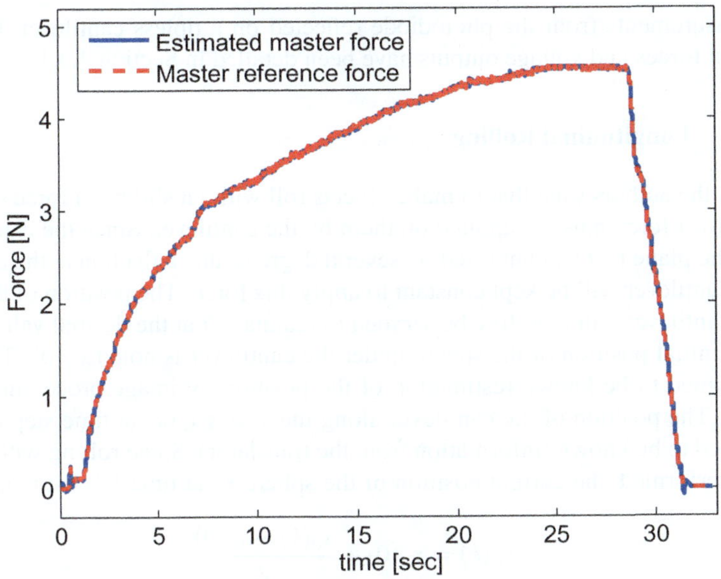

Fig. 4.27 Estimated and reference vertical forces on the master side for a 0.5 μm range teleoperated nanoindentation experiment on a PMMA substrate.

4.5 2-D Micro Teleoperation with Force Feedback

In previous sections, haptic feedback is designed to transmit interaction forces applied on cantilevers to users. A second approach is considered in this section. Haptic feedback does not intend anymore at transmitting nanoscale interactions, but at *helping users to perform a given task.* Force measurement is used to define information on the system state, such as the relative position of the object with respect to the tool. These indications are then sent to users through a 2D haptic feedback [38].

In this section, objects are moved by rolling using a tipless cantilever. This approach is a promising method to manipulate objects such as spheres and carbon nanotubes. It can also be used to release objects from a cantilever [39], or for indenting tasks [40]. Other manipulation modes, or different objects could have been chosen. Rolling is only one of the possible applications to illustrate the use of haptic feedback aiming at transmitting information on the system's state. Experiments are performed using *Force-Position* control, to benefit from the stability of this coupling over wide range of parameters.

4.5.1 Haptic Feedback Determination

During a rolling manipulation, the object is hiden by the tool, and is not visible on optical microscope images. Haptic feedback is designed to *transmit to users information about the position* P (x_p, y_p) *of the object with respect to the tool.* It is based on measurements from the photodiode reflected on a tipless cantilever. Relations between forces and voltage outputs have been detailed in Section 4.3.1.

4.5.1.1 Longitudinal Rolling

In [39], the authors state that to make objects roll without sliding, a force similar to the pull-off force must be applied on them by the cantilever. Since the cantilever is not in the plane (x, O, y) but tilted by several degrees, the deflection at the extremity of the cantilever will be kept constant to apply this force. The position of the origin of the cantilever will therefore be servoed to maintain it at the desired value.

The initial position of the sphere under the cantilever is noted $x_p(0)$. This point is presumed to be known (estimation of the position by image processing for example). The position of the cantilever along the x axis $x_c(k)$ at time step k, is also presumed to be known (information from the translator). Since rolling without sliding is performed, the current position of the sphere x_p at time k is estimated from:

$$x_p(k) = x_p(0) + \frac{x_c(k) - x_c(0)}{2} \tag{4.74}$$

For longitudinal rolling, the sphere is initially positioned at the extremity of the probe, in its middle line (point E $(x_{max}, 0)$. The position along the y axis is kept constant and is such that the cantilever experiences no torsion: $\phi(L,0) = 0$. The cantilever is then moved along the x axis, to make the sphere roll. To provide users with information about the position of the object being manipulated, a modified force f_{sm}^x is defined such that its value is minimum when the sphere is at the extremity of the cantilever and it increases when it get closer to its origin: users must apply a force to take the sphere away from the extremity of the cantilever. This force is derived from the normal force f_s^z applied by the sphere on the cantilever given by Equation (4.59) as seen in Section 4.3.1.2:

$$f_s^z(x_p, y_p) = k_z(x_p)\zeta(x_p, y_p) \tag{4.75}$$

However, as explained in Section 4.3.1.2, $\delta(L, y_p)$ is not available, since the laser is reflected in point (L,0). The force f_{sm}^x is thus computed as:

$$f_{sm}^x(x_p, y_p) = \frac{2L^3}{x_p^2(3L - x_p)} K_p(L)\zeta(L,0) \tag{4.76}$$

where $\zeta(L,0)$ is measured using the photodiode.

The profile of the force is plotted on Fig. 4.28(a) (as stated above, the position of the origin of the cantilever is servoed so that $\delta(L,0)$ is constant).

This force is indeed minimal at the extremity of the cantilever. So that this point represents a stable equilibrium position for users, an offset will be used, in order to render a force equals to zero.

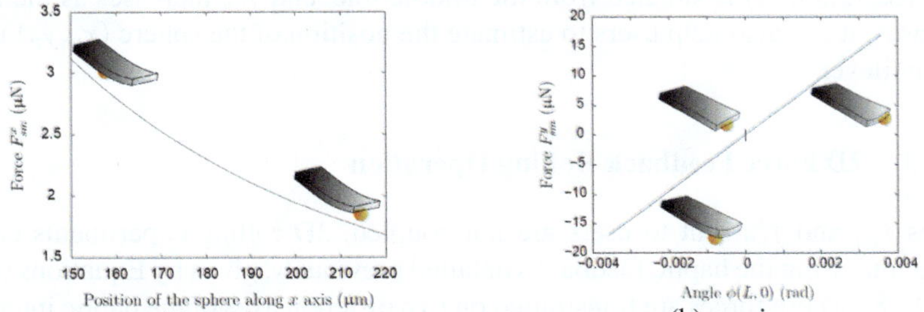

(a) x axis (b) y axis

Fig. 4.28 Profile of the force that will be sent to the user

4.5.1.2 Lateral Rolling

For lateral rolling, the position along the x axis is kept constant, while the cantilever is moved along the y axis so that the sphere rolls. While performing rolling tasks, it is important to keep the sphere in the middle line of the cantilever, so that it does not loose contact on a lateral side of the beam. To help users maintain the sphere on the middle line, a force f^y_{sm} is sent to them via the haptic interface. It is chosen to correspond to the force that should be applied with the lever arm $\|y_p\|$ to maintain the cantilever horizontal[10], $\frac{M_t(x_p,y_p)}{\|y_p\|}$. According to Equations (4.64) and (4.65) determined in Section 4.3.1.2, the moment, which would correspond to the rotation angle is expressed by:

$$M_t(x_p,y_p) = k_{t_\phi}(x_p)\left[\frac{f^z_s(x_p,y_p)}{2EI_x}y_p^2 + \phi(x_p,0)\right] \tag{4.77}$$

However, according to (4.77), the variation of this force is not easily interpretable. Therefore, this expression is linearized:

$$f^y_{sm}(x_p,y_p) = \frac{M_t(L,0)}{\frac{w}{2}} = \frac{k_{t_\phi}(L)}{\frac{w}{2}}\phi(L,0) \tag{4.78}$$

where $\phi(L,0)$ is the measure of the rotation angle given by the photodiode outputs.

The variations of the force f^y_{sm} are plotted on Fig. 4.28(b). Depending on the sign of $\phi(L,0)$, this force is either positive or negative. The higher $\phi(L,0)$ is, the higher is the magnitude of the force. If sent back to the user through a haptic interface, this force will indeed tend to keep the sphere on the middle line of the cantilever (which is the point where the rotation angle is nil). Since the angle increases the further

[10] Since the origin of the y axis is defined on the middle line of the cantilever, $\|y_p\|$ corresponds to the distance along the y axis between point P and the midle line.

the force $f_s^z(x_p, y_p)$ is situated from the middle line, and f_{sm}^y increases as the angle increases, it can also help users to estimate the position of the sphere (x_p, y_p) under the cantilever.

4.5.1.3 2D Force Feedback Rolling Operation

Forces f_{sm}^x and f_{sm}^y sent to users are not coupled. 2D rolling experiments can be carried out using the haptic feedbacks detailed previously, given by Equations (4.76) and (4.78). These forces are transmitted on two different axes of the haptic interface.

4.5.2 Experimental Results

The strategy described in Section 4.5.1 is evaluated with rolling experiments using glass spheres (radius: 25 μm).

4.5.2.1 Longitudinal Rolling

A rolling experiment along the x axis of the cantilever is presented. The vertical position of the probe is servoed in order to maintain a constant deflection at its extremity ($\delta(L,0) = 0.7$ μm , which corresponds to a force of 1.7 μN when the sphere is at point E).

During the experiment, the user makes the sphere roll from the point $x = x_{max} = L$ (extremity of the cantilever) to a point closer to the origin of the cantilever (noted x_{min}), and then back to a point situated between x_{min} and x_{max}, noted x_{int} (for intermediate). The user produces small oscillations by moving the sphere back and forth around this position x_{int}. Results are plotted on Fig. 4.29(a). The position of the sphere along the x axis, as well as the haptic forces f_m^x are represented.

Numerical values used for the controller gains are chosen according to Table 4.2. They are detailed in the caption of Fig. 4.29(a), which presents experimental results. According to the velocity scaling factor α_p^x used, a displacement of 1 cm in the haptic workspace corresponds to a displacement of 0.2 μm for the cantilever. During this experiment, the probe is moved by around 40 μm, which implies a displacement of 2 m of the haptic handle. Due to the limited workspace of the master device (around 35 cm for the x axis), a function to reset the position of the haptic device without moving the probe is used. The steps that can be seen on the plot of the position along the x axis are due to these instants when the user resets the position of the haptic device (and therefore keeps the position of the sphere constant). As a consequence, the force sent back to the user is constant during this operation[11].

As predicted in Section 4.5.1.1 (Figure 4.28(a)), and according to experimental results in Fig. 4.29(a), the haptic force increases as the sphere gets further from the extremity of the cantilever. For $x = x_{max}$, the force is nil. The operator has to make an effort to move the sphere further from this equilibrium position. The force felt by

[11] This remark is also valid for Fig. 4.29(b).

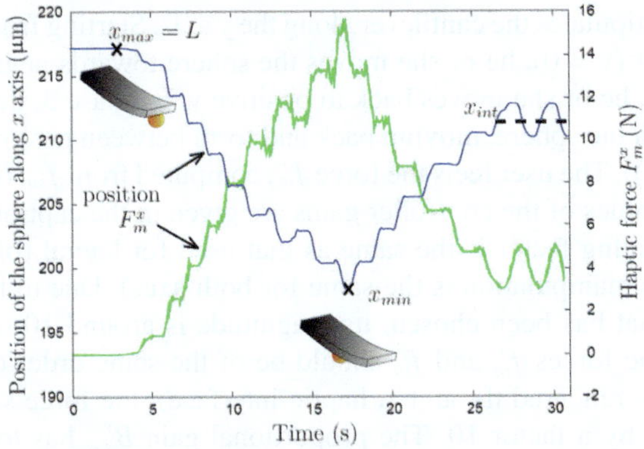

(a) Longitudinal rolling experiment. Parameters used: $\alpha_F^x = 25 \cdot 10^6$, $\alpha_p^x = 0.05 \cdot 10^6$, $K_{fp}^x = 100\ \mathrm{N \cdot m^{-1}}$, $B_{fp}^x = 0.1\ \mathrm{N \cdot s \cdot m^{-1}}$, $G_n^x = 825\ \mathrm{m \cdot N^{-1} \cdot s^{-1}}$

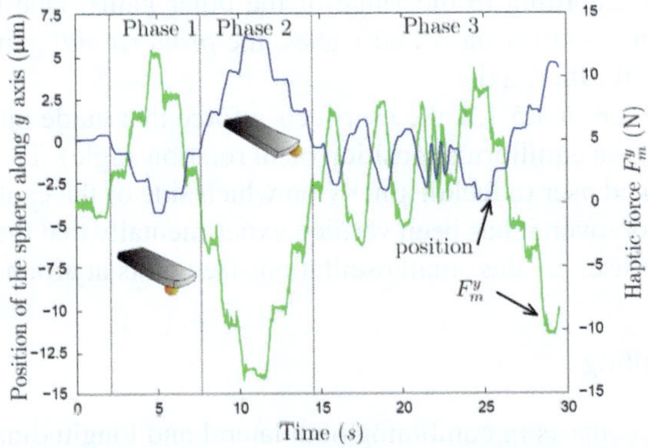

(b) Lateral rolling experiment. Parameters used: $\alpha_F^y = 2 \cdot 10^6$, $\alpha_p^y = 0.05 \cdot 10^6$, $K_{fp}^y = 100\ \mathrm{N \cdot m^{-1}}$, $B_{fp}^y = 1.5\ \mathrm{N \cdot s \cdot m^{-1}}$, $G_n^y = 19.8\ \mathrm{m \cdot N^{-1} \cdot s^{-1}}$

Fig. 4.29 Positions and forces during longitudinal and lateral rolling experiments

users also helps them to evaluate the distance between sphere and point $x = x_{max}$, as the magnitude of the force increases when the distance increases.

4.5.2.2 Lateral Rolling

As in the case of longitudinal rolling, the probe's vertical deflection is kept constant by servoing the vertical position of the origin of the cantilever. The x position is also kept constant.

The user manipulates the cantilever along the y axis. Starting from the middle line of the cantilever ($y = 0$), he or she moves the sphere towards negative y (phase 1). During phase 2, he or she moves back to positive y. In phase 3, he or she performs oscillations with the sphere, moving back and forth between positive and negative y (see Fig. 4.29(b)). The user feels the force f_m^y, computed from f_{sm}^y (Equation (4.78)).

Numerical values of the controller gains are given in the caption of Fig. 4.29(b). The velocity scaling factor is the same as that used for lateral rolling (the desired precision of the manipulation is the same for both axes). Due to the expression of the force f_{sm}^y that has been chosen, its magnitude is around 10 times higher than f_{sm}^x. Since haptic forces f_m^x and f_m^y should be of the same order of magnitude (so that they can be rendered through a haptic interface), the force scaling factor has been decreased by a factor 10. The proportional gain B_{fp}^y has to be increased to introduce damping into the system. This damping enables to limit oscillations due to the variation between positive and negative forces that arise when the sphere moves from the left to the right side of the cantilever, and *vice-versa*. As for lateral rolling, G_n^y is computed according to the value of the other gains. Due to the difference between the gains used for the x and y axes, the *proportional* gain of the y axis is smaller than that for the x axis.

As stated in Section 4.5.1.2, the user feels a force that guide him or her to bring the cantilever into an equilibrium position (a nil rotation angle). The forces sent back enable an untrained user to feel distinctly on which side of the cantilever the sphere is positioned. Moreover, it has been verified experimentally that the system remains stable even if the user creates small oscillations (see plots at around $t = 22s$).

4.5.2.3 2D Rolling

This experiment consists in combining both lateral and longitudinal rolling, to perform a 2D rolling task with force feedback. This is possible since the signals used are independent: there is no coupling between the f_{sm}^x and f_{sm}^y forces computed. Therefore they can be rendered to users using two axes of the haptic device.

As for lateral and longitudinal rollings the deflection $\delta(L,0)$ is constant. The movement performed by the operator in the plane (xOy) is depicted on Fig. 4.30(a). The shape of the movement is roughly circular, but noticeable differences have been made to better distinguish the corresponding forces sent to the user. The movement has been repeated three times in order to verify that the forces felt by the operator are constant for a given position. The gains used are the same as those of previous experiments.

The forces sent back to the user while performing this experiment are represented on Fig. 4.30(b), where the lateral force is plotted against the longitudinal force. Since the user chose to perform these circles without reaching the extremity of the cantilever, the haptic force along the x axis is never nil. For a given position, the operator feels the same force, as highlighted by the characteristic points P_1, P_2 and P_3. This haptic feedback helps the operator to estimate the position of the sphere under the cantilever.

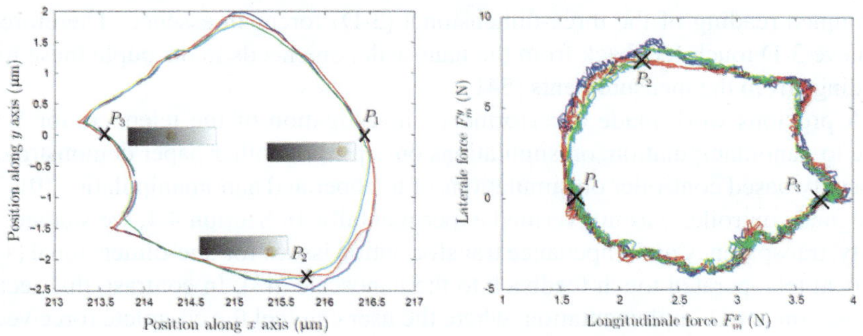

(a) Positions of the sphere on the plane (xOy)

(b) Haptic force along x and y axes

Fig. 4.30 Positions and forces on the (xOy) plane during the 2D rolling experiment

4.5.3 Conclusion

During these experiments, haptic forces provide users with comprehensible information of the system state, since they have to make an effort to take the sphere further from the defined equilibrium position (point E). This is possible thanks to the appropriate choice of the forces sent to users, and the stability performances of the *Force-Position* control which allows to render these forces sufficiently amplified (so that operators can feel them) while remaining stable. Using the same setup, other strategies to provide force information during a manipulation task can be considered.

It has been highlighted that due to the limited force measurements provided by AFM, strategies had to be used to provide comprehensive information on the system state. The same problem arrise while attempting to provide users with haptic feedback of nanoscale interaction forces. This issue is addressed in the next section, using a model of the substrate to decouple AFM force measurements and provide 3D haptic feedback.

4.6 3-D Teleoperated Touch Feedback Using AFM

In this section, an AFM is employed as the nanomanipulator to realize a convenient interface with nanoscale phenomena in terms of a versatile teleoperation system. With its real-time force feedback, the AFM can be helpful in the manipulation of soft, fragile, or complex objects such as biological specimens. Since the AFM cannot provide real-time visual feedback, studies that augment teleoperation with a virtual reality environment have been performed. Using well-established small-scale contact mechanics, the operator can gain additional feedback of the nanoenvironment from this augmented virtual reality [33, 41, 42, 43]. The most important limitation of AFM as the nanomanipulator, however, is that it cannot measure all three forces acting on the tip simultaneously. Rather, it gives two force measurements, a coupled reading of the three-dimensional (3-D) forces in essence. Therefore, to achieve 3-D touch feedback from the nanoscale, one needs to decouple these force readings from the measurements [54].

A previous work made a performance investigation of the teleoperation interface to nanomanipulation, on simulations only [33]. Another paper demonstrated a passivity based controller on simulations of teleoperated nanomanipulation [9]. The designed controller was not verified experimentally. In Section 4.4, we studied stability, transparency and impedance transformation issues for one-dimensional (1-D) vertical teleoperated touch feedback to the nanoscale [43]. In contrast, this section focuses on a 3-D implementation, where the user can feel the complete force vector acting on the tip [54]. A decoupling procedure is proposed to this end. After decoupling, and passivity control for stability, the operator will have the opportunity to feel the frictional properties of materials via the bilateral link. The proposed system allows for an enhanced teleoperation experience in the nanoscale by utilizing control theory and mathematical transformations.

The decoupling method using local surface slopes and an empirical friction model is discussed in Section 4.6.1. Section 4.6.2 improves the friction model to an adaptive one using an adaptive parameter observer. Stability of the bilateral control system is reinforced by a passivity based controller, designed in Section 4.2. In Section 4.6.3, the performance of the overall system is validated on experimental results via touching experiments with 3-D force feedback, where the tip is brought into contact with the surface and moved in arbitrary directions by the operator. Results on flat and non-flat, hard and soft substrates are demonstrated.

4.6.1 Three-Dimensional Force Decoupling

In modern AFMs, two force readings can be simultaneously taken as

$$F^* = V_{A-B}\frac{K_p}{s_n} = F_z + 2\frac{L_{tip}}{L}F_y \tag{4.79}$$

$$F_x = V_{LFM}\frac{k_l}{s_l}, \tag{4.80}$$

where K_p and k_l are the normal and lateral stiffness values of the cantilever, respectively; s_n and s_l are the normal and lateral sensitivity values of the optical measurement system, respectively; and L and L_{tip} are the length of the cantilever and the tip, respectively. In these equations, F^* is a coupled reading of the vertical force (F_z) and the longitudinal force (F_y), while the lateral force (F_x) can be directly measured.

This coupled force reading, at first sight, seems to render the system unobservable and make it impossible to build a controller. Nevertheless, augmenting theory with experiments and confining ourselves to non-changing surfaces, it is possible to avoid this issue as detailed below.

One previous method of force decoupling found in the literature involves the estimation of the longitudinal force using a tangent relation with the lateral force [41]. While this relation is correct, it is not general to various conditions. Most obviously, the tangent of the motion angle on a flat surface gets infinite for longitudinal motions, at which point the lateral force should yield values close to but not equal to zero due to noise. For this case, the longitudinal force would become undefined for a zero reading of lateral force, and very large (or infinite) for small perturbations.

As a solution one can apply a low-pass filter to the calculated longitudinal forces in the tangent-based method. For a low enough cut-off frequency, the jumps in the estimated values would be removed, with the cost of a high lag, and hence, inaccurate results. In comparison, the decoupling method proposed here does not suffer from such problems, recovers the tangent relation utilized in the previous works, and is general enough to be applicable to flat and non-flat, hard and soft substrates in a variety of experimental conditions.

To be able to decouple F_y and F_z from F^* for generating 3D touch feedback from the nanoscale, we propose to utilize the geometry of contact between the AFM tip and the sample. As the tip moves in contact with the surface, it will induce two forces; namely normal and frictional forces. Using a well-established friction force model that is applicable to nanoscale tribology, one can relate all three forces to these two main forces acting on the tip.

As the plane segment that the tip is in contact with has a certain orientation, to use a friction model on this surface, a perspective transformation of the tip motion on the surface should be performed. Taking the equation of the plane segment simply as:

$$z = s_x x + s_y y + z_o, \tag{4.81}$$

where $s_x \in \Re$ and $s_y \in \Re$ are the local slopes of the surface in x and y directions, respectively; and $z_o \in \Re$ is a z-offset; the normal vector of the surface around the contact is

$$\mathbf{n} = (-s_x, -s_y, 1). \tag{4.82}$$

When the unit vector in the y-axis, written as $\hat{\mathbf{j}} = (0, 1, 0)$, where $\hat{}$ shows that the vector is normalized, is projected on this surface, the resulting vector becomes

$$\mathbf{v} = \hat{\mathbf{j}} - (\hat{\mathbf{j}} \cdot \hat{\mathbf{n}})\hat{\mathbf{n}}. \tag{4.83}$$

The third orthogonal vector, then, is the cross-product of the other two vectors as

$$\hat{\mathbf{u}} = \hat{\mathbf{v}} \times \hat{\mathbf{n}}. \tag{4.84}$$

Using these three unit vectors, a transformation of forces (and positions) can be written as

$$\begin{pmatrix} F_x \\ F_y \\ F_z \end{pmatrix} = \begin{bmatrix} \hat{\mathbf{u}} \\ \hat{\mathbf{v}} \\ \hat{\mathbf{n}} \end{bmatrix}^T \begin{pmatrix} F_{fu} \\ F_{fv} \\ F_n \end{pmatrix}. \tag{4.85}$$

Here, $F_{fu} \in \Re$ and $F_{fv} \in \Re$ are the friction force components in u and v directions, respectively, and $F_n \in \Re$ is the normal force. What obviously follows is that $F_{fu} = F_f \cos\theta_m$ and $F_{fv} = F_f \sin\theta_m$ depend on the same friction force (F_f) and projected motion angle (θ_m) on the surface.

Friction force (F_f) in the macroscale depends linearly on the applied normal load (F_n) with a friction constant (μ) as

$$F_f = \mu F_n. \tag{4.86}$$

However, it has been shown that this behavior of friction changes in the micro/nanoscale, where friction force has an additional linear dependency on the contact area [44] as

$$F_f = \mu F_n + \tau A, \tag{4.87}$$

where τ is the interfacial shear stress and $A = \pi a^2$ is the contact area with a as the contact radius. Contact mechanics models relate contact radius to the applied load. This relation, combined with (4.87), gives a direct explicit relation between F_n and F_f. At this point, all the parameters of this relation can be calibrated to be able to solve for the three orthogonal force values during the experiment.

A second possibility, however, is to employ an empirical approach and avoid the cumbersome calibration process. Contact mechanics models generally define a cubic relation between the contact radius and normal force (*i.e.* $a \propto F_n^{1/3}$). Disregarding physical meanings, then, a simplification can be made as [44],

$$F_f = C_0 + C_1 F_n + C_2 F_n^{2/3}. \tag{4.88}$$

Note that, this mathematical model captures the linear and nonlinear nature of a general physical friction model given in (4.87). The linear term ($C_1 F_n$) directly reflects the μF_n component, while the friction due to the shear strength in the contact area is represented by the nonlinear equation: $C_0 + C_2 F_n^{2/3}$. The offset in this nonlinear equation is due to the finite contact area that occurs for zero normal load. The empirical model does not include dynamic effects of friction; such as stick-slip, etc but rather reflects a general frictional behavior. If desired, however, one can utilize a complex model for an AFM system with a better signal-to-noise ratio, due to the modularity of the overall teleoperation system.

Using (4.88), it is possible to have a single initial calibration experiment that gives all the data necessary to fit a curve and find the parameters C_i. This initial experiment should include a lateral motion of the tip $\theta_m = 0$ sliding on the surface with increasing pressure. As $F_{fv} = 0$ and $F_{fu} = F_f$, the force transformation in (4.85) and (4.79) reduce to

$$F_x = u_1 F_f + n_1 F_n \tag{4.89}$$

$$F^* = (u_3 + C u_2) F_f + (n_3 + C n_2) F_n, \tag{4.90}$$

where u_i and n_i are the ith elements of the corresponding unit vectors, for $C = \frac{2L_{tip}}{L}$. Using F_x and F^* readings from the AFM, F_f and F_n can be solved at each point and a linear curve fit to the polynomial friction function given in (4.88) can be made. Moreover, for a flat surface, this task is simplified as the transformation matrix in (4.85) becomes an identity matrix. Results of friction force model calibration experiments on three substrates, namely silicon, mica and poly-dimethylsiloxane (PDMS) are given in Fig. 4.31. In this figure, it is clearly seen that linear and nonlinear behavior of friction in the nanoscale is appropriately captured by the model.

This friction model is used to solve for normal forces using a root finder and after this point other components of the force vector can be solved for, giving complete force feedback ability for bilateral teleoperation.

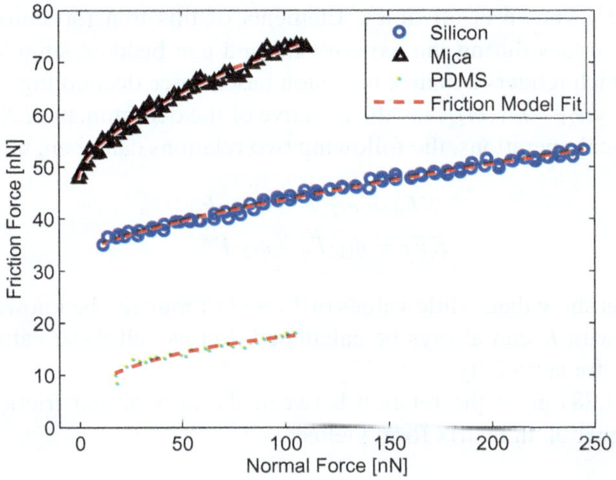

Fig. 4.31 Experimental friction calibration data on flat silicon, mica and PDMS substrates.

From [8], one can obtain conditions of instability for this particular passivity controller. For the system to remain active for all time, there needs to be a single condition–that the velocity of the haptic end-effector (or the tip) needs to change signs at each iteration. As a consequence, for the parameter observer, or the friction model to cause unstable operation; they need to cause high frequency vibrations in the tip position. First, the friction model itself is passive (friction acts against the motion direction, at all times). Secondly, the parameter estimator adapts the friction model, but does not allow it to become active (cause negative friction force). A possibility is vibrations in the force measurements due to noise, possible friction dynamics, etc. In this case, force vibrations would translate to oscillations in tip position as well. The rate of such mechanical vibrations, however, would be much slower than the operating bandwidth of the passivity controller (1 kHz). Not to mention, we are filtering our AFM signals with a 60 Hz low-pass filter.

4.6.2 Adaptive Friction Model

One drawback of the procedure outlined above is that it implicitly assumes a homogeneous surface, with constant frictional properties. To remove that assumption, one needs to build a parameter observer, which can update the friction model in real-time.

To this end, we utilized an adaptive observer design [45]. In its most general case, relation between the two forces acting on the tip and the measured forces by the optical system can be written as:

$$\begin{pmatrix} F^* \\ F_x \end{pmatrix} = \begin{bmatrix} m_{11} & m_{12} \\ m_{21} & m_{22} \end{bmatrix} \begin{pmatrix} F_f \\ F_n \end{pmatrix}, \tag{4.91}$$

where m_{ij} are the elements of a general force transformation matrix $\mathbf{M} \in \Re^{2\times 2}$ for $i, j \in \{1,2\}$, known at all times. Elements of this transformation matrix can take arbitrary values during the experiment, and can become singular for certain conditions, which renders a matrix inversion based force decoupling impossible.

Let $K = m_{12}m_{21} - m_{11}m_{22}$, *i.e.* the negative of the determinant of $\mathbf{M}$. Using simple mathematical operations, the following two relations can, then, be written:

$$KF_n = m_{21}F^* - m_{11}F_x \tag{4.92}$$

$$KF_f = m_{12}F_x - m_{22}F^*. \tag{4.93}$$

These relations show that, while values of F_n and F_x may not be known at all times, their product with K can always be calculated. Let us call these values $K_n = KF_n$ and $K_f = KF_f$ for simplicity.

Equation (4.88) gives the relation between the normal and friction forces, rewriting this equation in matrix form yields:

$$F_f = \begin{pmatrix} C_o & C_1 & C_2 \end{pmatrix} \begin{pmatrix} 1 \\ F_n \\ F_n^{2/3} \end{pmatrix}, \tag{4.94}$$

which immediately suggests that

$$K_f = \underbrace{\begin{pmatrix} C_o & C_1 & C_2 \end{pmatrix}}_{\mathbf{p}} \underbrace{\begin{pmatrix} K \\ K_n \\ K^{1/3}K_n^{2/3} \end{pmatrix}}_{\phi}. \tag{4.95}$$

In this final equation, only the parameters C_i, $i \in \{0,1,2\}$ are unknown, to be observed by the adaptive observer in what follows.

Defining $\mathbf{p}^* = \left(C_o^* \; C_1^* \; C_2^*\right)$ as the observed parameter array, the following equation is used for the parameter observer:

$$\hat{K_f} = \mathbf{p}^* \phi, \tag{4.96}$$

where $\hat{K_f}$ is the observer output or the estimated value of K_f. The observer error is

$$e_f = K_f - \hat{K_f} = K_f - \mathbf{p}^* \phi \tag{4.97}$$

with a residue

$$J(e_f) = e_f^2/2. \tag{4.98}$$

The last part of the adaptive observer design involves defining a dynamic equation that satisfies the Lyapunov stability condition as

$$\dot{C}_i^* = -\gamma_i \frac{\partial J}{\partial C_i^*}, \tag{4.99}$$

for $C_i^*(0) = C_{i0}$, $i \in \{0,1,2\}$ and γ_i is a proportional gain to tune the observer.

Note that, the mentioned adaptive parameter observer design assumes the parameters not to be varying with time at a fast rate. It may be argued that due to tip-wear, relation of contact area to normal load will change over time. Nevertheless, this change is too slow to be problematic.

An obvious way to compare the results of decoupling is to plot the measured and estimated values of F_x on the same plot. One such comparison is given in Fig. 4.32, where a heterogeneous frictional behavior is simulated on a flat PDMS substrate, such that a region on the surface gives a higher friction value. The static model is initially calibrated on the normal frictional region. The adaptive model starts with $C_{i0} = 0$, with $\gamma_i = 100$. As seen in this figure, the parameter observer successfully adapts the parameter of the friction model to adjust for the changes in the measurements.

4.6.3 *Experimental Results*

4.6.3.1 Validation of Decoupling Procedure on Flat Surfaces

In this section, a validation of the decoupling procedure, proposed in this work, is made on flat substrates. For this demonstration, surface slopes are removed (*i.e.* normal direction of the surface coincides with the z direction of the cantilever), the AFM servo is turned off and the tip is pressed vertically onto the surface for a certain amount. The vertical position of the tip is fixed afterwards. This allows the user to move the tip via the haptic device in a horizontal plane, exciting roughly a constant normal force with variations of less than 10 nN due to small errors in slope correction. In consequence, an approximately constant friction force is generated during motion. For a flat surface, the decoupled forces in the horizontal plane (F_x and F_y) should be the cosine and sine components of this friction force, respectively.

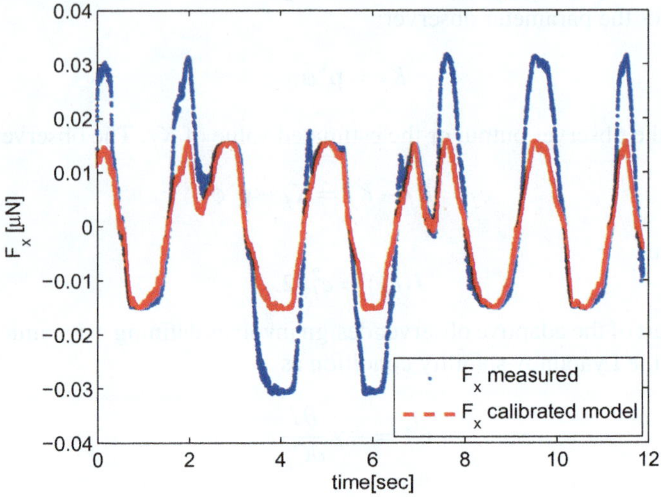

(a)

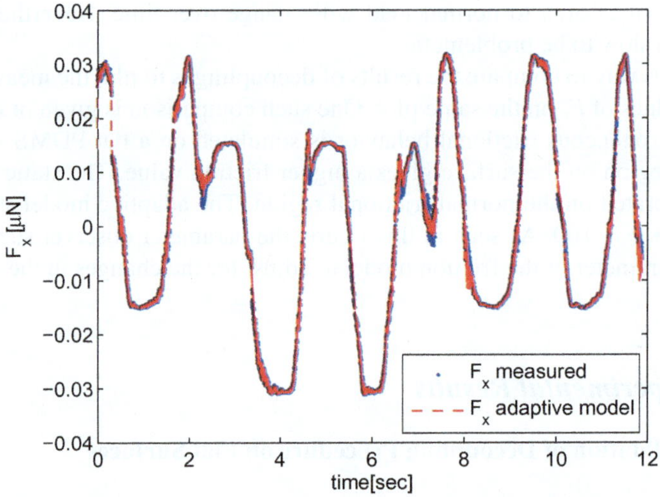

(b)

Fig. 4.32 Comparison of the measured and estimated lateral force (F_x) values using (a) a calibrated and (b) an adaptive friction force model on a flat heterogeneous PDMS surface.

Three experiments were done on flat silicon, mica, and PDMS substrates. The used friction model is calibrated for each surface, right before the experiments and shown in Fig. 4.31. Figures 4.33 and 4.35 display the results of decoupling for the two substrates, respectively. Forces in all three dimensions are effectively estimated online with the algorithm. As expected, lateral (F_x) and longitudinal (F_y) forces acting on the AFM cantilever are complimenting each other to achieve an approximately constant friction force resisting the horizontal motion. Normal (F_z) forces are also approximately constant, as expected.

For validation, the lateral force (F_x) in the decoupled model output is compared with its measured counterpart from the AFM. A comparison on the coupled force (F^*) does not give much insight since it is largely dominated by the normal force ($F_n = F_z$), which is generally about an order of magnitude larger than the frictional forces in the horizontal plane. Figures 4.34 and 4.36 compare the estimated lateral force values with the measured ones. As seen in these figures, the calibrated model does a good job in estimating the friction force value. However, the measured signal tends to have more variations due to noise or possibly small roughness effects.

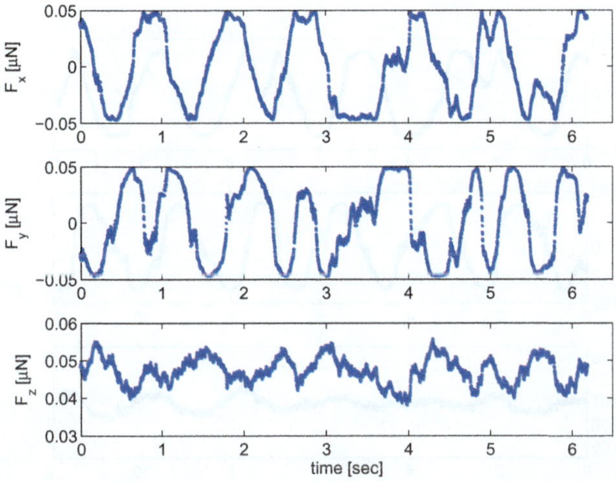

Fig. 4.33 3-D decoupled forces on the slave side for a flat mica substrate.

4.6.3.2 Three-Dimensional Force Feedback from Non-flat Surfaces

This section shows results of bilateral teleoperation for 3-D tip motion on two non-flat surfaces, namely the back side of a silicon wafer, and a silicone rubber substrate. Silicone rubber is molded on the back side of a silicon wafer to transfer a similar roughness. These two substrates are chosen to show the results of the designed

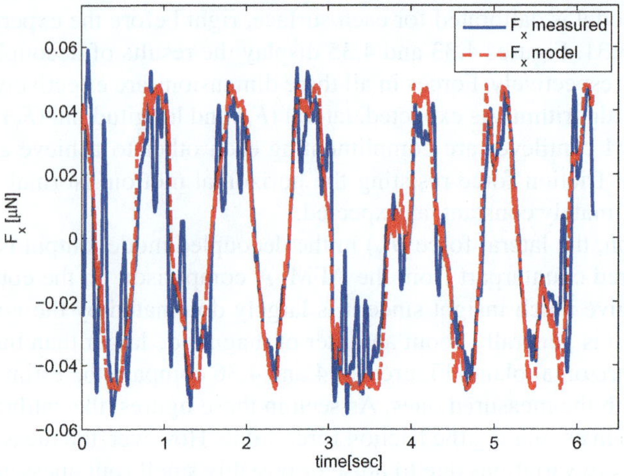

Fig. 4.34 Comparison of estimated and measured F_x values for a flat mica substrate.

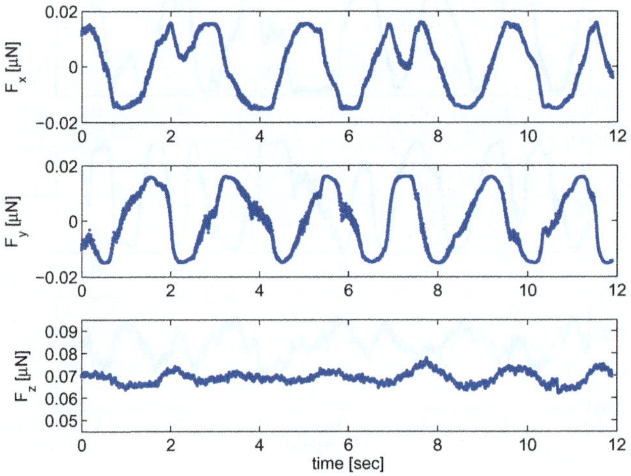

Fig. 4.35 3-D decoupled forces on the slave side for a flat PDMS substrate.

system for both hard and soft surfaces. Figures 4.37 and 4.39 show non-contact AFM images of the surfaces. Measurement of local surface slopes are made on these images, using OpenCV image processing library. The adaptive friction force model is utilized in these experiments.

Figures 4.38 and 4.40 show the force references to the haptic device since the haptic device used in our experiments does not have force sensors to measure forces exerted on the operator's hand.

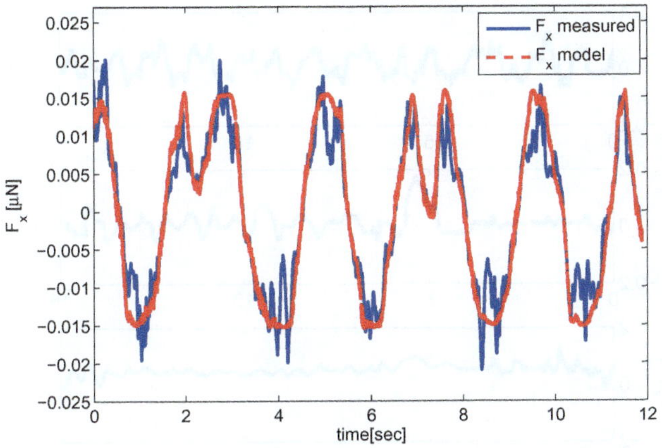

Fig. 4.36 Comparison of estimated and measured F_x values for a flat PDMS substrate.

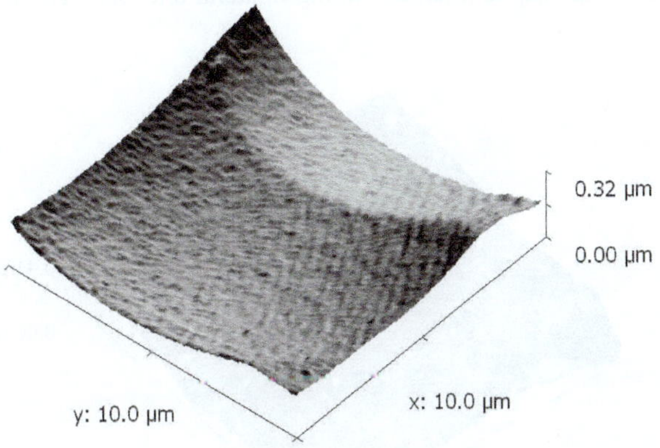

Fig. 4.37 10 μm $\times$ 10 μm non-contact AFM image of the back side of a silicon wafer.

A comparison of estimated and measured signals from the AFM is made in Figs. 4.41 and 4.43 for F_x and Figs. 4.42 and 4.44 for F^*. As seen in these figures, estimated signals track the measured ones accurately.

It should be noted that during these experiments, the passivity controller almost never kicks in. This is expected since friction is an entirely passive phenomenon, working against the tip motion on the surface. The most basic result from the experiments is that the proposed system is versatile enough to work on a wide range of materials, hard and soft. As expected, soft materials yield lower frictional forces. Since friction forces help stabilize the position of the tip on the surface, these materials would yield worse results for a possible trajectory following task. One may increase the force scaling factor to help the operator get a better feeling of the

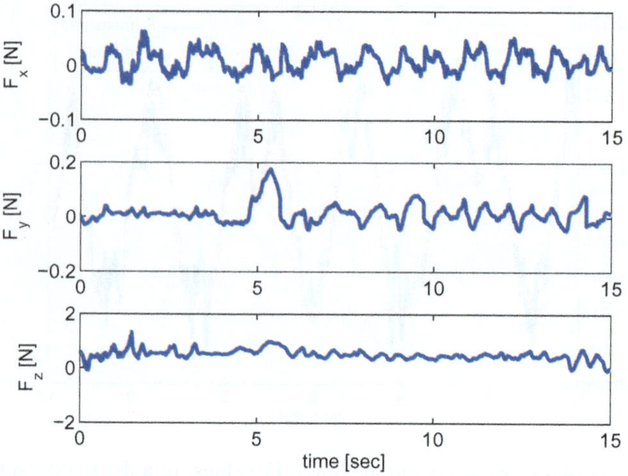

Fig. 4.38 Forces of the haptic interface during bilateral teleoperation on the back side of a silicon wafer.

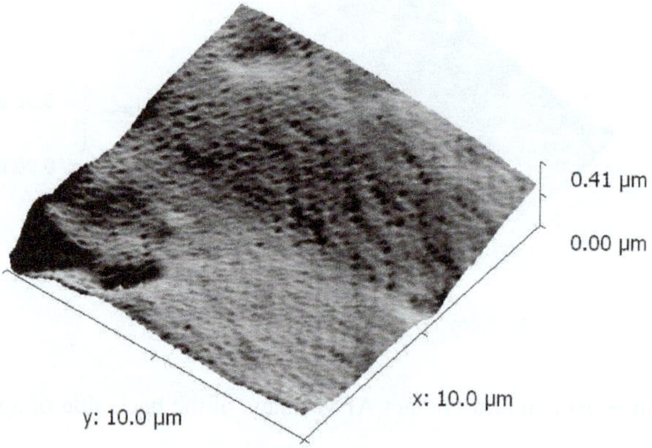

Fig. 4.39 10 μm × 10 μm non-contact AFM image of a silicone rubber substrate, molded on the back side of a silicon wafer to transfer roughness.

environment. Nevertheless, that study is out of the scope of this study and might be investigated in a future work.

The adaptive parameter observer makes the job of bilateral teleoperation easier, bypassing an initial calibration routine. Approaching the problem from another aspect, however, means that one can actually dictate the perceived behavior of the substrate and make it behave differently by simply picking a set of frictional parameters of a different material. Moreover, a heterogeneous frictional behavior or a composite substrate can also be simulated as previously realized in Fig. 4.32.

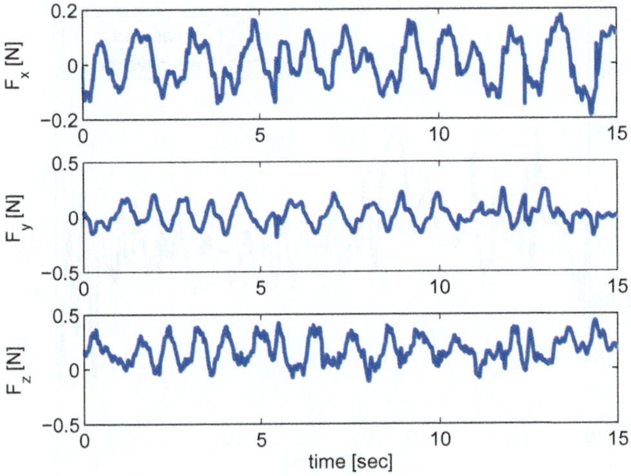

Fig. 4.40 Forces of the haptic interface during bilateral teleoperation on a silicone rubber substrate.

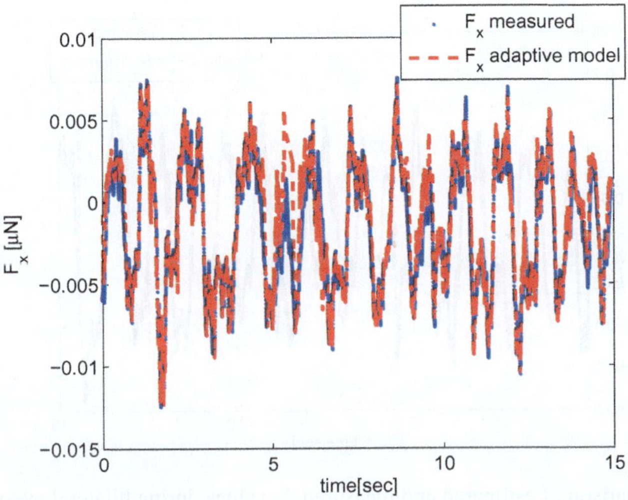

Fig. 4.41 Comparison of estimated and measured F_x values during bilateral teleoperation on the back side of a silicon wafer.

To further underline some limitations of the mentioned procedure, first, the interaction of the tip with the substrate should remain in the elastic regime, which means that no permanent changes in the environment are allowed. This limitation comes from the fact that force measurements are coupled and the decoupling procedure must include an additional knowledge of the surface topography. Since there is no real-time visual feedback during teleoperation, changes on the substrate should be

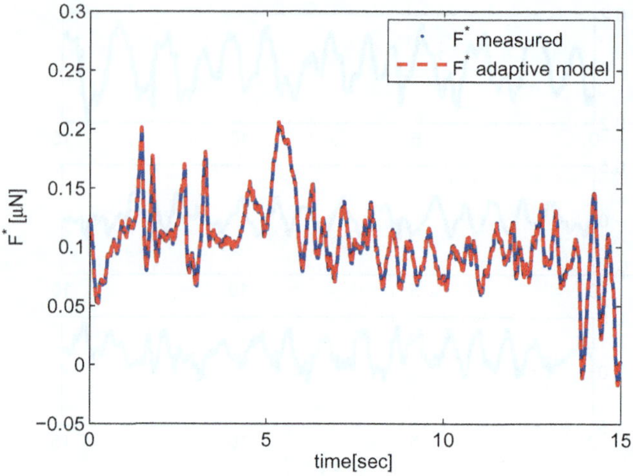

Fig. 4.42 Comparison of estimated and measured F^* values during bilateral teleoperation on the back side of a silicon wafer.

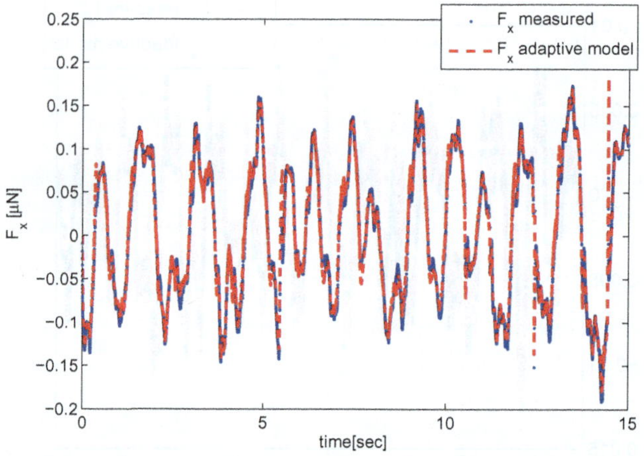

Fig. 4.43 Comparison of estimated and measured F_x values during bilateral teleoperation on a silicone rubber substrate.

avoided. This limitation can be lifted, if the change made can be accurately modeled. Secondly, from the same reasoning, any drift in the environment is also unacceptable. If one can estimate drift for sufficiently long intervals [46], this limitation can also be lifted and experiments can be started without the need to wait for the drift velocities to reduce to acceptable values.

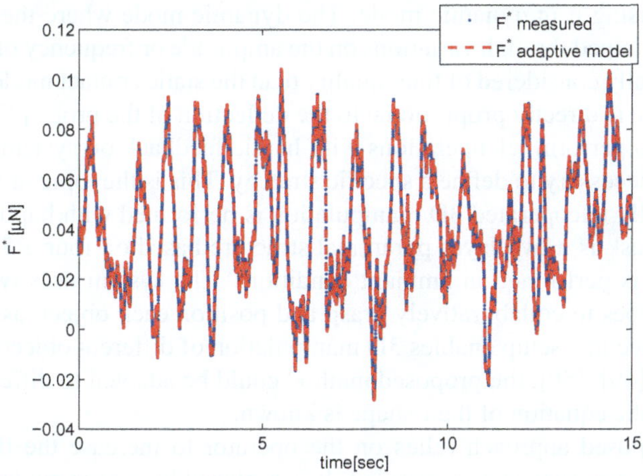

Fig. 4.44 Comparison of estimated and measured F^* values during bilateral teleoperation on a silicone rubber substrate.

4.6.4 Conclusion

In this section, a stable bilateral controller for a teleoperated system with 3-D decoupled force feedback is demonstrated with an AFM on the slave side. 3-D force feedback is achieved using local surface slopes and an empirical friction model. Friction model parameters are updated online using an adaptive parameter observer. A modified passivity controller takes care of instabilities in bilateral control. In conclusion, a robust system has been developed that works with very soft to very hard substrates to provide touch feedback from the nanoscale. Future works include teleoperated nanolithography and teleoperated manipulation of nanoobjects on a substrate using an AFM probe with a 3-D force feedback.

4.7 Haptic Teleoperation for 3D Microassembly of Spherical Objects

In the previous section, 3D teleoperation was proposed to give operators opportunity to feel the frictional properties of materials. This enables users to get a deeper knowledge of physical characteristics of objects. In this section, 3D teleoperation is considered to *help users perform microassemblies* [47]. 3D manipulations are performed to build 3D micro structures. These experiments are performed using the two tip manipulator detailed in Chapter 3.

Most of the teleoperation systems with haptic feedback use a single AFM cantilever and static measurement. Consequently, only one controllable contact point is available to the user. Using the AFM in contact mode implies also some limitations compared to using it in dynamic mode. The dynamic mode where the force measurement is obtained through variations on the amplitude or frequency of a vibrating probe is generally considered of finer quality than the static contact mode, where the measured force is directly proportional to the deflection of the probe [48].

However, to perform teleoperations with haptic feedback on systems which are complex it is necessary to define a specific strategy. This is the topic of this section, on which a fully teleoperated 3D manipulation is performed with haptic feedback. The selected task is a two layer pyramidal structure based on four $\varnothing 5 \pm 1\,\mu\text{m}$ microspheres. It is performed in ambiant conditions. The system uses two independent AFM probes to collaboratively grasp and position each object, as reported in Chapter 3. Since this setup enables 3D manipulation of different objects with force measurement [49], [50], the proposed method could be adapted to different objects provided that the equation of their shape is known.

As the proposed approach relies on the operator to increase the flexibility of the system, the pre-scan step is avoided. It is replaced by a user guided initial exploration. This exploration allows for online calculation of virtual guides, helping the operator to correctly align the dual-tip gripper with respect to the manipulated object, even in the case of poor visual accuracy. Additionally, different feedback schemes are presented for pick-and-place of microspheres. The first one provides information on the measured interaction forces directly while the second one assists the user in improving dexterity and avoiding collision. These feedbacks are proposed for the same task, but they correspond to different applications (comprehension of physical phenomena or safe and efficient manipulation).

4.7.1 Experimental Setup

The generic AFM micromanipulation platform used in ISIR is detailed in Section 4.3. However, some modifications must be made to enable the simulatenous use of two tips. Detailed specifications of the two tips manipulation setup are discussed in [51]. A brief summary is given here. The micromanipulation platform is depicted in Fig. 4.45. The AFM gripper is equipped with an optical microscope, and two sets of nanopositioning devices and optical levers to coordinate two AFM cantilevers with protruding tips (namely, Tip I and Tip II) facing each other, forming a dual-tip gripper.

Each cantilever disposes of its own optical lever, comprising a laser source and a four-quadrant photodiode. Cantilevers can be used in two different modes: tapping and static. For the tapping mode, a piezoceramic excites each probe at its natural frequency. The amplitude of the resulting oscillations is measured through the variations of the voltage output on the photodiode:

$$A = \beta \cdot \Delta V \tag{4.100}$$

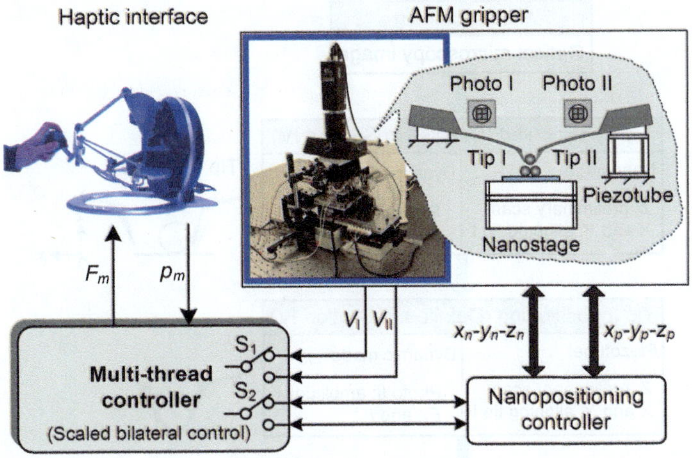

Fig. 4.45 AFM gripper based telemicromanipulation system: (*left*) The haptic device providing a user interface to control the 3D microassembly with real time haptic feedback; (*right*) The dual-probe gripper comprises two AFM cantilevers with protruding tips for pick-and-place micromanipulation.

where A is the amplitude measurement, $\beta = 10^{-6}\mathrm{m}\cdot\mathrm{V}^{-1}$ is a calibrated conversion factor and ΔV is the differential voltage response of the photodiode.

In static mode, the normal force applied on the cantilever F is measured directly from the output voltage of the photodiode as detailed in Section 4.3.1:

$$F_s^z = \frac{K_p}{s_n}\Delta V \tag{4.101}$$

where $K_p = 2.8\mathrm{N}\cdot\mathrm{m}^{-1}$ is the normal stiffness of the cantilever, and $1/s_n = 8\cdot10^{-7}\mathrm{m}\cdot\mathrm{V}^{-1}$ is the sensitivity of the optical levers.

An Omega haptic interface, manufactured by Force Dimension is provided for intuitive user control of the manipulator. The user manipulates the handle and the resulting position p_m is scaled down to be used to control the actuators (nanostage and piezotube). The haptic force F_m sent to the user through the haptic interface is based on measurements from the two photodiodes (V_I and V_{II}). As represented by the switches S_1 and S_2 in Fig. 4.45, different translators and feedbacks are used at each step of the microassembly. Next sections detail the use of the haptic interface to interactively perform a microassembly task.

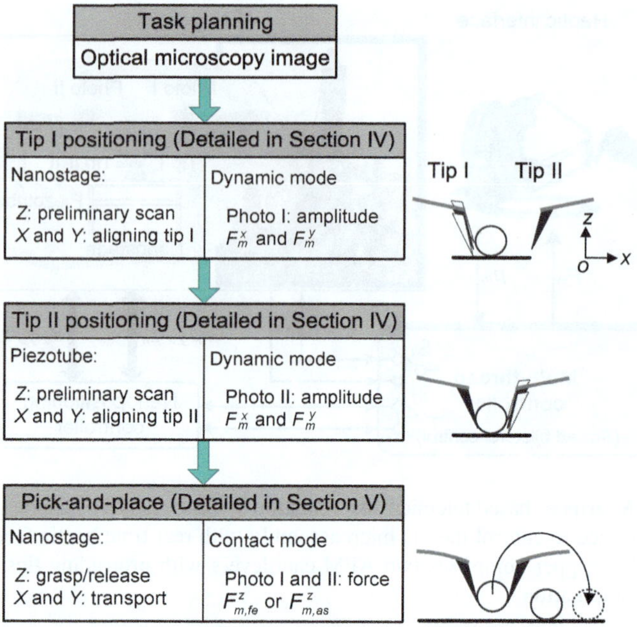

Fig. 4.46 Steps of dual-tip pick-and-place manipulation. The manipulation area and a coarse positioning of the tips are determined using the optical microscope. The operator then sequentially places each tip on both sides of the object using the Omega, controlling respectively the sample holder through the nanostage and Tip II through the piezotube. In both steps haptic feedback is provided based on the amplitude variations of each tip in tapping mode. Once the object is held between the two tips, the lift-off and placing down are achieved by haptic control of the sample holder. Contact mode measurements are used to provide the force feedback.

4.7.2 3D Microassembly Protocol

At microscales complex strategies must be used to manipulate objects [52]. Dual-tip manipulation has certain advantages over other 3D manipulation techniques, as detailed in Chapter 3. However, these advantages come at the cost of augmented complexity of the overall manipulation process. Different delicate steps are required to position each tip correctly and to coordinate the pick-and-place of the object. An overview of the complete manipulation scenario is depicted in Fig. 4.46.

In order to grasp the object between two tips, each contact point has to be aligned with the center of the manipulated sphere. Given the relative sizes of manipulated objects, AFM cantilevers and their protrudent tips, a vision-based control scheme under an optical microscope does not provide sufficient resolution and precision. An initial AFM scan would give additional information for correct positioning, but that is a time-consuming step and comes at the risk of disturbing the manipulation scene.

The approach proposed here is based on a user-driven exploration of the manipulated object. Haptic feedback allows the operator to feel when he/she touches the object while freely exploring the manipulation area. Note that during this operation the vertical position of the probes are constrained to a few micrometers above the substrate and the operator controls only the horizontal motion. The data recorded during this exploration is processed online and generates a virtual guide to pull the user to the optimum contact point. The user "feels" and sequentially adjusts the contact force for both tips, ensuring an adequate grip on the object. In the third phase of the manipulation, both grippers are immobilized on both sides of the object and the operator controls the motion of the sample holder, while still receiving haptic feedback calculated from the output of the two probes.

The choice of the particle to be manipulated is made using a top view optical microscope. The coarse positioning of the tips is also performed using this visual feedback. They are moved using the micropositioning modules (manual and motorized stages). The positioning at the correct height is achieved by automated detection of the substrate. The user then selects the task to be realized using a user interface. This protocol could be simplified by automated transitions between these steps. In that case, the completion of the pick-and-place operation would be determined by the user's decision, to give him or her the possibility of picking up again the sphere to bring it elsewhere.

Tip-alignment phases, including the haptic feedback and virtual guide generation and pick-and-place phases with two different haptic schemes are detailed in the following. They are carried out in ambient conditions, at 20°C and relative humidity of 48%.

4.7.3 *Assisted Gripper Alignment*

The alignment of each tip is a user-driven process. The operator moves the tip while receiving haptic feedback derived from amplitude measurements of the AFM probe. During the initial exploration and prior to the generation of virtual guides, the haptic feedback is only on the x axis (Figure 4.47(a)). As the operator manually scans the surface of the to-be-manipulated object, the data are recorded to reconstruct its shape and create the virtual guide. This virtual guide generates the haptic feedback along the y axis, pulling the tip to the calculated grasp line y_0, parallel to the x axis, and crossing the sphere's center.

4.7.3.1 Tapping Mode Measurements

In tapping mode, each probe is excited at its natural frequency. At constant z position above the substrate and away from objects, this results in oscillations at constant amplitude, noted as A_0. While approaching an object, starting from a few hundreds of nanometers, the tip contacts the object intermittently and the amplitude A_t decreases until a minimum value A_{CP} is reached at full-contact between the tip and the object.

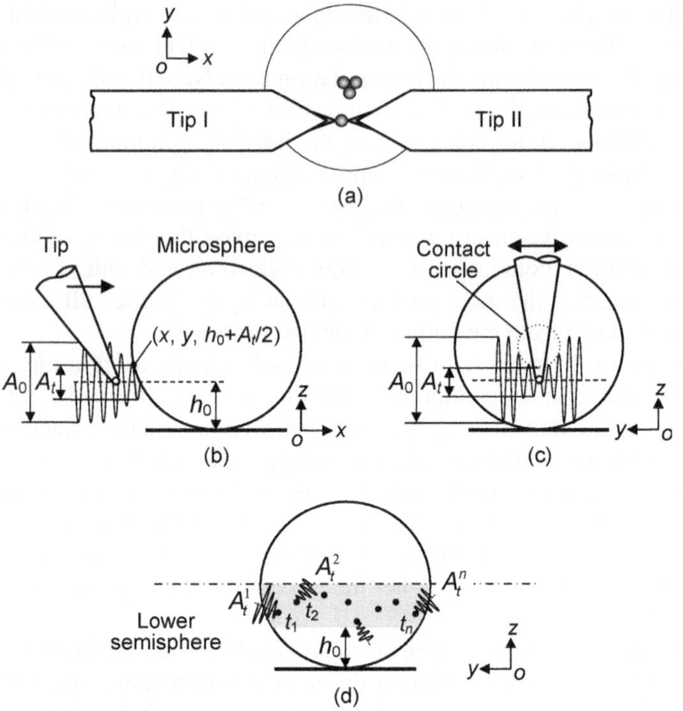

Fig. 4.47 A schematic diagram of haptic exploration by local scan of the lower semi-microsphere using a oscillating cantilever: (a) Top view of desired grasp configuration; (b) Front view shows the tip tapping the microsphere while approaching on the x-axis; (c) Side view shows the tip tapping the microsphere when scanning on the y-axis with a fixed x position; (d) n contact points recorded with random x_t^i and y_t^i positions with matching amplitudes A_t^i.

Figure 4.47 illustrates the principle of object detection from amplitude variations. The tip is first set to a given z position above the substrate h_0. This step is achieved in an initial phase and the user controls the motion only in the (x, y) plane parallel to the substrate. While the tip moves on the grasp direction of the gripper, the x axis, amplitude decreases until contact (Figure 4.47(b)). On the y axis, perpendicularl to the grasp direction, both tips must be aligned with the center of the sphere. This matches the minimum of amplitude along the y axis, at a fixed x position (Figure 4.47(c)).

4.7.3.2 Haptic Feedback for Tip Alignment

To align the gripper with the sphere and bring it to contact the visual feedback from the optical microscope does not provide sufficient resolution. Haptic force aims to compensate for this lack of visual feedback. The haptic coupling is the *DFF*

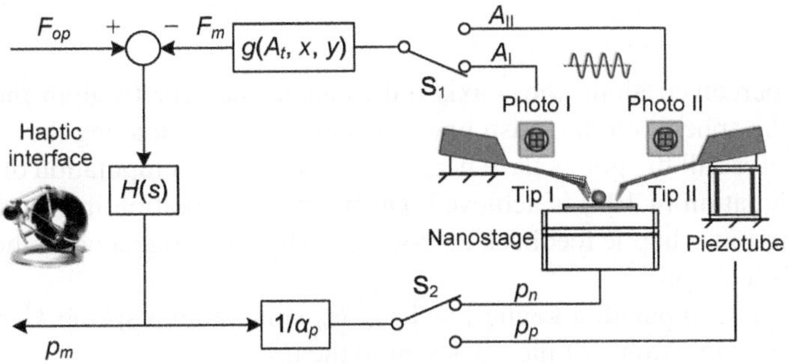

Fig. 4.48 *DFF* haptic coupling for dual-tip gripper alignment. The user manipulates the actuators by setting the position of the haptic device. Haptic feedback is derived from amplitude measurements. Depending on the considered tip, the switches S_1 and S_2 enable the user to manipulate the nanostage or the piezotube, and accordingly receive the amplitude measurement from Photodiode I or II.

control scheme, modified to take into account constraints due to the used setup (see Fig. 4.48). In particular, each tip is sequentially aligned on the grasp line and brought to contact. Haptic feedback is determined based on amplitude of oscillations.

x axis

The haptic feedback along the x axis should provide the following information:

- force null when the tip is far from the object
- increasing force as the tip approaches the object
- increasing force as the tip applies a force on the object

According to the variation of amplitude described previously, the following haptic feedback F_m^x is proposed and satisfies above stated requirements:

$$F_m^x = \begin{cases} -\alpha_a(A - A_0) & \text{if } A > A_{CP} \\ -\alpha_a(A - A_0) + k_h^x(x - x_{CP}) & \text{else} \end{cases} \tag{4.102}$$

where α_a is a scaling factor. x_{CP} is the contact point position. The amplitude A_0 is measured at the beginning of the experiment while the tip oscillates at its natural frequency. Two cases are distinguished:

- before contact (first equation): as the amplitude is decreasing while the tip approaches the object, an increasing repulsive force is sent to the user so that he or she is aware of the presence of the object
- while in contact (second equation): a spring k_h^x between the position of the contact point x_{CP} and the current position of the tip x is added to the feedback of the first equation. It simulates the force applied by the tip to the sphere. The contact point location x_{CP} is set when the amplitude measurement reaches A_{CP} at full contact. A_{CP} is an arbitrary threshold, set according to the conclusions given in [49].

y axis

The force perceived along the y axis must enable the user to align the tips with respect to the sphere, on the grasp line. The haptic feedback along the y axis is not available before all the points have been recorded and the computation of the virtual guide by Equation (4.106) is achieved. During this exploration in search of the y_0 position, the x axis haptic feedback is provided to the user so that he or she perceives the sphere's location.

When y_0 is computed, a haptic feedback F_m^y simulating a spring k_h^y between y_0 and the current position y of the tip is sent to the user:

$$F_m^y = k_h^y(y - y_0) \tag{4.103}$$

4.7.3.3 Virtual Guide Generation

During the initial exploration in tapping mode, n contact points (x_t^i, y_t^i), with their matching amplitudes A_t^i $(i = 1..n)$ are collected (Figure 4.47). The contact position data are acquired only if the actual amplitude A_t is in the $[15\%A_0, 70\%A_0]$ interval to avoid false positives. In order to define the z_t^i coordinate for each contact point (x_t^i, y_t^i), the approximation $z_t^i = A_t^i/2$ is proposed. This is a relative position since the cantilever is oscillating around the z position h_0 set manually. The calculation of the z_0 coordinate of the sphere is thus relative to h_0, and is not accurately known. However, as the only parameter useful for haptic feedback is the y_0 coordinate, this approximation is acceptable.

Figure 4.47(d) represents the n points $(t_1, ...t_n)$ recorded during the exploration process. These points are used to reconstruct the shape of the manipulated sphere, calculate the grasp line, and provide the haptic feedback along the y axis. With prior knowledge of the shape of the object, and the n recorded points, the sphere can be reconstructed from the surface equation:

$$(x - x_0)^2 + (y - y_0)^2 + (z - z_0)^2 = R^2 \tag{4.104}$$

where R is the radius of the sphere, and x_0, y_0, and z_0 are the coordinates of its center. This can be written as:

$$x^2 + y^2 + z^2 - C_a x - C_b y - C_c z + C_d = 0 \tag{4.105}$$

where: $C_a = 2x_0$, $C_b = 2y_0$, $C_c = 2z_0$, and $C_d = x_0^2 + y_0^2 + z_0^2 - R^2$. Finding the coefficients C_a, C_b, C_c, and C_d allows us to define the coordinates at the center of the microsphere and its radius. To do so, the equation which best fits the n recorded points is determined using a least mean square algorithm. The solution of the following system of equations gives the coordinates of the sphere's center:

$$\begin{bmatrix} x_t^1 & y_t^1 & z_t^1 & -1 \\ x_t^2 & y_t^2 & z_t^2 & -1 \\ & & & \\ x_t^n & y_t^n & z_t^n & -1 \end{bmatrix} \begin{bmatrix} C_a \\ C_b \\ C_c \\ C_d \end{bmatrix} = \begin{bmatrix} (x_t^1)^2 + (y_t^1)^2 + (z_t^1)^2 \\ (x_t^2)^2 + (y_t^2)^2 + (z_t^2)^2 \\ \cdots \\ (x_t^n)^2 + (y_t^n)^2 + (z_t^n)^2 \end{bmatrix} \tag{4.106}$$

The coefficients C_a, C_b, C_c and C_d can be deduced from: $\left[C_a\ C_b\ C_c\ C_d\right]^T = M^{-1}.Y$ where ^{T} is for the matrix transposition and:

$$M = \begin{bmatrix} \Sigma(x_t^i)^2 & \Sigma x_t^i y_t^i & \Sigma x_t^i z_t^i & -\Sigma x_t^i \\ \Sigma x_t^i y_t^i & \Sigma(y_t^i)^2 & \Sigma y_t^i z_t^i & -\Sigma y_t^i \\ \Sigma x_t^i z_t^i & \Sigma y_t^i z_t^i & \Sigma(z_t^i)^2 & -\Sigma z_t^i \\ -\Sigma x_t^i & -\Sigma y_t^i & -\Sigma z_t^i & n \end{bmatrix}$$

$$Y = \begin{bmatrix} \Sigma x_t^i[(x_t^i)^2 + (y_t^i)^2 + (z_t^i)^2] \\ \Sigma y_t^i[(x_t^i)^2 + (y_t^i)^2 + (z_t^i)^2] \\ \Sigma z_t^i[(x_t^i)^2 + (y_t^i)^2 + (z_t^i)^2] \\ -\Sigma[(x_t^i)^2 + (y_t^i)^2 + (z_t^i)^2] \end{bmatrix}$$

The position of the grasping point along the y axis is then computed as $y_0 = \frac{C_b}{2}$. Note that as all the points are on the same side of the sphere along the x axis ($x < 0$ for Tip I and inversely for Tip II), the calculated x_0 coordinate may be inaccurate. However, as stated above, the only parameter used for virtual guide is y_0.

The generation of this virtual guide depends on the number of points n and their respective positions. An empirical analysis to define a minimum value for n and the distribution of the points recorded is presented below along with experimental analysis.

4.7.3.4 Experimental Validation of Tip Alignment

Manipulated objects are microspheres, with a diameter of $4-6\mu$m (Figure 4.49). Before aligning the tips with the spheres, they are first positioned manually at the correct vertical position (around $500-600$nm) above the substrate[12]. Each tip is then sequentially positioned by the operator at each side of the object. For Tip I, the user actually controls the nanostage transporting the sample holder. As the nanostage includes a closed-loop position controller, the Omega supplies directly set-point values for its motion. For the alignment of Tip II, the piezotube actuator is used. As this actuator is in open-loop, it does not provide accurate positioning. However, coupling the piezotube with a haptic interface is equivalent to a closed-loop force-feedback scheme with the operator as the controller. All experimental data presented below are acquired using the nanostage and provide accurate position information.

[12] For this step, solutions based on teleoperation with haptic feedback are detailled in Section 4.4 for approach-retract experiments.

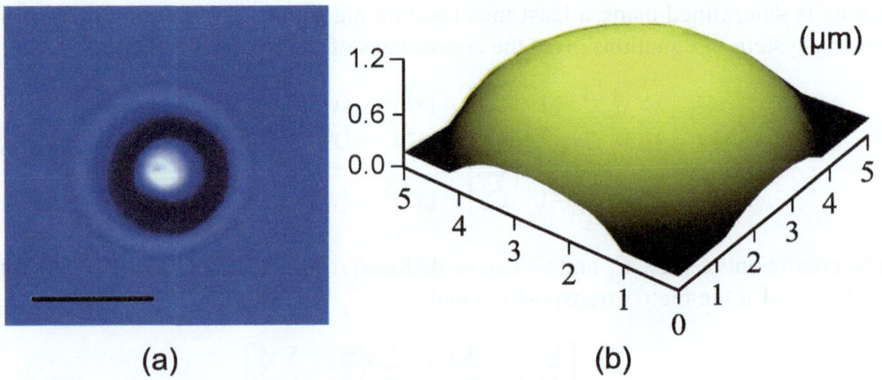

Fig. 4.49 (a) An optical image of a microsphere under 100× optical magnification (the scale bar represents 5μm). (b) An AFM image scan on the microsphere.

The piezotube, although lacking precise position measurements, gives qualitatively similar results.

x-Axis Haptic Feedback

Experimental results acquired while moving the tip along the x axis and contacting the microsphere are depicted in Fig. 4.50. The position of the tip is represented in Fig. 4.50(a), the amplitude measurement is depicted in Fig. 4.50(b) and the haptic feedback in Fig. 4.50(c). In area **1**, the tip is away from the sphere and the feedback is null. In area **2**, the user distinctly perceives the haptic feedback as the tip approaches the sphere and intermittent contact starts. An additional feedback is transmitted when an effort is applied by the cantilever on the sphere in area **3**, increasing the sensation of stiffness.

Compared to using direct force measurement from a cantilever in static mode, the tapping mode amplitude measurement provides a better sensitivity on the x axis. In static mode, as the measurement direction is almost aligned with the probes' length, the equivalent stiffness is extremely high compared to K_p on the z axis. Hence, a static detection on the x axis would only occur when a quite important force is already applied on the object. In contrast, the use of the tapping mode allows earlier detection of the object as only intermittent contact with the object produces a detectable signal. This allows users to be aware of the object's presence and prevents their involuntarily pushing.

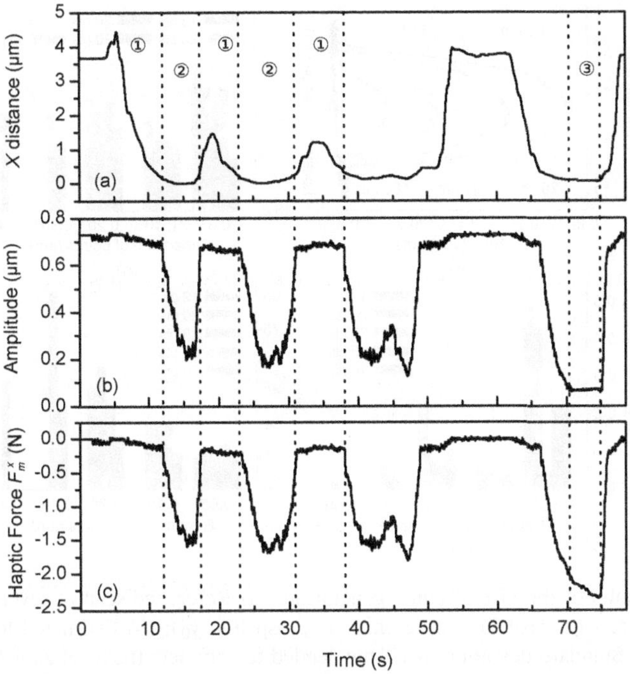

Fig. 4.50 Haptic feedback along the x axis while exploring the to-be-manipulated sphere: (a) Tip position; (b) Measured tip oscillation amplitude; (c) Force sent to the user. For the haptic feedback, the coefficients are set to: $\alpha_a = 3 \cdot 10^6 \mathrm{N.m}^{-1}$, $k_h^x = 10 \cdot 10^6 \mathrm{N.m}^{-1}$ and $a_{CP} = 0.1\,\mu\mathrm{m}$.

Generation of the Virtual Guide

As stated above, the preliminary exploration provides the data points to generate the virtual guide and the associated y axis feedback. The influence of the two parameters - the number of points n used in Equation (4.106), and the minimum distance between two points (noted d) - is experimentally explored in order to optimize the virtual guide. Table 4.7 summarizes the different trials. Each experiment (for a given (n,d) couple) is repeated 5 times.

Table 4.7 Parameters used for virtual guide generation

		Minimum distance d between two points (μm)				
		0.05	0.1	0.2	0.3	0.4
Number of points n	12	x	x	x	x	x
	20	x	x	x	x	
	25	x	x	x	x	
	30	x	x			
	40	x				

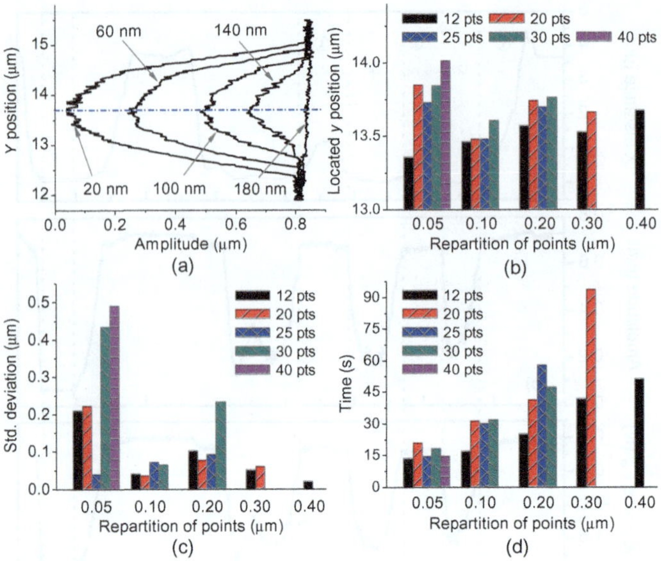

Fig. 4.51 Results of the virtual guide generation: (a) Reference scan at several x positions from the sphere (the dash line represents the grasp line y_0); (b) Estimated location of the grasp line; (c) Standard deviation; (d) Time needed to complete the localization of y_0.

Figure 4.51 compares a reference AFM scan and y_0 obtained by shape reconstruction, for the considered (n,d) couples. The standard deviation is displayed in Fig. 4.51(c). As all these results are user dependent - since users are free to choose any trajectory for the initial exploration - they should be treated only qualitatively.

The first observation is that increasing the number of points n has little impact on the accuracy of the virtual guide if these points are close to each other. Moreover, if n is set too high, more points on the edges of the semi-sphere are required. As the position data at these locations are less accurate, the standard deviation is higher. On the other hand, setting a minimum value for d forces data points to be more evenly distributed on the surface and leads to a better estimation (Figure 4.52) as supported by the decrease of the standard deviation.

Figure 4.51 also shows that except for small d values, increasing the number of points highly increases the time-cost of the virtual guide generation. Since only the hemisphere facing the tip is accessible and the tip is constrained in the vertical direction, setting d to a minimum value limits the maximum number of points that can be acquired.

Choosing $d = 0.3\mu m$ and $n = 12$ is a good trade-off between precision of the results and time-cost of the guide generation. These values are selected for the following manipulations.

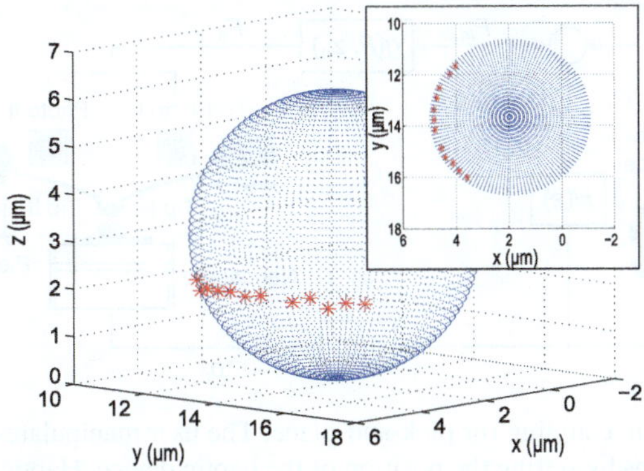

Fig. 4.52 Sphere reconstructed based on $n = 12$ and $d = 0.4\mu$m. The red dots represent the points acquired during the exploration step.

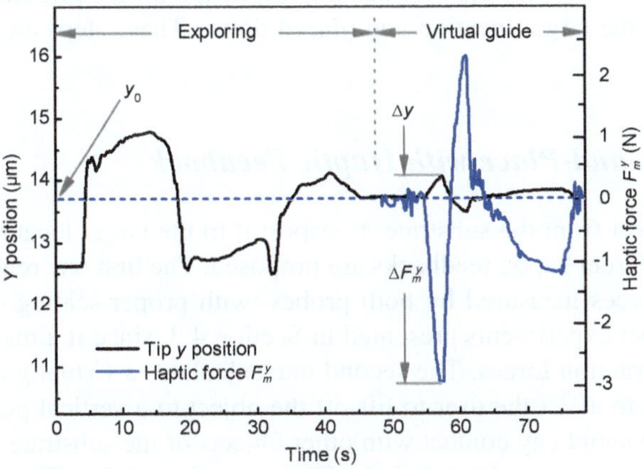

Fig. 4.53 Haptic feedback along the y axis while aligning the tip with the sphere: position of the tip and force computed with $k_h^y = 10 \cdot 10^6 \text{N.m}^{-1}$.

y-Axis Haptic Feedback

The y axis feedback is effective as soon as the virtual guide is generated. Its value is calculated using Equation (4.103). Figure 4.53 represents the force perceived by the user. The position of the tip as well as y_0 are also given. The haptic feedback on the y axis helps the user to align the tip with respect to the sphere, as y_0 is at the equilibrium point of the virtual spring.

Precise positioning is achieved since contact information is transmitted to the user through the x axis of the haptic device F_m^x, and alignment is ensured thanks to the haptic force F_m^y.

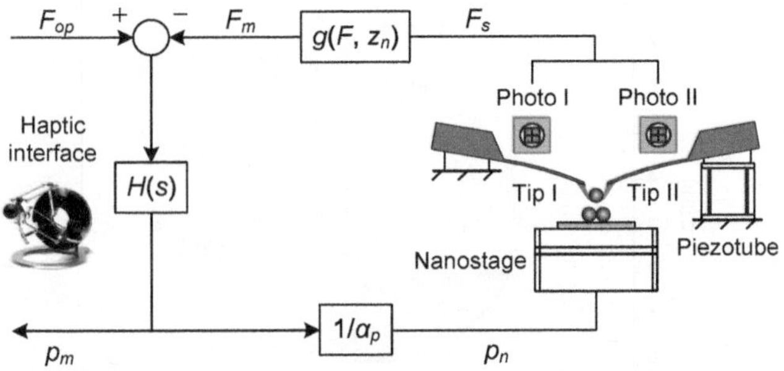

Fig. 4.54 *DFF* haptic coupling for pick-and-place. The user manipulates the nanostage with respect to the two tips by setting the position of the haptic device. Haptic feedback is derived from force measurement from the sum of the two photodiodes' outputs.

When the object is grapsed between the two tips, it must be lifted from substrate, transported to the target location and placed down. These steps are detailed in the following.

4.7.4 Pick-and-Place with Haptic Feedback

To lift the object from the substrate, transport it to the target location, and place it down, two different haptic feedbacks are proposed. The first one renders to the user directly the forces measured by both probes, with proper scaling. It is similar to approach retract experiments presented in Section 4.3, since it aims at transmitting nanoscale interaction forces. The second one calculates a virtual guide using these measurements to assist the user to lift-off the object to a vertical position set sufficiently high to avoid any contact with other objects or the substrate. It also ensures that the placing down is voluntary. In both cases, as depicted in Fig. 4.54, the Omega haptic device is used for position control of the sample holder through the nanostage, while the two tips holding the microsphere are immobilized. The force data are obtained in static mode, from the deflection of each probe measured directly on photodiodes using Equation (4.101).

Haptic feedback is rendered along the vertical ascending z axis. Hence, a positive value results in a force pushing the haptic handle upwards, away from the substrate; it is hence called "repulsive", while a negative value pulls the handle downward towards the substrate ("attractive").

4.7.4.1 Haptic Feedback of Nanoscale Interactions

This first haptic feedback returns to the user the nanoscale interactions of the pick-and-place operation as faithfully as possible. Results are similar to approach retract experiments presented in Section 4.3, since in both cases nanoscale interaction forces are transmitted. In particular, snap-in and pull-off phenomena are present.

Haptic feedback is synthesized from force responses of Tip I and Tip II. As detailed in [49], adhesive forces f_{ao} between the sphere and the substrate can be estimated as:

$$f_{ao} = f_{s,I} + f_{s,II} \tag{4.107}$$

where $f_{s,I}$ (resp. $f_{s,II}$) is the force applied to Cantilever I (resp. II). Hence, the haptic force rendered to the user is computed as:

$$\begin{aligned} F^{z}_{m,fe} &= \alpha_F (f_{ao} - f_0) \\ &= \alpha_F [(f_{s,I} - f_{s,I_0}) + (f_{s,II} - f_{s,II_0})] \end{aligned} \tag{4.108}$$

where $f_0 = f_{s,I_0} + f_{s,II_0}$ is the force measured when the tips are holding the sphere before lift-off and it is naturally proportional to the grasping force applied by the tips to the object. Removing this offset allows the user to discard the grasping force which is not useful for pick-and-place. Moreover, in the case where the grasped object is lost hazardously during the lift-off, the measured forces F_I and F_{II} will fall back to zero and Equation (4.108) will give a negative value, pulling back the probes to the substrate.

A force amplification factor α_F is used to scale the measured forces and the haptic force sent to the user. The nominal value used here is $\alpha_F = 2.0 \cdot 10^6$. This coefficient is set considering the magnitude of nanoscale interactions that should be felt by the user (in particular the pull-off force). Detailed discussion on the definition of this parameter can be found in Section 4.3.

Figure 4.55 represents the haptic feedback during a pick-and-place operation of a 5μm sphere from a glass substrate and the insert depicts forces measured from probes. The curve's starting point is the contact state between the microsphere and the substrate. As the nanostage moves down (hence the object held by the tips is lifted), probes are bent down measuring negative forces (inset i). During the pickup, when the nanostage position reaches around −900 nm, the microsphere pulls off the substrate with a minimum force of −1125 nm overcoming the adhesion. Note that after the pull-off, the measured force falls to −550 nm, and not to the pre-pick-up null value (inset ii). Actually, as the tip/object contact points are in the lower hemisphere, during the lift-off the object slides slightly down, increasing the grasping force[13]. During the transport phase, a change on the force can be noted. This is again due to the sliding of the sphere in the gripper. This hypothesis is backed up by the approx. 0.2 μm difference seen between pick-up and touch down positions along the z axis.

---

[13] Both pick-up and unintentional loss of a sphere yield a negative force feedback until the pull-off (either the pull-off of the sphere from the substrate, or the tips from the sphere). After the pull-off, in case of picking-up the sphere, the force feedback remains negative due to the grasping force. In case of an unintentional loss of the sphere, the cantilevers will go back to their neutral position, i.e. with no bending. In that case, the haptic feedback sent to the user will become positive since the zero haptic feedback corresponds to the initial grasping force applied on the sphere. The user will thus be able to distinguish the pick-up and the unintentional loss of the sphere.

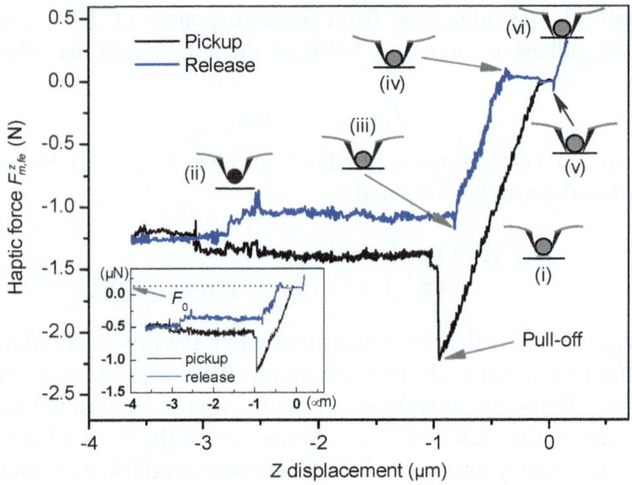

Fig. 4.55 Haptic force and measured normal force responses from both microcantilevers during the pick-and-place manipulation of a microsphere: (i) Pick-up occurs; (ii) The microsphere is detached from substrate after the pull-off; (iii) The microsphere snaps into the substrate; (iv) The gripper/microsphere pulls-off; (v) The gripper snaps into the substrate; (vi) Manipulation ends with slight bending of the microcantilevers.

During the placing down operation, the microsphere snaps-in the substrate (inset iii). As the object is pushed to the substrate, between (iii) and (iv) the contact force compensates the grasping force, until the tips pull off the sphere (inset iv) and slide down from the object to the substrate (inset v). At this point, it is sufficient to move apart both tips along the y direction to release the sphere from the gripper and achieve the operation. Note that since the contact area at object/tip interface is much smaller than at the object/substrate interface due to the sharp tips used, the problem of the object adhering to a probe is limited.

4.7.4.2 Haptic Feedback Providing Assistance

The haptic scheme presented above allows the user to feel the nanoscale interactions, especially the adhesion and the well-known pull-off phenomena. However, it is arguable if this feedback has a positive effect on manipulation dexterity. It may be more interesting to conceal these effects of which the operator is unfamiliar with and to replace the haptic feedback with a virtual guide. An assistive haptic feedback, with a positive effect on the dexterity, should fulfill following requirements:

- the user does not have to use great effort to lift the sphere off
- the sphere should be kept at a given z position during transport to avoid contact
- the sphere should be placed down only voluntarily, not because the operator touches the substrate accidentally
- while placing down the sphere, the effort applied should be as little as possible to enable an easy release of the tips. Sufficient haptic information should be provided so the user can effectively feel that the placing down has been achieved.

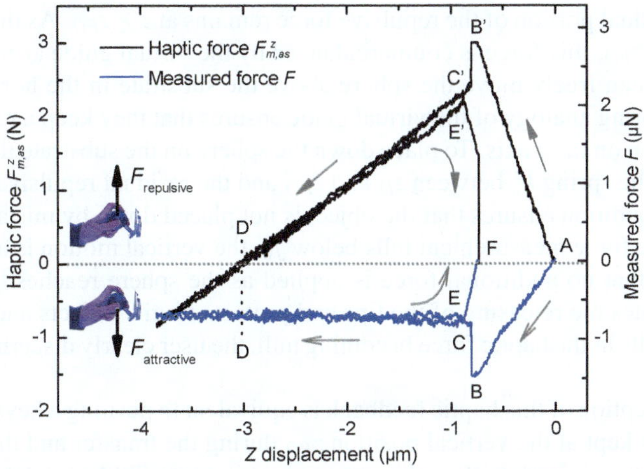

Fig. 4.56 Haptic feedback assistance for pick-and-place operation: (A) The pick-and-place operation begins, the sphere is on the substrate; (B-C) The sphere pulls-off; (D) The sphere is maintained above the substrate at a given vertical position; (E) The sphere is moved toward the substrate; (F) The sphere is placed down on the substrate. For haptic feedback, the parameters are set to: $\alpha_F = 2.5 \cdot 10^6$, $k^z_h = 1 \cdot 10^6$ N.m^{-1}. Forces higher than 3 N are truncated.

The proposed approach is based on the use of the opposite of the measured interaction force as haptic feedback. As such, for example, the pull-off force will result in a positive force on the haptic handle, actually pushing the held object away from the substrate. The expected perception is comparable to releasing a pressed keyboard button.

As previously, the haptic feedback is computed from force measurement from both photodiodes. To fulfill the requirements stated above, the output of the photodiode is converted into the haptic force $f^z_{m,as}$:

$$F^z_{m,as} = \begin{cases} -\alpha_F (f_{ao} - f_0) & \text{if } z < z_{PO} \\ -\alpha_F (f_{ao} - f_0) + k^z_h (z - z_{PO}) & \text{else} \end{cases} \quad (4.109)$$

where α_F is a force amplification factor, f_{ao} is the force measured from the photodiode, f_0 is the grasping force as above; z is the position on the vertical axis and z_{PO} is the vertical position of nanostage corresponding to the end of pull-off phenomenon. Its value is detected online during the manipulation from the sudden drop in force measurement; k^z_h is the stiffness of the virtual spring of the haptic guide which will effectively pull the object above z_{PO} and keep it at a constant vertical position.

The result of this feedback scheme is depicted in Fig. 4.56 for a pick-and-place operation on the substrate. A repulsive force, proportional to the opposite of the measured adhesive forces, assists the user to lift the object during the pull-off. Immediately after the pull-off phenomena, the measured force does not fall back to its initial value, due to the sliding of the sphere between the tips as explained above.

Thus, a residual portion of the repulsive force remains at $z \geq z_{PO}$. As the spring k_h^z is activated at z_{PO}, this force is counterbalanced by the virtual guide at z_D, above z_{PO}. Then users can freely move the sphere above the substrate in the horizontal plane while the spring analogy of the virtual guide ensures that they keep a relatively constant position on the z axis. To place down the sphere on the substrate the user has to counteract the spring k_h^z between z_D and z_{PO} and the residual repulsive force below z_{PO}. This condition ensures that the object is not placed down by mistake.

When the force measurement falls below f_0, the vertical motion is automatically stopped so that no additional force is applied as the sphere reaches the substrate. This facilitates the releasing of the tips and protects fragile objects and the gripper. As this results in the haptic force becoming null, the user clearly discerns the placing down.

The perception of this haptic feedback is equivalent to pushing a keyboard button: the object is kept at the vertical position z_{PO} during the transfer and the user has to voluntarily push it back to the substrate to place it down. Table 4.8 details each step of the pick-and-place operation using this virtual guide with phase transitions and user perception.

Table 4.8 Assistance based on haptic feedback for pick-and-place operation

Step	**Haptic feedback**	**Position of the tips: Eq. used**
A: Lift-off begins.	f_0 is acquired, $f_{m,as}^z$ is null.	$z < z_{PO}$: 1
A-B: The sphere is pulled to overcome adhesive forces.	$f_{m,as}^z$ is repulsive and increases: it helps the user lifting the sphere.	$z < z_{PO}$: 1
B-C: Pull-off happens.	The pull-off is detected, the position z_{PO} is acquiried, the feedback switches from Eq. 1 to Eq. 2.	$z = z_{PO}$: $1 \rightarrow 2$
C-D: The sphere is detached from the substrate.	$f_{m,as}^z$ is repulsive, the spring k_h^z starts to counterbalance it.	$z > z_{PO}$: 2
D: The sphere is manipulated at around z_D.	$f_{m,as}^z = 0$: the virtual spring counterbalances the residual repulsive force at z_D ($> z_{PO}$).	$z = z_D$: 2
D-E: Placing down operation begins.	$f_{m,as}^z$ is repulsive and increases: the user pushes down the sphere towards the substrate.	$z_D > z > z_{PO}$: 2
E: The sphere reaches to z_{PO}.	The feedback switches from Eq. 2 to Eq. 1.	$z = z_{PO}$: $2 \rightarrow 1$
E-F: The sphere is moved towards the substrate.	$f_{m,as}^z$ decreases.	$z < z_{PO}$: 1
F: The sphere is placed down: $f = f_0$.	The vertical motion is stopped. $f_{m,as}^z = 0$	$z < z_{PO}$: 1

Using this scheme the pick-and-place operation is made easier and safer as the haptic feedback helps the user to perform the given steps correctly.

4.7.4.3 Comparison of Two Haptic Feedback Methods

Two haptic feedback methods are presented in this paper for pick-and-place operation. However, the resulting feedback in the users is different (see Table 4.9). The haptic feedback of nanoscale interactions aims at transmitting physical phenomena. The user directly feels the adhesive forces, the pull-off phenomena, and the contact with the substrate. It improves the understanding of these interactions. Haptic feedback for assistance is not designed to be related to the physical forces but to facilitate the manipulation operation. The choice of the method will thus depend on user needs and the specificities of the task.

Table 4.9 Comparison of haptic feedback methods for pick-and-place operations

Step	Direct haptic feedback of nanoscale interactions	Haptic feedback providing manipulation assistance
The sphere is lifted-off.	The haptic feedback is attractive, and corresponds to adhesive forces.	The haptic feedback is repulsive to help the user to lift the sphere.
The sphere is above the substrate.	A residual haptic force attracts the user. It corresponds to the slide down of the sphere in the gripper.	A spring attracts the user to an equilibrium position.
The sphere is placed-down.	The haptic feedback is attractive and decreases. It is null when the sphere is on the substrate.	The haptic feedback is repulsive. The forces the user must counteract ensure a voluntary placing-down. The feedback is null when the sphere is on the substrate.

4.7.5 Construction of a Two-Layer Pyramid

In order to validate 3-D manipulation capabilities of the haptic system, a microstructure is built. The example of a two-layer pyramid composed of four nylon microspheres with diameter of 4-6μm is chosen [53]. Microspheres were deposited on a freshly cleaned glass substrate. An area of interest for the experiments was selected under an optical microscope with 20× optical magnification. Figure 4.57 shows the selected area and the inset shows the assembly sequence. After a coarse positioning under the optical microscope, the total manipulation, which includes the alignment of the gripper and the pick-and-place operation, takes less than five minutes per sphere. Since this manipulation is based on the operator, this time mainly depends on users' skills. It is given only as an indication here.

The 3D micropyramid is built using pick-and-place manipulation. Figure 4.58 shows the 3-D micromanipulation process. Figure 4.58(a) and Figure 4.58(b) are

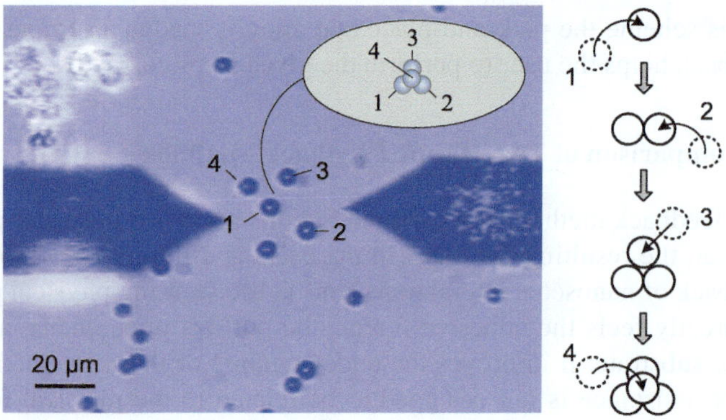

Fig. 4.57 Task planning of the micropyramid assembled by four microspheres. The right-side diagram shows the assembly protocol, in which microspheres 1, 2, and 4 are assembled by pick-and-place manipulation while microsphere 3 is pushed to its target position.

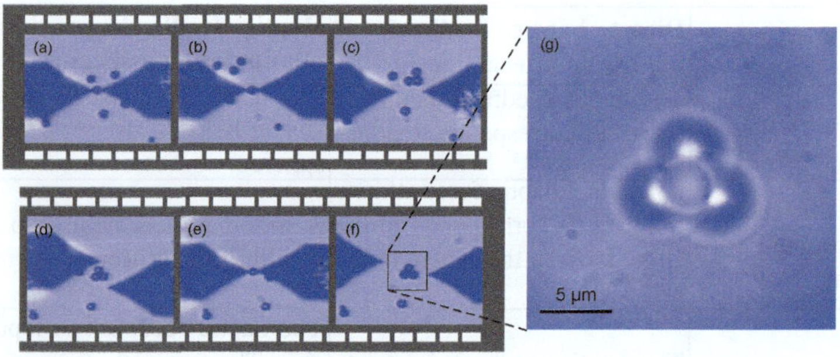

Fig. 4.58 Teleoperated 3-D microassembly demonstration of a micropyramid: (a–d) Four photos intercepted from the assembly process of the first layer of the micropyramid; (e and f) Assembly process of the second layer (the fourth microsphere) of the micropyramid. The top view photos (a)–(f) are captured under magnification of 20×. (g) Microassembly result magnified 100×.

captured when the first sphere is picked and placed. Figure 4.58(c) shows the second sphere manipulated and released at its final position. The first layer is finished after the third microsphere has been manipulated by pushing (Figure 4.58(d)). Figure 4.58(e) and Figure 4.58(f) describe the transport of the last microsphere which completes the assembly. The ultimate result is shown in Figure 4.58(f), and with a magnification of ×100 in Figure 4.58(g).

Figure 4.59 shows the haptic feedback of nanoscale interactions during the manipulation of the fourth sphere. As this sphere is put back above the others, the placing down occurs $\sim 4\mu m$ above the initial position.

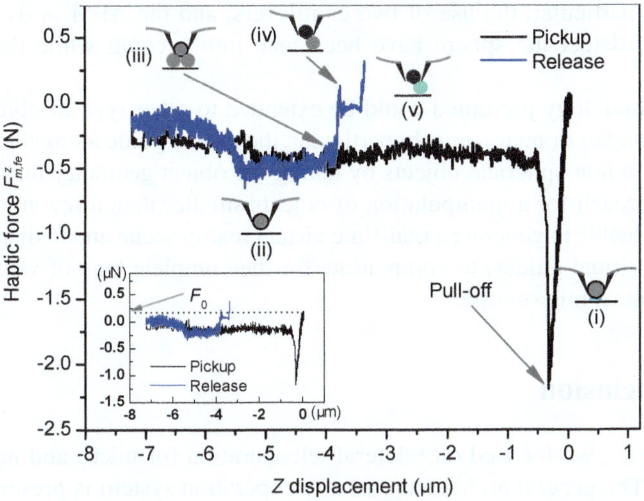

Fig. 4.59 Haptic feedback during the microassembly operation of the fourth microsphere: (i) Pick-up occurs; (ii) The microsphere is detached from the substrate; (iii) The microsphere snaps into the first layer of microspheres; (iv) The gripper/sphere pulls-off; (v) Grippers snap into the first layer of microspheres.

In the case where assistive feedback is used for manipulation of the fourth sphere, two points should be considered:

- The given vertical position at which the sphere is maintained with zero force feedback must be higher than the first layer of the spheres. This is adjusted by changing the value of the stiffness k^z_h. To increase the z position above the substrate, this stiffness should be decreased.
- While placing down the sphere, the z position will not become lower than z_{PO}. The force feedback will not decrease before placing down the sphere. When the measured force will be such that $f \leq f_0$ (the sphere is deposited onto the pyramid), the vertical motion of the sphere will be stopped and the force feedback will become null.

In these circumstances assistive haptic feedback produces results similar to the ones presented in Fig. 4.56. Moreover, if the height at which the sphere should be placed is known, Equation 4.109 can be adjusted by tuning k^z_h so that haptic feedback is identical to the one transmitted when the sphere is placed down on the substrate.

4.7.6 Remarks

Despite the complexity of the micromanipulation system, a 3D manipulation has been performed using teleoperation with haptic feedback, based on a specific strategy. In particular, the use of two cantilevers, and the AFM in dynamic mode to precisely detect the sphere have been take into account while designing this strategy.

The methodology presented could be extended to other type of objects, such as carbon nanotubes or nanowires. In particular, the virtual guide along the y axis could be adapted to non-spherical objects by taking the object geometry into account. To use this approach for a manipulation of objects smaller than a few micrometers, it might be valuable to generate a real-time virtual reality scene and to display it in 3D, along with virtual guides, to compensate for the complete lack of visual feedback from the optical microsocpe.

4.8 Conclusion

In this chapter, we focused on bilateral teleoperation for micro and nanoscale experiments. The general architecture of a teleoperation system is presented, as well as common criteria to analyse the performance of a control scheme, such as transparency, stability, and quality of the teleoperation system.

First, three control schemes are detailed, analyzed and compared. A theoretical evaluation of their performance is conducted. An experimental comparison of these control schemes is performed on 1D teleoperation tasks. Two custom AFM-based micro/nanomanipulation platforms, which enable teleoperation, are described. Advantages and drawbacks of AFM measurements are detailed. It is shown that AFM systems generate only two force measurement, which necessitates the augmentation of theoretical models to reconstruct 3D force measurements for 3D haptic feedback.

Next, 2D and 3D teleoperation experiments have been performed on the described platforms. They enable both a better understanding of nanoscale interactions, such as frictionnal effects, and higher efficiency in manipulation tasks, such as 3D assembly of microspheres. The missing force information is reconstructed either based on frictional models, or on the use of AFM in dynamic mode.

Finally, complex experiments are performed using teleoperation, based on both adequate control schemes, and models or strategies to derive efficient haptic feedback from AFM measurements. These studies demonstrate the feasibility of complicated experiments using teleoperation systems, and the associated benefits in terms of both efficiency and understanding of nanoscale phenomena.

What is detailed in this chapter is a first step toward haptic feedback nanomanipulation. Prior to using it, though, a real-time virtual reality scene should be generated and displayed in 3D, along with virtual guides, to compensate for the lack of visual feedback. The use of other tools, such as tweezers, should also be considered. However, they usually lack force measurements. Vision might be another solution to derive haptic feedback.

4.9 Nomenclature

Haptic:

- F_{op}, F_m, F_s: User force, haptic force rendering and force on the slave side, respectively
- $H(s)$: Haptic device transfer function
- α_a, α_F, α_p: Amplitude, force and displacement scaling factor, respectively
- M_h, B_h: Inertia, damping of the haptic device

Force measurements:

- K_s, K_p: Surface, probe stiffness, respectively
- K_{eq}: equivalent stiffness which comprises the probe and the substrate stiffnesses
- V_{A-B}, V_{LFM}: Photodiode voltage output for a vertical, horizontal deflection respectively
- s_n, s_l: normal, lateral sensitivity values of optical measurement system respectively
- V_I, V_{II}: Voltage output of Photodiode I, II
- A_I, A_{II}: Amplitude output of Photodiode I, II
- ζ, ϕ: Deflexion, angle of rotation of the cantilever

References

1. Fearing, R.S.: IEEE/RSJ Intelligent Robots and Systems Conference, pp. 212–217 (1995) 145
2. Sitti, M.: IEEE Robotics and Automation Magazine 14(1), 53 (2007) 145
3. Hokayem, P.F., Spong, M.W.: Automatica 42, 2035 (2006) 145, 148
4. Yokokohji, Y., Yoshikawa, T.: IEEE Int. Workshop on Intelligent Robots and Systems, pp. 355–362 (1990) 145, 146
5. Cavusoglu, M.C., Sherman, A., Tendick, F.: IEEE Trans. Robot. Automat. 18(4), 641 (2002) 145, 146, 173
6. Hogan, N.: IEEE Int. Conf. Robotics and Automation, vol. 3, pp. 1626–1631 (1989) 145, 146, 151
7. Hollis, R.L., Salcudean, S., Abraham, D.W.: IEEE Int. Conf. Micro Electro Mechanical Systems, pp. 115–119 (1990) 146
8. Ryu, J.H., Kim, Y.S., Hannaford, B.: IEEE Trans. on Robotics 20(4), 772 (2004) 146, 165, 201
9. Kim, S., Sitti, M.: IEEE Trans. on Automation Sci. and Eng. 3(3), 240 (2006) 146, 151, 198
10. Chopra, N., Berestesky, P., Spong, M.W.: IEEE Trans. on Control Systems Technology 16-2, 304 (2008) 146
11. Berestesky, P., Chopra, N., Spong, M.W.: IEEE Int. Conf. Robotics and Automation, pp. 4557–4564 (2004) 146
12. Secchi, C., Stramigioli, S., Fantuzzi, C.: IEEE Int. Conf. Robotics and Automation, pp. 3290–3295 (2003) 146
13. Boukhnifer, M., Ferreira, A.: Robotics and Autonomous Systems 54, 601 (2006) 146
14. Sitti, M., Hashimoto, H.: IEEE/ASME Trans. Mechatron. 8(2), 287 (2003) 146, 166

15. Ferreira, A., Mavroidis, C.: IEEE Robotics and Automation Magazine 13(3), 78 (2006) 146
16. Millet, G., Haliyo, S., Régnier, S., Hayward, V., pp. 273 –278 (2009) 147
17. Diolaiti, N., Niemeyer, G., Barbagli, F., Salisbury, J.: IEEE Transactions on Robotics 22(2), 256 (2006) 147
18. Lawrence, D.: IEEE Transactions on Robotics and Automation 9(5), 624 (1993) 148
19. Hannaford, B., Ryu, J.H.: IEEE Transactions on Robotics and Automation 18(1), 1 (2002) 151
20. Artigas, J., Preusche, C., Hirzinger, G., Melchiorri, B.G.C.: Third Joint Eurohaptics Conference and Symposium on Haptic Interfaces for Virtual Environment and Teleoperator Systems, pp. 488–493 (2009) 151
21. Adams, R., Hannaford, B.: IEEE Transactions on Robotics and Automation 15(3), 465 (1999) 152
22. Llewellyn, F.: Proceedings of the IRE 40(3), 271 (1952) 152
23. Micaelli, A.: Tlopration et tlrobotique (Herms Science 2002), ch. 6. Asservissements et lois de couplage en tlopration (2002) 152
24. Haddadi, A., Hashtrudi-zaad, K.: Third Joint Eurohaptics Conference and Symposium on Haptic Interfaces for Virtual Environment and Teleoperator Systems, pp. 220–225 (2009) 152
25. Bolopion, A., Cagneau, B., Haliyo, S., Régnier, S.: Journal of Micro - Nano Mechatronics 4(4), 145 (2009) 155
26. Goethals, P., Gersem, G.D., Sette, M., Reynaerts, D., Brussel, H.V.: Proceedings of the Second Joint EuroHaptics Conference and Symposium on Haptic Interfaces for Virtual Environment and Teleoperator Systems (2007) 155
27. Gil, J.J., Avello, A., Rubio, A., Flórez, J.: IEEE Transactions on Control Systems Technology 12(4), 583 (2004) 162, 163
28. Lee, S., Song, J.: IEEE/ASME Trans. on Mechatronics 6(1), 58 (2001) 170
29. Onal, C.D., Sumer, B., Sitti, M.: Review of Scientific Instruments 79, 103706 (2008) 172
30. Maugis, D.: J. of Colloid Interf. Sci. 150(11), 243 (1992) 180
31. Sumer, B., Sitti, M.: J. Adh. Scie. Techn. 22, 481 (2008) 185
32. Steinfeld, A., Fong, T., Kaber, D., Lewis, M., Scholts, J., Schultz, A., Goodrich, M.: Human Robot Interaction, pp. 33–40 (2006) 185
33. Vogl, W., Ma, B.K.L., Sitti, M.: IEEE Trans. on Nanotechnology 5(4), 397 (2006) 186, 187, 197, 198
34. Vettiger, P., Cross, G., Despont, M., Drechsler, U., Drig, U., Gotsmann, B., Hberle, W., Lantz, M.A., Rothuizen, H.E., Stutz, R., Binnig, G.K.: IEEE Trans. on Nanotechnology 1(1), 39 (2002) 188
35. Eleftheriou, E., Antonakopoulos, T., Binnig, G.K., Cherubini, G., Despont, M., Dholakia, A., Drig, U., Lantz, M.A., Pozidis, H., Rothuizen, H.E., Vettiger, P.: IEEE Trans. on Magnetics 39(2), 938 (2003) 188
36. Fisher-Cribbs, A.C.: Mechanical Engineering Series: Nanoindentation. Springer, Heidelberg (2002) 188
37. Liu, C., Lee, S., Sung, L., Nguyen, T.: J. Appl. Phys. 100(3), 033503 (2006) 189
38. Bolopion, A., Cagneau, B., Régnier, S.: Proceedings of the IEEE International Conference on Intelligent Robots and Systems, pp. 3265–3270 (2009) 191
39. Haliyo, S., Dionnet, F., Régnier, S.: International Journal of Micromechatronics 3(2), 75 (2006) 191, 192
40. Tafazzoli, A., Pawashe, C., Sitti, M.: Proceedings of the IEEE International Conference on Robotics and Automation, pp. 263–268 (2006) 191

41. Li, G., Xi, N., Yu, M., Fung, W.: IEEE/ASME Trans. on Mechatronics 9(2), 358 (2004) 197, 199
42. Marliere, S., Urma, D., Florens, J., Marchi, F.: EuroHaptics, pp. 246–252 (2004) 197
43. Onal, C.D., Sitti, M.: The International Journal of Robotics Research 28(4), 484 (2009) 179, 197, 198
44. Bhushan, B.: Handbook of Micro/Nano Tribology, 2nd edn. CRC Press, Boca Raton (1999) 200
45. Ioannou, P., Sun, J.: Robust Adaptive Control. Prentice Hall, Englewood Cliffs (1996) 202
46. Krohs, F., Onal, C.D., Sitti, M., Fatikow, S.: ASME Journal of Dynamic Systems, Measurement, and Control (2008) (in press) 210
47. Bolopion, A., Xie, H., Haliyo, S., Régnier, S.: IEEE/ASME Transactions on Mechatronics (2010) (to appear) 211
48. Sitti, M.: Proceedings of the IEEE Conference on Nanotechnology, pp. 75–80 (2001) 212
49. Xie, H., Régnier, S.: Journal of Micromechanics and Microengineering 19(7), 075009 (p. 9) (2009) 212, 217, 225
50. Xie, H., Haliyo, S., Régnier, S.: Nanotechnology 20, 215301 (p. 9) (2009) 212
51. Xie, H., Régnier, S.: IEEE/ASME Transactions on Mechatronics 16(2), 266–276 (2011) 212
52. Savia, M., Koivo, H.: IEEE/ASME Transactions on Mechatronics 14(4), 504 (2009) 214
53. Saito, S., Miyazaki, H.T., Sato, T.: Journal of Robotics and Mechatronics 14(3), 227 (2002) 229
54. Onal, C.D., Sitti, M.: Teleoperated 3-D force feedback from the nanoscale with an atomic force microscope. IEEE Transactions on Nanotechnology 9, 46–54 (2010) 198

Chapter 5
Automated Control of AFM Based Nanomanipulation

Fabricating microstructures has been investigated for over a decade. Self-sustained electromechanical systems in these small scales are becoming more and more of a necessity to make further scientific explorations in physics, biology, and chemistry as current systems have many limitations in their interaction with small scale phenomena due to their size. Also, building miniature agents will one day enable massively parallel operations such as distributed sensory data, fast manipulations of small scale materials for repair operations, or smart materials that can change shape or material properties to interact with large scale objects efficiently. These devices might one day replace current designs even on the macro scale.

Micromanipulation implies precise interactions with microscale objects for pulling, pushing, cutting, indenting, picking and placing, etc. In this context, submicron or nanometer scale resolution positioning is imperative.

The physics at small scales is very different than what one is used to in macro scale. Different kinds of forces scale differently when dimensions are scaled down linearly. Inertial (volumetric) forces, such as weight, that dominate in macro-scales are negligible with respect to areal (adhesive) and peripheral (capillary) forces, especially at scales smaller than 10 μm [1, 2, 3].

One method of producing microstructures is utilizing micro/nanoelectromechanical systems (MEMS/NEMS) monolithic fabrication processes, which have a top-down approach to the problem. These processes typically start with a block of material and move down to achieve the final product with generally some chemical and/or optical lithography routines. The most important drawback of this approach is that, unfortunately not every 3D geometry can be achieved with it. This fact is summarized in the definition of MEMS/NEMS fabrication processes, which are said to be $2\frac{1}{2}$D processes. Also, materials that can be used with them are limited.

Even though these approaches can be considered as a subset of micromanipulation, what is generally understood from the term is moving sub-units, such as molecules, particles, rods, etc. to achieve the final product as an assembly with a bottom-up approach. Potentially, micromanipulation could achieve any 2D or 3D structure imaginable. Also, it does not necessarily have limits on material to be used. However, in today's technology, it suffers mainly from the lack of a general method

H. Xie et al.: Atomic Force Microscopy Based Nanorobotics, STAR 71, pp. 237–311.
springerlink.com

with autonomous operation capability at 2D and three-dimensional (3D) scales [4]. This chapter focuses on how AFM could be used to automatically manipulate micro and nano-objects.

The first section will describe the principle of the automated two-dimensional micromanipulation using AFM based robotic system. The second will focus on three-dimensional automated micromanipulation using a nanotip gripper with multi-feedback and how force sensing could be useful for automated micromanipulation. The third section is devoted to three-dimensional nanomanipulation using a robotic AFM system with two probes. The subsection IV will introduce the concept of parallel imaging/manipulation force microscopy, in which the probes are alternately used to scan or to manipulate nanoobjects. The section V will describe in full details this concept by focusing on high speed imaging and nanomanipulation.

5.1 Automated Two-Dimensional Micromanipulation

This section describes a 2D autonomous micromanipulation control system using a pushing based mechanical pushing or pulling method. An atomic force microscope (AFM) probe tip is utilized to push and pull 4.5 μm diameter polystyrene (PS) particles on a flat glass substrate in 2D. The work is generalizable to different materials, sizes and geometries. Also, an iterative sliding mode parameter estimator that satisfies the Lyapunov stability criterion is used to observe the parameters of the transformation between the image coordinates and world coordinates. Sliding mode control (SMC) is a robust control technique with many important features including insensitivity to matched parameter variations and disturbance rejection. SMC simply works by constraining the plant on a predefined manifold $\sigma = 0$ in the state space and therefore reducing the order of the motion [15, 16, 18]. The gist of SMC is selecting a manifold such that control objectives are realized on or with the help of it using a reaching condition such as the Lyapunov stability criterion to confine the motion on the manifold as utilized in this work.

The inclusion of the sliding mode estimator enables one to convert references in image coordinates to world coordinates and also to extrapolate the positions between the two frames for a larger bandwidth and hence, smoother and faster motions without overshoot.

The organization of this section is as follows. In subsection I, the problem of manipulating microparticles autonomously to form predefined patterns and assemblies are defined with the experimental setup. In subsection II, the image processing procedure to detect particles in real-time is explained. In subsection III an iterative discrete sliding mode parameter estimator is designed to find the transformation between the world-frame and the image-frame. In subsection IV, the microparticle manipulation procedure is defined as pushing/pulling with an AFM probe tip. In subsection V, the algorithm to order manipulation of individual microparticles to form a user-defined pattern is explained.

5.1.1 Problem Definition

The experimental setup is designed as shown in Fig. 5.1 for autonomous pushing or pulling of microparticles using an AFM probe tip. A glass slide carrying the particles on the bottom side is attached to a nanometer precision *XY* piezoelectric stage (Queensgate NPS-XYZ-100A, with effective *x-y-z* range of 100 μm $\times$ 100 μm $\times$ 15 μm with $\pm$ 5 nm closed-loop precision). The stage has its own closed-loop PID controller using capacitive sensors and driver that takes position inputs from a PC with a D/A card and moves to those positions rapidly. The *z*-axis is extracted from the piezoelectric stage and attached to another, manual stage. An AFM probe is attached to this z-stage upside-down, so it can touch the bottom side of the substrate and hence, the particles. This inverted probe setup is required not to obstruct the images of microparticles during micromanipulation for real-time visual servoing based control. This setup is placed under a top-view reflective type optical microscope (Nikon Eclipse L200) with a video camera (Dage-MTI DC-330), connected to a frame-grabber on the PC to close the loop. A constraint in this inverted configuration is the necessity of a transparent substrate.

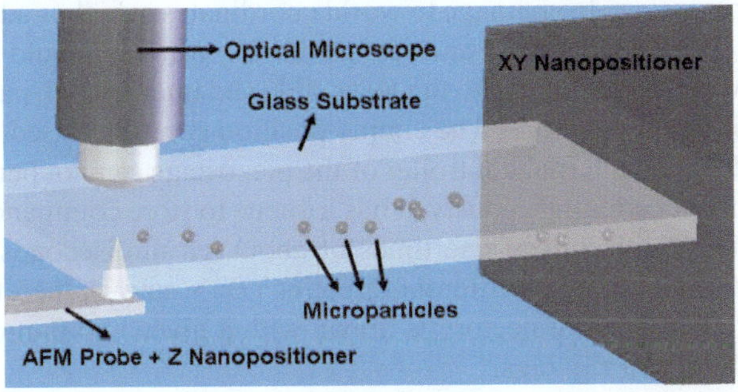

Fig. 5.1 Structure of the autonomous micromanipulation system using an AFM probe for pushing/pulling microparticles on a glass substrate using a top-view optical microscope visual feedback.

Particles are deposited on the glass slide in a deionized water solution that is passively evaporated. The setup can easily be used in an inverted microscope system by depositing the particles on the top-side of a glass slide and touching with the probe from the top.

As mentioned briefly before, the dominance of inertial and gravitational forces degrade and diminish as objects scale down to micron size. Manipulated particles are chosen to be 4.5 μm in diameter. Therefore, adhesive forces play the most important role in the manipulation process. As these forces are geometry, material type and environmental condition dependent, a pick and place kind of approach is hard to achieve due to release problems. However, using an AFM probe with a very sharp tip (Ultrasharp NSC12/50 non-contact probe made by Micromasch Inc.), adhesive forces are unable to cause permanent stiction. In theory, adhesive forces between the substrate and particles should be much larger than those of the tip-particle interaction. However, this theory assumes that contact with the object is conducted with only the tip, which is not the case in practice. Instead, particles contact the side of the tip and the area of contact is much larger. Still, for the selected tip, it is experimentally verified that repeatable release can be achieved for a vertical motion and adhesion can stick the particle to the tip for slow horizontal motions in a quasi-static equilibrium. This enables pushing and pulling of particles by the tip, while no permanent stiction occurs. The effective forces while manipulating particles horizontally or releasing/coming into contact with particles vertically for the success of a manipulation task is investigated in subsection 5.1.4 and depicted in Fig. 5.3. The relationships of net horizontal and vertical forces to the cone half angle θ and tip substrate separation L, while pulling and releasing the particles are shown in Fig. 5.4 and 5.5.

For an autonomous micromanipulation system with visual feedback, there are three main issues to be addressed. First, a fast image processing algorithm should be utilized to detect particle positions. A gradient based circle detection algorithm is used for this as described in subsection 5.1.2. Second, the transformation between the world-frame and the image-frame should be found to convert the reference commands to the piezostage from image to world coordinates, which is addressed in subsection 5.1.3. Third, an accurate and fast position controller should be designed to move the AFM probe tip relative to the glass slide and interact with particles to move them into their target positions. The simple position controller used is explained in subsection 5.1.4. The internal controller of the piezostage can not perform positioning in different axes simultaneeously, it has a queue to store commands and perform them sequentially with typical rise times of about ten milliseconds. As the references are sent sequentially in different directions (x, y, and z), therefore to achieve a smooth motion, a linear trajectory is drawn with a given maximum velocity with negligible incremental zig-zags.

5.1.2 Image Processing

The first challenge and the starting point of the work is the real-time detection of particles using optical microscope images. Since an image based controller will be utilized to ensure the mechanical manipulation of particles by pushing and pulling with an AFM probe tip to their respective target positions, it is imperative to process each frame as it is fed to the frame-grabber and update positions accordingly. OpenCV library is used to handle all image processing tasks.

Spherical PS particles (Banglabs Inc.) 4.5 μm in diameter are used for all experiments. PS particles are chosen to be manipulated since they can be used in optical microdevice prototyping applications, are hydrophobic, which removes the effects of capillary forces, and are commercially available in wide range of sizes. Under the microscope, any spherical object, and hence the particles, have circular shapes. Many circle detection schemes are proposed in the image processing literature. Among them, Hough transform is the most common algorithm that is known to be robust under noisy conditions [26]. In general, it is possible to detect any arbitrary shape that can be quantized with parameters [27]. Although it has many favorable properties, most crucial problem of the Hough transform is the fact that it is slow and hard to include in a real-time process. There have been some improvements to the generalized Hough transform, such as adding gradient direction information or probabilistic calculations [28]. However, even with these improvements, the speed of the process is still not enough to process every frame repeatably.

Algorithm 1. Detection of Particles

1: Calculate the gradient image.
- Calculate the x and y gradients using the Sobel operator.
- Calculate the magnitude and direction of gradients at each pixel.

2: Utilize an adaptive threshold, Th, according to the maximum gradient magnitude, $\max(gr)$, such that $Th = K\max(gr)$, $K \in [0,1]$.
3: Search for pairs of gradient vectors with the two properties mentioned above.
4: Utilize a voting mechanism to avoid false positives.
5: Search for circles with the predefined particle radius.
6: Disregard circles closer to each other than a specified threshold value.

Utilizing the gradient directions of a grayscale image, Rad *et al.* [29] proposed a new algorithm to detect circles quickly. The algorithm is based upon the two reasonable assumptions that for each circle,

1. There will be pairs of gradient vectors with opposite directions.
2. The slope of the line that connects the two base points of these vectors would be about the same as the slope of the first vector.

The details of this particle detection algorithm are given in Algorithm 1.

This algorithm can detect particles whether or not the cantilever is at the back of them, or the tip is close to them with the assumption that initially, particles are not in contact. This application of the mentioned circle detection algorithm takes about 0.03 sec for a 320 × 240 pixels frame on average on a non real-time environment, which allows every frame to be processed.

5.1.3 Parameter Estimation

Before focusing on the pushing/pulling of the particles, it is necessary to discuss calibration of the camera. One issue of visually servoed micromanipulation attempts is the calibration of length scale between the image-frame and the world-frame. As the end-effector moves according to references in the real-frame, a calibration procedure is typically the first step to map the corresponding positions between the two frames. A transformation or even a simple look-up table can be used for the mentioned mapping. Since using or generating look-up tables is time and resource consuming, and using purely mathematical procedures might yield inaccurate results, an iterative discrete sliding-mode parameter observer is proposed to converge to the solution in this study.

The mapping between the two frames is simply a linear (affine) transformation and a translation. This mapping can represent any 3D orientation of the real-frame as projected onto the image-frame. Assuming that a piezo-stage is fastened carefully under the microscope, the elements of the transformation matrix $\mathbf{A} \in \Re^{2\times 2}$ and the translation vector $\mathbf{b} \in \Re^{2\times 1}$ would be constant. The resulting transformation operation could be described as

$$\underbrace{\begin{pmatrix} x^i \\ y^i \end{pmatrix}}_{\mathbf{x}_i} = \underbrace{\begin{bmatrix} a_{11} & a_{12} \\ a_{21} & a_{22} \end{bmatrix}}_{\mathbf{A}} \underbrace{\begin{pmatrix} x^r \\ y^r \end{pmatrix}}_{\mathbf{x}_r} + \underbrace{\begin{pmatrix} b_1 \\ b_2 \end{pmatrix}}_{\mathbf{b}} \tag{5.1}$$

where $\mathbf{x}_i \in \Re^{2\times 1}$ and $\mathbf{x}_r \in \Re^{2\times 1}$ are the state (position) vectors in image and real frames, respectively.

This transformation could be converted to parameter space as

$$\mathbf{x}_i = \underbrace{\begin{bmatrix} x^r & y^r & 0 & 0 & 1 & 0 \\ 0 & 0 & x^r & y^r & 0 & 1 \end{bmatrix}}_{\mathbf{K}} \underbrace{\begin{pmatrix} a_{11} \\ a_{12} \\ a_{21} \\ a_{22} \\ b_1 \\ b_2 \end{pmatrix}}_{\mathbf{p}} \tag{5.2}$$

where $\mathbf{p} \in \Re^{6\times 1}$ is the parameter vector, to be estimated. One important requirement for the matrix $\mathbf{K}$ is that $\text{rank}(\mathbf{K}) = m = 6$ for the proposed sliding-mode estimator to converge to a solution. This means that the number of states should at least be equal to the number of parameters and cannot be true with its current definition. Assuming that the transformation is constant (for at least two iterations), this matrix could be redefined to become a nonsingular (square) matrix.

The parameter estimator is, as mentioned, an iterative one. This means that discrete values are obtained for the input and output vectors iteratively with a similar approach to the controller previously designed in [16] and [17]. Simply combining the current values with the two previous values, (5.2) becomes,

$$\mathbf{x}_i = \underbrace{\begin{bmatrix} x^r_k & y^r_k & 0 & 0 & 1 & 0 \\ 0 & 0 & x^r_k & y^r_k & 0 & 1 \\ x^r_{k-1} & y^r_{k-1} & 0 & 0 & 1 & 0 \\ 0 & 0 & x^r_{k-1} & y^r_{k-1} & 0 & 1 \\ x^r_{k-2} & y^r_{k-2} & 0 & 0 & 1 & 0 \\ 0 & 0 & x^r_{k-2} & y^r_{k-2} & 0 & 1 \end{bmatrix}}_{\mathbf{K}} \underbrace{\begin{pmatrix} a_{11} \\ a_{12} \\ a_{21} \\ a_{22} \\ b_1 \\ b_2 \end{pmatrix}}_{\mathbf{p}} \tag{5.3}$$

redefining image state vector as

$$\mathbf{x}_i = \left(x^i_k \; y^i_k \; x^i_{k-1} \; y^i_{k-1} \; x^i_{k-2} \; y^i_{k-2}\right)^T. \tag{5.4}$$

Now that the necessary conditions are held for the estimator to work, its working principle could be further explained. Defining the estimated parameters as $\mathbf{u} \in \Re^{6\times 1}$ and the image state vector according to this estimation as $\hat{\mathbf{x}}_i = \mathbf{K}\mathbf{u}$, estimation error becomes

$$\mathbf{e} = \mathbf{K}(\mathbf{p} - \mathbf{u}). \tag{5.5}$$

Choosing the sliding mode variable $\sigma \in \Re^m$ as

$$\sigma = \mathbf{e}, \tag{5.6}$$

for det($\mathbf{K}$) $\neq 0$, for stability, a positive definite Lyapunov function of the form

$$v(\sigma) = \frac{\sigma^T \sigma}{2}, \tag{5.7}$$

is used, with the derivative

$$\dot{v}(\sigma) = \sigma^T \dot{\sigma}. \tag{5.8}$$

If the control function is designed such that

$$\dot{\sigma} + \mathbf{D}\sigma = 0, \tag{5.9}$$

for positive definite symmetric matrix $\mathbf{D} \in \Re^{m\times m}$, the Lyapunov function derivative becomes a negative-definite function as

$$\dot{v}(\sigma) = -\sigma^T \mathbf{D}\sigma, \tag{5.10}$$

which satisfies the Lyapunov stability criterion.

From (5.5) and (5.6), sliding mode variable is

$$\sigma = \mathbf{K}(\underbrace{\mathbf{p}}_{\mathbf{u}_{eq}} - \mathbf{u}). \tag{5.11}$$

Here $\mathbf{u}_{eq} \in \Re^m$ is the equivalent control defined by Utkin [15], which makes $\sigma = 0$. Solving (5.11) for $\mathbf{u}_{eq}$ yields

$$\mathbf{u}_{eq}(t) = \mathbf{u}(t) + \mathbf{K}^{-1}\sigma. \tag{5.12}$$

Putting (5.11) in (5.9),

$$\dot{\sigma} + \mathbf{D}\mathbf{K}(\mathbf{u}_{eq}(t) - \mathbf{u}(t)) = 0, \tag{5.13}$$

the only unknown that prevents the calculation of $\mathbf{u}$ is $\mathbf{u}_{eq}$, which is hard to calculate. However, since it is a smooth function, an approximation could be made by using the previous time step value of the $\mathbf{u}$ in (5.12) such that

$$\mathbf{u}_{eq}(t) \approx \mathbf{u}(t - \Delta t) + \mathbf{K}^{-1}\sigma. \tag{5.14}$$

Using this approximation in (5.13) and solving for current **u** gives

$$\mathbf{u}(t) = \mathbf{u}(t-\Delta t) + [\mathbf{DK}]^{-1}(\dot{\sigma} + \mathbf{D}\sigma)|_{t-\Delta t}, \quad (5.15)$$

or in discrete-time

$$\mathbf{u}_k = \mathbf{u}_{k-1} + [\mathbf{DK}]^{-1}(\dot{\sigma} + \mathbf{D}\sigma)|_{k-1}. \quad (5.16)$$

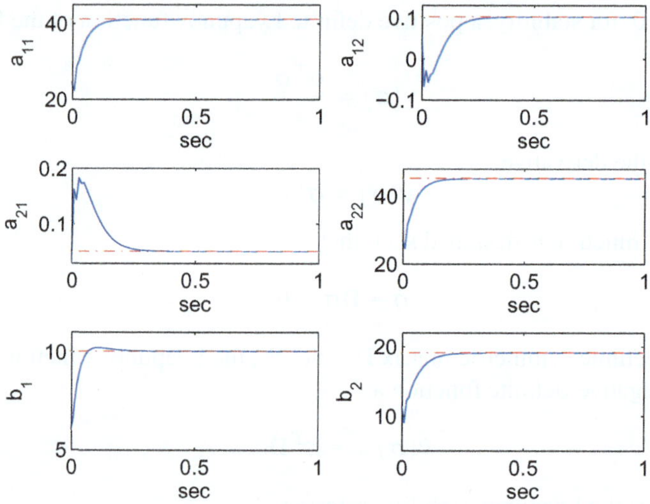

Fig. 5.2 Simulation results for the sliding mode iterative parameter estimator.

The effectiveness of the proposed observer is demonstrated on simulation results in Fig. 5.2. The experimental results are in line with these results as similar convergence curves are achieved.

5.1.4 Controlled Pushing and Pulling

As the AFM probe tip radius is tiny (around 20 nm) with respect to the size of the particles to be manipulated by pulling/pushing, the contact area with the particle is much smaller than particle substrate contact. This leads to smaller adhesive forces as the contact area is reduced. However, while pulling (or pushing) horizontally on the glass slide, the force that opposes adhesive forces is the particle-substrate friction force, which can not be enough to prevent the particle from sticking near the probe tip. Moreover, it is a small probability to ensure contact with only the tip of the nanoprobe, which has a length of about 20-25 μm. And contact with the side of the probe increases the contact area, and hence, the adhesion.

A theoretical analysis on using pulling for the manipulation of microparticles is realized in this subsection. Since the spinning friction is much lower than rolling and sliding frictions, even a small alignment error results in the spherical particle to spin (and roll) around the tip [21]. This causes an instability and makes it harder to control a merely pushing based manipulation attempt. In contrast, pulling provides a stable alternative that simplifies the trajectory of the tip with respect to the slide.

As the motion speed will be very small, the system can be considered to be in quasi-static equilibrium written as

$$F_A^T \cos\theta - F_F^T \sin\theta = F_F^S, \tag{5.17}$$

$$F_A^T \sin\theta + F_F^T \cos\theta = F_A^S. \tag{5.18}$$

where the subscripts denote adhesive (A) and frictional (F) forces and the superscripts indicate forces at the interface of the particle with the tip (T) and substrate (S). Angle θ is the angle of the tip shape with respect to the vertical axis. For a conical tip, this angle is invariant to the direction. Effective forces in the system while moving horizontally, touching the particle (i.e. pushing or pulling) and vertically (i.e. coming into contact or releasing) are depicted in Fig. 5.3. According to this geometrical interpretation, for particles to be pushed/pulled into their target positions, net horizontal force exerted on the particle by the probe should be larger than the frictional force exerted by the glass slide.

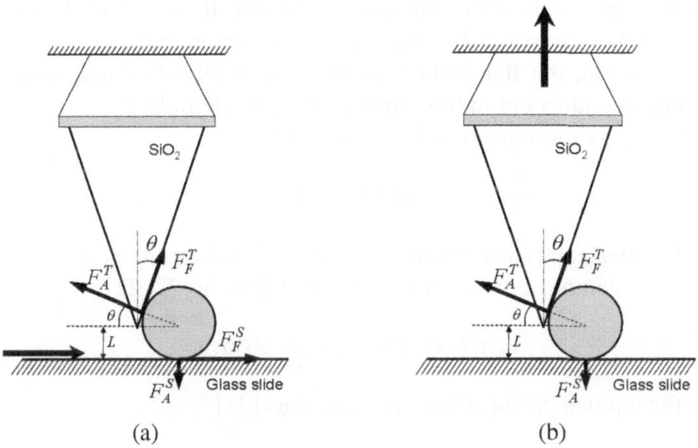

Fig. 5.3 Effective forces during (a) particle pulling with the substrate motion and (b) releasing particles with the probe motion.

For the particles to adhere to the tip, the vector sum of the horizontal components of adhesive and frictional forces on the tip should be larger than the frictional force on the glass slide. Also, for the particle to be released from the tip after a successful positioning, the vectoral sum of the vertical components of the adhesive and frictional forces on the tip should be smaller than the adhesive force on the glass slide. This can be further explained with the following equations.

$$F_A^T \cos\theta - F_F^{T^{\max}} \sin\theta > F_F^{S^{\max}}, \tag{5.19}$$

$$F_A^T \sin\theta + F_F^{T^{\max}} \cos\theta < F_A^{S^{\max}}. \tag{5.20}$$

Adhesive force (pull-off force) is related to the equivalent radius R between two contacting spheres with radii r_1, r_2 and the relative surface energy $\Delta\gamma$ according to Johnson-Kendall-Roberts (JKR) elastic contact mechanics model as [30, 21, 31]

$$F_A = \frac{3}{2}\pi\Delta\gamma R. \tag{5.21}$$

$\Delta\gamma$ for PS-SiO_2 interface is about $\Delta\gamma \approx 2\sqrt{\gamma_{PS}\gamma_{SiO_2}} = 0.16$ J/m^2. Material properties are given in Table 5.1. Equivalent radius can be found from,

$$\frac{1}{R} = \frac{1}{r_1} + \frac{1}{r_2}. \tag{5.22}$$

Here, as the tip-particle interaction is a cone-sphere elastic contact, the cone is approximated to a sphere with radius equal to the radius of the level curve (circle) at the contact with the particle as

$$r_c = r_t + (r_p(1+\sin\theta) - L)\tan\theta, \tag{5.23}$$

where r_c is the cone radius at the contact, $r_t = 20\ nm$ is the radius of the tip, $r_p = 2.25\ \mu m$ is the particle radius and L is the tip sample separation. Also at the particle-substrate interaction, the flat surface could be considered as a sphere of infinite radius to yield an equivalent radius equal to the particle radius.

Defining the maximum frictional forces simply as

$$F_F^{\max} = \tau A, \tag{5.24}$$

where τ is the interfacial shear stress and $A = \pi r^{*2}$ is the contact area with r^* as the contact radius. τ can be approximately determined as in [32, 21]

$$\tau \approx G/29 \cong 28.34 \ \texttt{MPa}, \tag{5.25}$$

where G is the equivalent shear modulus given as [21]

$$G = \left(\frac{2-\nu_1^2}{G_1} + \frac{2-\nu_2^2}{G_2}\right)^{-1} \cong 0.71 \ \texttt{GPa}, \tag{5.26}$$

where $G_i = \frac{E_i}{2(1+\nu_i)}$ is the shear modulus, E_i is the Young's modulus, and ν_i is the Poisson's ratio for $i = 1, 2$. For calculating A, its radius can be determined from JKR contact radius equation [30]:

$$r^* = \left[\frac{R}{K}\left(N + 3\pi R\Delta\gamma + \sqrt{6\pi R\Delta\gamma N + (3\pi R\Delta\gamma)^2}\right)\right]^{1/3}, \tag{5.27}$$

for N being the net normal force. K is the equivalent Young's modulus of the two spheres and defined as

$$K = \left[\frac{3}{4} \left(\frac{1-v_1^2}{E_1} + \frac{1-v_2^2}{E_2} \right) \right]^{-1} \cong 4.37 \text{ GPa}. \tag{5.28}$$

While trying to pull an object moving horizontally, the net vertical (normal) force N on the particle by the glass slide could be found simply as

$$N = -F_A^T \sin\theta - F_F^T \cos\theta. \tag{5.29}$$

Putting this force in (5.24), the maximum frictional force on the slide could be determined. Drawing $F_A^T cos\theta - F_F^{T^{max}} sin\theta - F_F^{S^{max}}$ with respect to θ, and L, the effects of the tip cone half angle and tip-sample separation on pulling 4.5 μm PS particles are investigated theoretically in Fig. 5.4 and 5.5, top plots.

Also, for successful release of the particle by moving vertically, $F_A^T sin\theta + F_F^{T^{max}} cos\theta - F_A^S$ should be smaller than zero. This value is also demonstrated according to tip cone half angle and tip-sample separation in Fig. 5.4 and 5.5, bottom plots, respectively. Analysis is made until $L = 2.25$ μm for convenience (i.e. contact angle with the particle is equal to the (acute) tip half-angle as shown in Fig. 5.3).

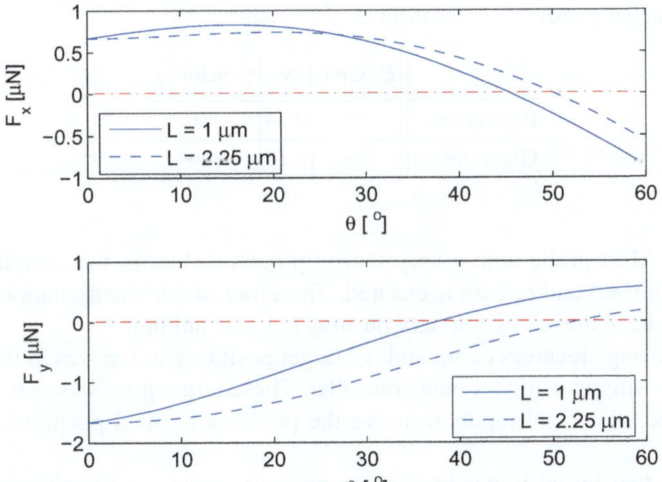

Fig. 5.4 Theoretical net horizontal and vertical forces upon pulling (top plot) and releasing (bottom plot), recpectively, according to cone half angle θ. For successful pulling, horizontal forces should be positive (i.e. in pulling direction) and for successful release, vertical forces should be negative (i.e. in releasing direction).

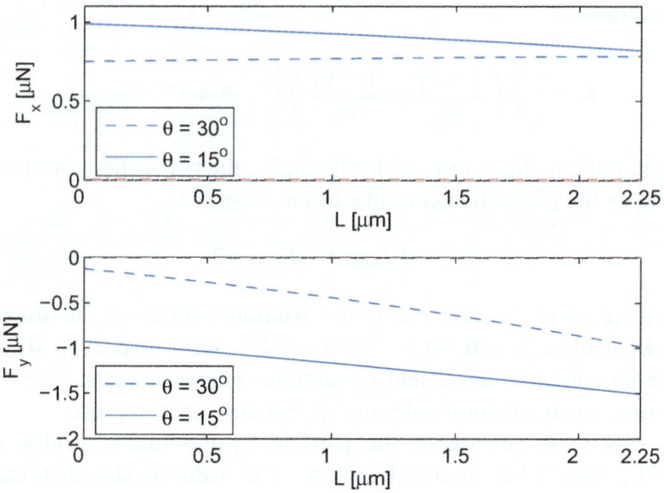

Fig. 5.5 Theoretical net horizontal and vertical forces upon pulling (top plot) and releasing (bottom plot), respectively, according to tip distance from substrate *L*. For successful pulling, horizontal forces should be positive (i.e. in pulling direction) and for successful release, vertical forces should be negative (i.e. in releasing direction).

Table 5.1 Material properties for the theoretical analysis

	E (GPa)	ν	γ (mJ/m^2)
Polystyrene	3.8	0.34	50
Glass (SiO_2)	73	0.17	160

Using an AFM probe with a very sharp tip moved close to the sample, pulling force is maximized and release is ensured. Therefore, stable manipulation of micro particles can be achieved by not only pushing but also pulling.

The piezo-stage receives commands of target positions and moves to those specified target positions using its own controller. These target positions can, however, be considered as control inputs to move the particles to pixel positions in image coordinates.

There are two initial issues in positioning the particles accurately and moving them with the probe. First, even though the controller can run in about 3-5 kHz bandwidth, the images are updated in only 25 Hz, which, in turn, defines the speed of the feedback loop. This causes either large rise times or overshoots in positioning, both of which are unacceptable. To solve this problem, the parameter estimator designed in the previous subsection is utilized to extrapolate the data between the two frames and provide a rough estimation of the real-time position of the particle of interest.

The second problem is due to the nature of the setup. As the whole glass slide moves with the stage, target positions of the particle are not stationary, and should be updated by the program. One solution to this problem is using a landmark on the glass slide (another particle) inside the frame that is untouched by the probe as a reference. This approach simplifies the updates and initial experiments were carried with it as the one in Fig. 5.6.

However, this puts another constraint, such that there needs to be another particle present in the scene during manipulation. One other solution is using the estimated transformation between the world-frame and the image-frame. There are two sources of errors in this approach. First is the error in estimation, which is less than 2 pixels (0.6 μm). Second, the error caused by the pixel discretization, which is by definition, smaller than a pixel, but accumulates throughout the manipulation and make large errors, if not taken care of. However, in this work, pixel discretization was taken into account to remove that accumulation and the constraint of having another particle on the screen all the time is lifted. The experiment in Fig. 5.9 was conducted using this approach.

Also, having to have another particle detected in the scene all the time was causing the manipulation to be slower than possible not to lose the reference particle outside the region of interest described in subsection 5.1.2. An individual manipulation was taking a little more than one minute with this approach. However, with the second method, this time is reduced to fifteen seconds on average. The former method was also eliminating the possibility of an assembly operation since once two particles contact each other, the gradient image is affected and a particle that is already assembled could not be detected and would not be able to be used as a reference. Using the estimated transformation, assembly of particles could be endouvered in subsection 5.1.6.

With these necessary additions to the manipulation algorithm, a simple controlled trajectory generation procedure can be devised to realize successful positioning of particles. As mentioned before, motion trajectories of the tip, relative to the glass slide are almost linear. First, a test is held to determine the vertical z-position of the probe that enable and eliminate contact between the particles and the probe. The difference in distance along z-axis between the two positions are more than the particle diameter. The algorithm decides whether to move the probe up or down depending on whether it wishes to push or pull the particle or not. First, the substrate is moved such that the tip is right behind the particle according to the line that connects the particle to its target position. Then the probe is moved up and target of the tip is defined to be the far edge of the target circle, since the particle could spin around the tip until it reaches there and be pulled to the target. Whenever the tip loses contact (i.e. distance between the two is larger than the radius) the whole process repeats. The algorithm exits when the position error is less than 2 pixels (0.6 μm). It has to be noted that, the controller does not behave differently for pushing or pulling particles.

Finally, to realize manipulation, it is imperative to move the slide accurately for the tip to reach its target stably. To ensure this, a velocity that varies linearly as a

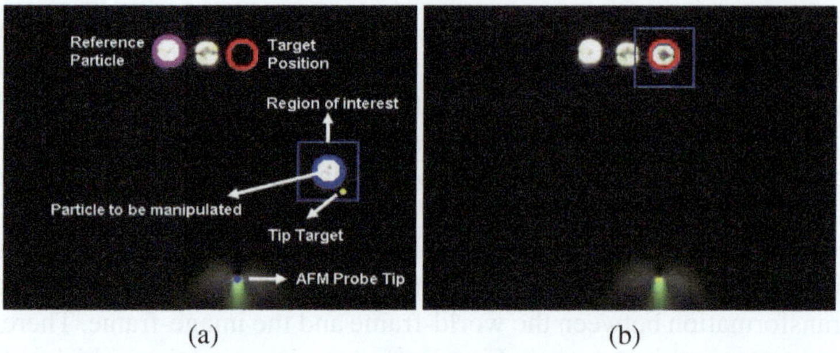

Fig. 5.6 Optical microscope top-view images of (a) before and (b) after controlled manipulation of a single 4.5 μm diameter PS particle to a target position autonomously.

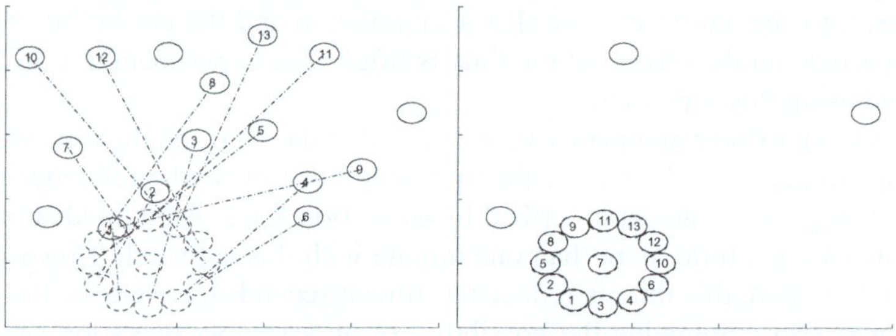

(a) Before manipulation (b) After manipulation

Fig. 5.7 Simulated assembly operation with a complicated target pattern using Algoritm 2. (Numbers depict the order of manipulation.)

fraction of the pixel distance between the current and target positions of the tip, is calculated so that it satisfies the Lyapunov stability criterion as,

$$\mathbf{v} = \dot{\mathbf{e}}_p = -\mathbf{D}\mathbf{e}_p, \tag{5.30}$$

for positive-definite $\mathbf{D}$, where $\mathbf{e}_p$ is the position error vector. Another small maximum velocity is defined to saturate the calculated Lyapunov velocity. From the saturated velocity in image coordinates, which has the opposite sign of the errors, x and y components of the next step are extracted and converted to world coordinates using the transformation in subsection 5.1.3. This motion, in very small increments, is run at around 3-5 kHz until the errors are less than 2 pixels (0.6 μm).

This vision-based algorithm is implemented on a PC. Initially, the user is asked for the tip position, the particle to be moved, the target position and the reference particle, as shown in Fig. 5.6. Then it runs the parameter estimator for 15-30 seconds to converge to a solution, while the stage is moved in a circular trajectory of 10 μm radius to eliminate any singularities. 10-15 data points (current and previous positions of a particle) are used to reduce any kind of disturbance or discretization errors. Then, the process explained above in detail is run until the particle is moved to its target position with an error smaller than 0.6 μm. No obstacle avoidance is implemented at this point.

To test for the repeatability and analyze the errors of positioning, a single particle is pushed/pulled in random distances to user specified random target positions 59 times. The average error and standard deviation in these successive manipulations are 0.49 and 0.14 μm, respectively.

5.1.5 *Trajectory Planning for Pattern Formation*

Since a single probe will be used for pushing/pulling particles in their respective references, generating non-linear trajectories is inconvenient and should be avoided for simplicity. The constraint in using linear trajectories is that one particle can not exist between the initial and final positions of another.

Algorithm 2. Pattern Formation Planning

```
Take a frame.
Detect all particles.
Generate all possible linear trajectories between each particle and target position.
Determine number of cuts on the generated trajectories by the particle and target positions.
for n = 1 → number of targets do
    Determine the particle to be manipulated and the target position it will be moved to.
    Update number of cuts.
end for
```

At this point, the first reasonable approach is using a distance-based planning, so that the nearest particle is pushed/pulled to each reference. However, in most initial and many final configurations, two of which are shown in Fig. 5.7 and 5.9, such a planner would fail since the placed particles would block others.

To overcome this problem, a metric that defines the problem of blockage caused by each particle is developed. The metric simply counts the number of blockages each particle and reference position causes to every possible trajectory. The particle to be pushed/pulled is the particle that cuts the most number of potential trajectories and the reference position it will be pushed/pulled to is the one that cuts the minimum number of potential trajectories. This approach would take the most "problematic" particle to the least interfering position. The algorithm for pattern formation planning is summarized in Algorithm 2.

The fact that the pattern formation planning algorithm minimizes the number of cuts on all possible trajectories can be interpreted as an optimization in terms of this blockage metric. Taking the particle blocking the most number of trajectories to the target position that blocks the least number of trajectories is to maximize the number of manipulations. The success of the manipulation (and assembly) of multiple particles with linear trajectories depends on this blockage metric, and hence, the algorithm yields a suboptimal solution to the problem to be complimented with the second algorithm described in the next subsection. With this approach, it becomes needless to implement more complex trajectories than simple linear ones, which in turn decreases the total manipulation time and distance travelled.

Also note that, assuming particles do not move by themselves this planning could be realized offline, before starting any manipulation, reducing the online processing burden.

5.1.6 Assembly

To realize an assembly of microparticles, by definition, there should be some attachment between them. This attachment can be in terms of chemical bonding, a bonding material that acts as a glue, or simply adhesive forces (primarily van der Waals). It has been experimentally verified that once the particles are brought into contact, they have a pretty strong and stable adhesion that keeps them together.

In fact, development on the assembly procedure in this work is bound to be slow due to this phenomenon. That is, it is hard to have many experiments to test and improve the procedure since once two particles get into contact, they are not easily separated and it was more reasonable to find new particles for further experiments most of the time.

The manipulation of a particle typically ends once it is pulled to its target position with an error less than a specified threshold, taken to be two pixels in this work. However, it is not always possible to pull all particles into an assembly since other particles can become obstacles to the linear trajectory connecting the initial and final positions of particle to be manipulated. This especially becomes a problem unless the number of particles are larger than the number of target positions and generally for the last manipulation.

The solution to this problem, without losing the simplicity of using linear trajectories, has been found to divide the manipulation for these inaccessible target positions into two sub-manipulations in linear trajectories. Another algorithm to search for a second target position that is accessible both from the initial and to the final position of the particle is included.

The second algorithm is activated once there is a target position inaccessible for every particle. It searches in circles of which, the target position is the center, radially increasing the radius until a suitable position is found. All the search is done offline before any manipulation is started, so it does not add any burden to the manipulation. This algorithm is depicted in Fig. 5.8. Also, since assembly operation needs more accuracy, another addition is a decrease in the velocity and it takes about 45 seconds per individual manipulation in an assembly operation.

Another problem might arise if a particle should be pulled to a position such that the tip gets in the middle of the current particle and another previously manipulated particle. This can occur if a closely packed target assembly (e.g. a 3×3 square made of 9 particles) is endeavored. However, even in a closely packed target assembly, this is a small probability as the tip radius is tiny with respect to the particle radius. Not to mention, even if the AFM tip gets in between two particles, there is a good possibility that the two particles would snap into contact once the tip is moved

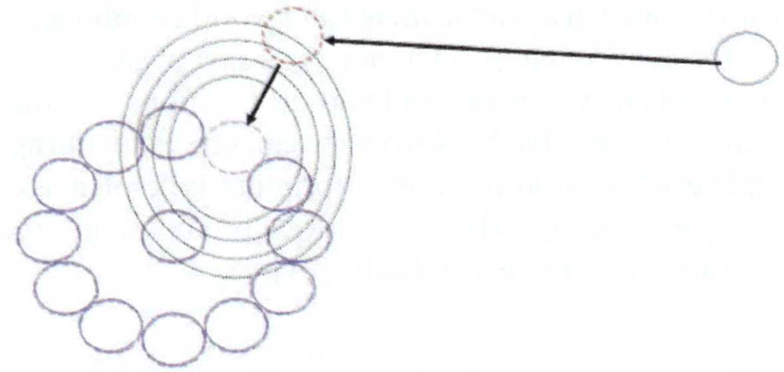

Fig. 5.8 The search algorithm for assembly. A suitable "first stop" is searched around the target position radially.

down. Therefore, in this work, possible errors due to this possibility are taken to be negligable.

Figure 5.9 shows an experimental result of the assembly procedure. As seen in this figure, particles are pulled into their target positions utilizing the pattern formation and assembly algorithms. The average error and standard deviation in the demonstrated manipulation in Fig. 5.9 are 0.64 and 0.44 μm, respectively. Movie of the complete assembly operation can be seen in [33].

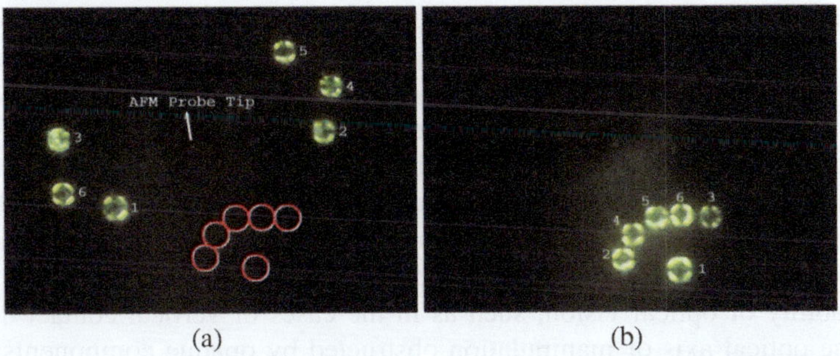

Fig. 5.9 Optical microscope top-view images of (a) before and (b) after a visually servoed manipulation of six 4.5 μm diameter PS particles into an assembly autonomously. (Numbers depict the order of manipulation.)

Autonomous manipulation of 4.5 μm PS particles with a single AFM probe based on visual information by an optical microscope is demonstrated with a positioning accuracy less than 0.64 μm in average. Smooth linear trajectories are generated for the motion of the probe relative to the substrate for speed and ease of operation. Multiple particles are successfully positioned with a task planner that chooses the order of manipulation based on a metric that defines the blockage of particles and target positions on the generated linear trajectories.

A globally stable iterative discrete sliding mode observer estimates the parameters of the transformation between the image and world coordinates. This estimation is used not only for conversion of reference positions of the piezoelectric stage to world coordinates, but an extrapolation of particle positions according to motions of the stage between two frames fed back from the camera. Potential applications could be particle template fabrication for contact printing type of maskless micro/nano fabrication techniques [25]; optoelectronic device prototyping; tribological and adhesion characterization of particles and cells [21].

5.2 Three-Dimensional Automated Micromanipulation Using a Nanotip Gripper with Multi-feedback

It is well known that pick-and-place is very important for 3-D microstructure fabrication since it is an indispensable step in the bottom-up building process. However, so far, there are few references in the literature that report mechanical pick-and-place manipulation of microobjects with feature sizes less than 10 μm, especially manipulation confined in air. The main difficulty in sufficiently completing such pick-and-place manipulation at this scale lies in fabricating a very sharp end-effector that is capable of smoothly releasing microobjects deposited on the substrate. Moreover, this end-effector has to provide enough grasping force to overcome strong adhesion forces [55, 56, 57] from the substrate as well as being capable of sensing and controlling interactions with the microobjects. Furthermore, compared with the manipulation of larger microobjects under an optical microscope, visual feedback at several microns more suffers from the shorter depth-of-focus and the narrower field-of-view of lenses with high magnifications, although different schemes or algorithms have been introduced on techniques of autofocusing [58, 59] and extending focus depth [61]. Compared with vision-based automated 2-D micromanipulation, automated 3-D micromanipulation at the scale of several microns to submicron scale is more challenging because of optical microscope's resolution limit (typically 200nm). Moreover, additional manipulation feedback is needed that is beyond the capability of optical vision, such as in the cases of vertical contact detection along the optical axis or manipulation obstructed by opaque components. Therefore, multi-feedback is of vital importance to achieve such accurate and stable 3-D micromanipulation at the scale of several microns to submicron scale.

In this section, an atomic force microscope (AFM) based 3-D micromanipulation system (3DMS) with a nanotip gripper is presented. The 3DMS mainly consists of two collaborative AFM cantilevers with protrudent tips. The nanotip gripper is constructed from these two tips that can be used to build 2-D micropatterns by pushing and pulling microobjects on a single surface, and more importantly, to achieve 3-D microstructures by pick-and-place involving grasp, pick-up, transport and release steps. We have used the 3DMS to fabricate five micropyramids with two layers by manipulating nylon microspheres with diameters of 3 μm - 4 μm. Compared with other means of 3-D micromanipulation in air, the developed 3DMS is more controllable due to its real-time interactive force sensing and process monitoring.

This section is organized as follows. Subsection 2 describes the prototype 3DMS design and presents its experimental setup. Subsection 3 introduces manipulation schemes for pick-and-place micromanipulation, including methods for grasping point searching and contact detection with amplitude feedback from dithering cantilevers, a 3-D manipulation protocol for microspheres and real-time force sensing. In subsection 4, we show the 3-D micromanipulation results of five micropyramids with two layers with each micropyramid being constructed from four microspheres.

5.2.1 System Configuration

As shown in figure 5.10(a) and (b), the proposed 3DMS is equipped with an optical microscope and two sets of similar devices commonly used in a conventional AFM, mainly including two cantilevers, two sets of nanopositioning devices and optical levers. The optical lever, typically composed of a laser and a quadrant photodiode, is used to detect the actions of each cantilever during micromanipulation. These two optical levers are arranged on two vertical planes due to the space limitation of the optical microscope, as seen in figure 5.10(b). The bottom inset of 5.10(c) shows that the nanotip gripper is constructed from two protrudent nanotips of AFM cantilevers I and II that are facing each other. With a mounting angle of 8°, each tip has a tilted angle about 70° on the side view, providing the nanotip gripper with a clamping angle $\theta = 40°$.

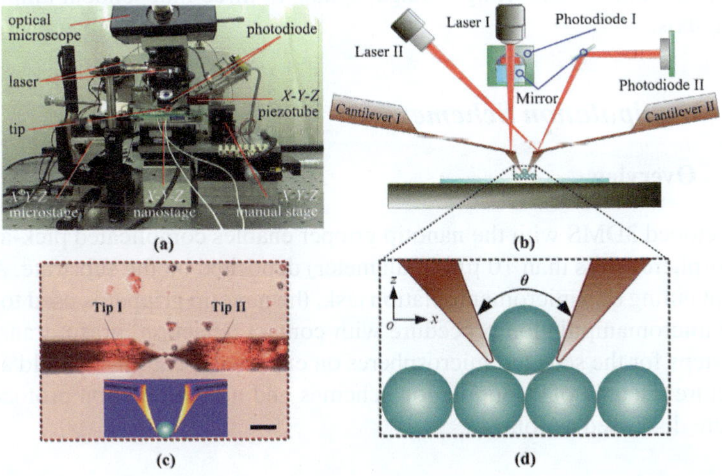

Fig. 5.10 Structure and manipulation schemes of the 3DMS. (a) A photo of the developed 3DMS. (b) System configuration of the 3DMS, which mainly consists of two sets of devices used in a commercially available AFM, including two cantilevers and the corresponding two sets of nanopositioning devices and optical levers. (c) A microscopic image captured during the pickup operation of a microsphere using the gripper constructed by tip I and tip II. The bottom inset shows a side view of the pick-and-place scheme with a gripper. The scale bar represents 20μm. (d) Close-up figure showing the scheme for grasping a microsphere with the nanotip gripper.

A typical pick-and-place micromanipulation scheme is depicted in figure 5.10(d), in which the nanotip gripper is used to pick up and place a microobject in its target position. The 3DMS configuration is described in detail as follows:

1. Cantilever I, which is immovable during the pick-and-place micromanipulation, is fixed on an X-Y-Z micropositioning stage for coarse positioning. The normal spring constant of cantilever I and the sensitivity of its optical lever are calibrated as 2.43N/m and 0.65nm/mV [61], respectively.
2. Cantilever II is actuated by an X-Y-Z piezotube (PI P-153.10H) for gripper opening and closing operations. The piezotube has a scan range of 10 μm × 10 μm × 10 μm and a sub-nanometer resolution, which is mounted on an X-Y-Z manual microstage. For accurate positioning on tip II, hysteresis of the piezotube is accurately compensated by the PI operator [62]. The normal spring constant of cantilever II and the sensitivity of its optical lever are calibrated as 2.48N/m and 0.58nm/mV [61], respectively.
3. An X-Y-Z closed-loop nanostage (MCL Nano-Bio2M on the X-Y axes, PI P-732.ZC on the Z-axis), with a maximum motion range of 50μm × 50μm × 10μm and a resolution of 0.1nm, is used to support and transport samples.
4. A data acquisition (NI 6289) card with a resolution of 18 bits in A/D transfer and a maximum sampling frequency of 625 kHz is used for high-speed capturing photodiode voltage output to estimate deflections on both cantilevers induced by force loading or resonant oscillation. The acquisition card is also used to actuate the piezotube by exporting voltage signals to three independent amplifiers for each axis.

5.2.2 *Manipulation Schemes*

5.2.2.1 Overview

The developed 3DMS with the nanotip gripper enables complicated pick-and-place of microspheres (less than 10 μm in diameter) deposited on the substrate. After trajectory planning of a micromanipulation task, the nanotip gripper is used to perform the 3-D micromanipulation procedure with contact detection, grasp, transport and release steps for the selected microspheres on expected locations to build a 3-D microstructure. Details of manipulation schemes and a manipulation protocol of the 3DMS are discussed as follows.

5.2.2.2 Configuration of the Two-Tip Gripper

Configurations of commonly used grippers with two parallel clamping jaws are shown in figure 5.11(a)-(c), which are different from the shape of the jaws or roughness of the contact surface. Figure 5.11(a) shows the most widely used configuration and the gripper seen in figure 5.11(b) is designed for holding a microobject more strongly. In order to decrease adhesion forces between the gripper and the microobject, a rough contact surface has been proposed [46] and, as shown in figure 5.11(c), a special tip end with tiny contact surface is adopted. Using these three types of commonly used microgrippers, microobjects with a feature size larger than 10 μm were successfully manipulated in various environments.

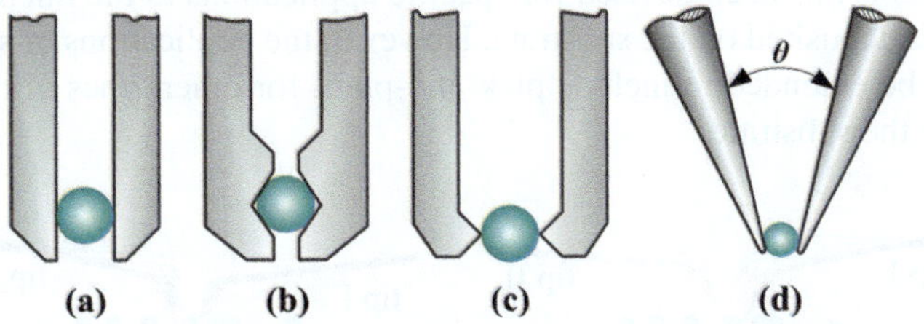

Fig. 5.11 Different types of microgripper configuration. (a) The most widely used microgripper which has parallel clamping jaws. (b) A gripper with a closed configuration is designed to hold microobjects more strongly in grasping operation. (c) A gripper with a tiny contact area is designed to reduce tip-microobject adhesion forces. (b) A gripper constructed from two AFM tips is adopted in the developed 3DMS. the relatively larger contact area between the gripper and the microobject [55, 57].

However, as the size of the microobject reduces to several microns or submicron, problems will arise with these conventional grippers:

1. Sticking phenomena becomes more severe due to
2. The tip diameters of the clamping jaws are comparable in size to the microobjects to be grasped. In this case, the clamping jaws are not geometrically sharp enough to pick up smaller microobjects deposited on the substrate.

Therefore, in order to pick up smaller microobjects without obstacles, such as the sticking phenomena, and enhance grasping ability to overcome the strong adhesion forces between the microobject and the substrate, a gripper with a new configuration is proposed in figure 5.11(d). This gripper is constructed from two AFM tips with the following benefits:

1. The AFM tip end is very tiny (typically around 10 nm in radius) with respect to the size of the microobject to be manipulated, which leads to smaller adhesive forces because the contact area with the microobject is much smaller than the microobject-substrate contact.
2. However, a smaller contact area also leads to weaker contact friction, which is proportional to the contact area at this scale [63]. In this case, the friction forces of the gripper-microobject contact, generally used for grasping operations by grippers with parallel clamping jaws, might not be strong enough to break the adhesion forces of the microobject-substrate contact. Fortunately, for the proposed nanotip gripper, as shown in figure 5.11(d), a positive clamping angle θ is configured making grasping stronger due to the contribution of repulsive forces from the tip-microobject contact, compounding with the tip-microobject friction forces to overcome the adhesion forces of the microobject-substrate contact.

5.2.3 *Protocol for Automated Pick-and-Place*

Microparticles and microspheres are being intensively investigated as significant experimental materials for micromanipulation and microassembly. Thus, one of the 3DMS protocols has been devised for specific applications to the microparticles or microspheres deposited on the substrate. However, the applications of such a protocol can also be extended to include pick-and-place for other types of microobjects dispersed on the substrate.

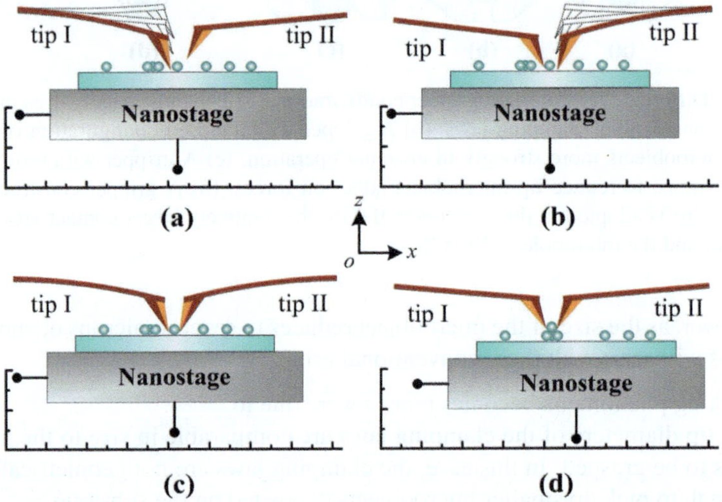

Fig. 5.12 A protocol for the pick-and-place manipulation of microspheres to build a pyramid. Four main steps are involved in a manipulation procedure. The dithering tip I is used to search for the grasping point by local scanning with amplitude feedback in step (a). In step (b), the grasping point between tip II and the microsphere is detected with the dithering tip II. A nanotip gripper is formed for the grasping operation as both tips make contact with the microsphere. The pick-and-place manipulation is achieved in step (c) by moving the nanostage on each axis. (d) A micropyramid is achieved as the microsphere is placed on the first microsphere layer.

As shown in figure figure 5.12, the pick-and-place procedure mainly includes following steps:

- System initialization. Firstly, each axis of the nanostage and the piezotube are set in a proper position, which will provide the pick-and-place manipulation with enough motion range on each axis.
- Task planning. Once the system initialization is ready, task planning is started with image processing of a global view containing all microspheres to be manipulated and the cantilever tips. In our method, the optical microscopy image is just used for coarse positioning of microobjects and the gripper, providing a distribution of the microspheres for task planning. Subsequently, fine positioning of the

microspheres is performed using amplitude or force feedback from the gripper. This will be discussed on the section 3.5.

- Contact with tip I. Figure 5.11(a) shows that tip I approaches a microsphere to make contact by moving the nanostage on the X-axis. A gap (typically 500nm – 800nm in our experiments) between tip I and the substrate should be kept during the approach to ensure contact with the very end of tip I. The actual grasping point and contact on the microsphere can be detected by the amplitude feedback. Methods for grasping point searching and contact detection will be discussed section 3.5.
- Contact with tip II. Similarly, as in the step depicted in figure 5.11(b), tip II approaches the microsphere by moving the piezotube on the X-axis. Once tip II and the microsphere are in contact, in figure 5.11(c), a nanotip gripper is configured for a grasping operation.
- Pick-and-place. Once the gripper is constructed, as shown in figure 5.11(d), the microsphere is picked up, transported and placed by moving the nanostage on each axis with a proper displacement that depends on the diameter of the microsphere and its destination. The complete 3-D micromanipulation procedure is monitored by the real time force sensing.

5.2.4 Task Planning

Unlike trajectory planning for 2-D pushing/pulling micromanipulation on a singe surface, the blockage problem in trajectory planning can be eliminated in pick-and-place micromanipulation, since the microsphere can be picked up and lifted to a height exceeding that of the microobjects on the substrate before being transported. An algorithm, for this case, based on the shortest path solution, is proposed for linear trajectory planning on 3-D microstructure formations.

For the manipulation example shown in figure 5.13, the task planning is described as follows:

1. Capture a frame of gray microscopic images.
2. Detect the central position of each microsphere in the image space using the method adopted in [43].
3. Determine target positions and manipulation sequences according to the microstructure to be fabricated.
4. for n = 1 to N (number of targets)
    - Generate all possible linear trajectories between each microsphere and the target position tn (x, y) in the image space.
    - Select the microsphere with a central position on (x, y) that has the shortest path to the target position, as show in figure 5.13(a).

   end for
5. Transform the position sequences tn (x, y)/on (x, y) from the image space into Tn (x, y)/On (x, y) nanostage motion space for a actual motion planning of the micro-manipulation.

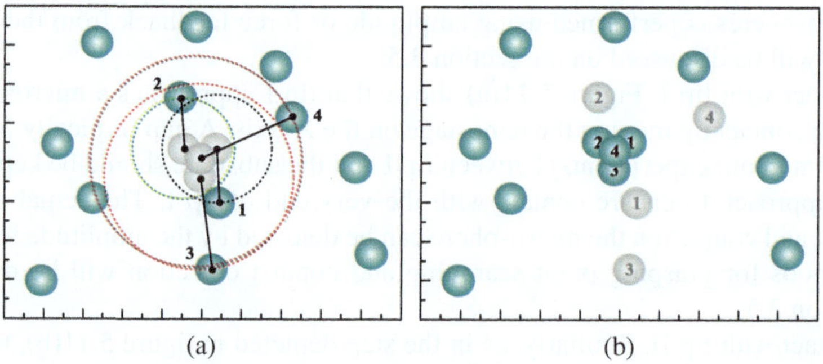

Fig. 5.13 Simulated 3-D assembly of a micropyramid constructed from four microspheres using a shortest path algorithm, in which the numbers refer to the order of the manipulation. (a) Before manipulation. (b) After manipulation

Once the task planning is completed, for each target position, the nanotip gripper is used to pick up and place the nearest microsphere. In this case, amplitude and force feedback on each microcantilever are used to detect grasp, pick-up and place operations instead of commonly used microscopic vision feedback, which will be discussed on the following sections.

5.2.5 Grasping Point Searching and Contact Detection

In the developed 3DMS, real-time amplitude or force feedback from the gripper is used to detect the interactions between the tip and the microobject.

As shown in figure 5.14(a)(a), the dithering cantilever with the first mode of its natural resonance is utilized to locate grasping points and detect the contact. In order to locate two grasping points distributed over the diameter of the contact circle on the microsphere, as shown in inset I of figure 5.14(a), the dithering cantilever sweeps along the Y-axis at a distance about half the microsphere's diameter as it is approaches the microsphere. When the tip laterally taps on the microsphere, the grasping point can be accurately located by searching for the minimum amplitude response of each single scan. A corresponding experimental result can be found in the figure 5.14(b), in which the tip laterally sweeps the microsphere within a range of 1.6 μm with a free oscillating amplitude of about 320 nm. Six different distances to the microsphere were tested from 120 nm to 20 nm with an interval of 20nm and ultimately, the grasping point is well located with an accuracy of 10 nm, which is greater than that achieved with the optical microscope.

From the scheme depicted in inset II of figure 5.14(a), amplitude feedback is also used for contact detection. In the experimental result shown in figure 5.14(c), contact between the tip and the microsphere is achieved as the amplitude reduces to a steady value close to zero. In addition, contact can be also detected by the normal force response from the gripper. As shown in figure 5.14(d), a full normal force response in an approach-retraction loop can be recognized by snap-in, contact and pull-off

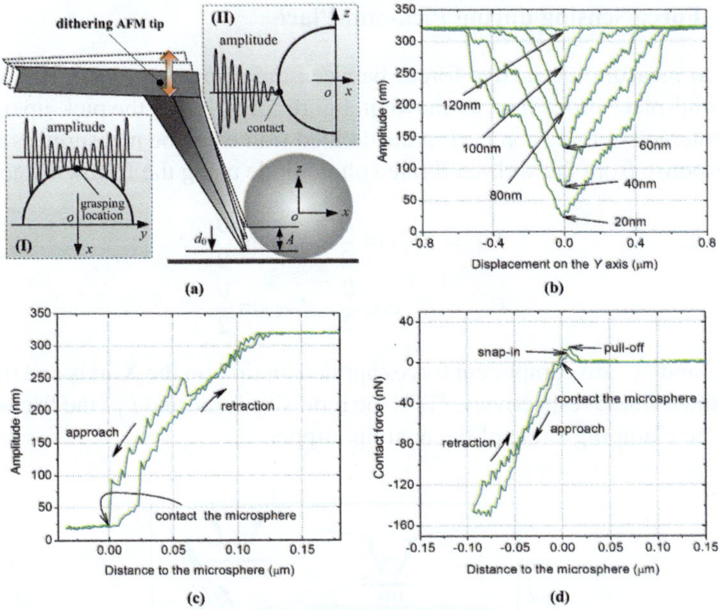

Fig. 5.14 Schemes and experimental results of grasping point searching and contact detection on a microsphere. (a) Schematic diagram of grasping point location (inset I) and contact detection (inset II) using the amplitude feedback of a dithering microcantilever with amplitude A. d0 is the distance between the tip and the substrate. (b) Amplitude responses of the microcantilever when it is sweeping along the Y-axis with different distance from the microsphere on the X-axis. Experimental results show that the locating accuracy of the grasping point can be as precise as 10 nm. (c) Amplitude responses of the cantilever when it is approaching the contact point on the microsphere. (d)As an alternative method, normal force responses could also be used for contact detection.

steps, usually in the presence of the tip-substrate contact. The interaction responses of the microcantilever described in figure 5.14(b)-(d) are sufficient to detect not only the contact between the tip and the microobject, but also the grasping state. In our experiments, as the contact between tip I and the microsphere is made, tip I retreats 10 nm -20 nm in order to make a tiny gap between tip I and the micro-sphere. This gap enables smart recognition of the grasping state as being tip II contact with the microsphere after a slight further push. Compared with operations under the optical microscope, the amplitude detecting method has two obvious advantages:

1. Grasping points and contact can be detected below the opaque components, more importantly, with a lever of accuracy that is far beyond the capability of the optical microscope.
2. Benefiting from AFM-based accurate force and amplitude measurement, the grasping points and contact can be successfully detected even with very weak interaction forces at the nano-Newton scale, protecting the fragile tips and the microobjects from damage during manipulation.

5.2.5.1 Force Sensing during Pick-and-Place

In order to measure interactive forces between the gripper and the microsphere in the pick-and-place procedure, as shown in figure 5.15, during the pick-up operation, the 3-D interactive forces on tip I in the defined frame can be measured as a normal signal response from the well-calibrated photodiode using the following equations:

$$F_{z1} = F_{f1}\cos\frac{\theta}{2} + F_{r1}\sin\frac{\theta}{2} \tag{5.31}$$

$$F_{x1} = F_{r1}\cos\frac{\theta}{2} - F_{f1}\sin\frac{\theta}{2} \tag{5.32}$$

where F_{z1} and F_{x1} are component forces applied on tip I on the X-axis and the Z-axis in the defined frame respectively, F_{r1} is the repulsive force and F_{f1} the friction force, and θ is the clamping angle of the nanotip gripper.

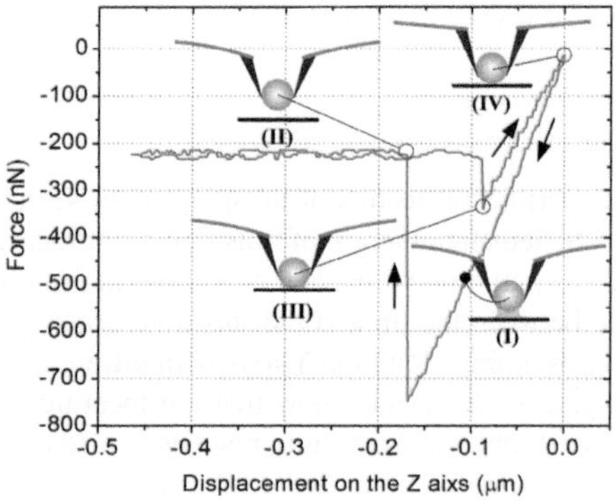

Fig. 5.15 Force simulation of a pick-up operation with the gripper, which has a clamping angle of 40°

For the sake of calculation convenience, the angular deflection of the cantilever is used because of its linear relation to the voltage output of the optical lever. The bending angular deflection on the free end of the cantilever comprises two parts: ϕ_{z1} and ϕ_{x1}, caused by F_{z1} and F_{x1}, respectively. These two parts can be calculated from:

$$\phi_{z1} = \frac{F_{z1}L^2}{2EI} \tag{5.33}$$

$$\phi_{x1} = \frac{F_{x1}Ll}{EI} - F_{f1}\sin\frac{\theta}{2} \tag{5.34}$$

where L is the effective length of the cantilever, l is the cantilever's tip height, E is Young's modulus of the silicon and I is the moment of inertia on the cantilever's cross-section. L and l are measured as 250μm and 9.5μm, respectively.

Here, we estimate $F_{f1} = F_{r1}$ using the assumption $\mu = 0.33$, which is one of experimental values obtained from an AFM lateral force calibration [64]. As the gripper has a clamping angle $\theta = 40^0$, from Equation (5.31)–(5.34), a simplified relationship between ϕ_{z1} and ϕ_{x1} can be estimated as:

$$\phi_{x1} = 0.089\phi_{z1} \tag{5.35}$$

Therefore, to simplify calculation of the adhesion forces F_{ao} between the microsphere and the substrate, the cantilever deflection caused from F_{x1} will be omitted in the following calculations. Therefore, F_{z1} can be easily estimated from the normal voltage output of the optical lever of tip I from [65]:

$$F_{z1} = \beta_1 \Delta V_1 \tag{5.36}$$

where β_1 is the normal force sensitivity; and ΔV_1 is the voltage response of the photodiode due to the force loading. A similar result can be also deduced on tip II. Once F_{z1} and F_{z2} are known, the adhesion force F_{ao} can be calculated from:

$$F_{ao} = F_{z1} + F_2 = \beta_1 \Delta V_1 + \beta_2 \Delta V_2 \tag{5.37}$$

where β_2 is the normal force sensitivity and ΔV_2 is the voltage response of the photodiode on tip II.

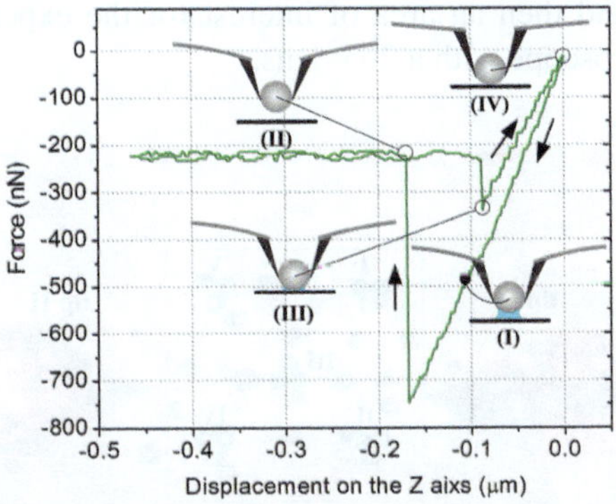

Fig. 5.16 Synthesized normal force responses from both microcantilevers during the pick-and-place manipulation of a microsphere. (I) Pick-up. (II) Pull-off. (III) Snap-in. (IV) Contact on the retraction branch.

Figure 5.16 shows a full force spectroscopy curve during the pick-and-place manipulation of a microsphere deposited on glass slide with an ambient temperature of 20°C and relative humidity of 38%. The force spectroscopy curve is synthesized from the force responses on tip I and tip II. The curve starts from the contact state between the microsphere and the substrate. As the nanostage is move down to pick up the microsphere, the cantilever is bent leading to negative forces. During the pickup, when the nanostage position reaches -170 nm, the micro-sphere pulls off the substrate with a pull-off force of 746nN. After the pull-off, the force curve returns to −220nN, and not to the initial force because of contributions of friction forces between the gripper and the microsphere. Inset I shows that the microsphere slides down to the substrate during the pick-up operation, which leads to bending deformations maintained by the friction on the nano-tips, as shown in inset II. During the retracting branch, earlier snap-in occurs at a distance of about 50 nm from the starting point, which indicates that the microsphereslides the same distance during the pick-up operation, as seen in inset III. Further retraction leads to a continued increase in the force response with a higher gradient than that of the pick-up operation until both the nanostage position and the magnitude of the normal force return to the initial grasping state in inset IV. Once such a force spectroscopy curve occurs during the pick-and-place manipulation, a stable grasping as well as a successful releasing operation can be validated.

5.2.6 Experimental Results and Discussion

5.2.6.1 Task Description

In order to validate 3-D automated manipulation capability of the developed 3DMS, nylon microspheres with diameter of 3μm − 4μm were manipulated to build 3-D microstructures in our experiments. The microspheres were deposited on a freshly cleaned glass slide and then an area of interest for the experiments was selected under an optical microscope with a 20× lens.

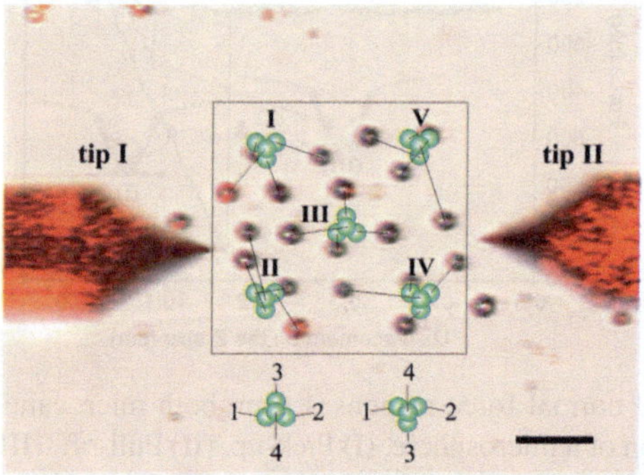

Fig. 5.17 An optical microscope image before the micromanipulation. Twenty microspheres with diameters of 3μm − 4μm will be manipulated to build five microsphere pyramids (labeled from I to V). The bottom insets show two types of assembly sequences depicted by numbers. The scale bar represents 15μm.

Figure 5.17 shows a plan image view of the selected area, which contains more than 24 microspheres, 20 of which separated in a 56 μm square frame are going to be manipulated to build five microsphere pyramids labeled by assembly sequences from I to V. Each pyramid is constructed from four microspheres with two layers. The bottom insets show two types of assembly sequences, depicted by numbers, for two different arrangements of the pyramids. Tip I and tip II, with a laser spot focused on each cantilever end, are located beside the manipulation area after system initialization, for the convenience of task planning. After task planning, a 3-D microassembly task is performed with the sequences predefined from the motion planning to build five micropyramids, shown as schematic structures by green microspheres.

5.2.6.2 Manipulation Results

Figure 5.18 shows the 3D micromanipulation process of the micropyramids. Figure 5.18(a) and (b) are intercepted when the first layer of the pyramid II and VI are assembled, respectively. The image shown in figure 5.18(c) is captured as the first layer of the five pyramids has been completed, in which twenty microspheres have been already placed in their reference positions with the pick-and-place operations. Once the first layer is ready, the remaining five microspheres are sequentially picked up and placed on the second-layer reference positions. Figure 5.18 9(d) and figure 5.18(e) describe the transporting process of the twenty-first and the last microsphere, respectively. The ultimate result is shown in figure 5.18(f). In addition, the assembly result is displayed more distinctly under the microscope with a magnification of 100×, as seen in figure 5.18(g).

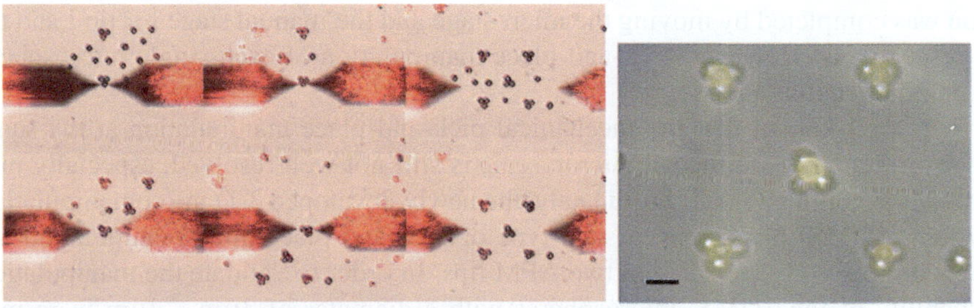

Fig. 5.18 3-D microassembly results of the micropyramids. (a) - (c) show three images intercepted from the assembly process of the first layer of the micropyramids. (d)-(f) depict the assembly process of the second layer of the micropyramids. The images (a) -(f) are captured under magnification of 20 ×. (g) The 3D microassembly results under magnification of 100×. The scale bar represents 5 μm.

The whole manipulation process was completed in 16 minutes excluding the time required for the user to relocate the gripper, so the average manipulation time of each microsphere is about 48 seconds, which mainly includes about 20s for microsphere grasping including the processes of grasping point searching and contact detection using amplitude feedback, 10s - 35s for microsphere release and, the remaining time for transport. The release time for the microspheres on the second layer is much longer than that for the first layer because substantial contact between microspheres is more difficult to achieve than the microsphere-substrate contact, especially the second-layer microspheres on pyramid II and pyramid IV, which have smaller diameters than the supporting micro-spheres. These two microspheres were successfully released after two release operations. Several aspects of the microassembly can be explained in detail as follows.

As mentioned above, the sharp AFM tip results in tiny adhesion forces between the gripper and microsphere. However, in order to achieve a reliable release operation, it is first necessary to make certain that the very end of the tip I is in contact with the microsphere by keeping a proper distance between the tips and the substrate in the grasping operation. Moreover, it is very important to ensure a full microsphere-substrate contact by waiting forseveral seconds before opening the nanotip gripper. In addition, the tips should keep dithering in their natural resonance during the whole pick-and-place manipulation procedure, especially in the release process, In order to reduce adhesion forces between the gripper and the microsphere attributable to the high inertial forces applied to the microsphere [65]. In our experiments, the dithering gripper did not fail once in 40 times of pick-and-place manipulations of microsphere deposited on the glass slide; in contrast, four microsphere release failures occurred using the non-dithering gripper. By applying schemes or strategies mentioned above, the sticking obstacle is estimated in the experiments. In addition, note that several interruptions occurred in user's nanotip gripper relocation operations during the whole microassembly of the five micropyramids due to the constraint of the limited nanostage motion range of 50 μm $\times$ 50 μm, which is less than the manipulation area of 56 μm $\times$ 56 μm. The gripper relocation was completed by moving the microstage and the manual stage for tip I and tip II respectively. However, pick-and-place manipulation of each single microsphere is definitely automated.

It is well known that the mechanical pick-and-place manipulation at the scale of several microns to the submicron scale is still not well resolved, especially manipulation confined in air. Fortunately, the newly developed 3-D micromanipulation system (3DMS) has achieved such type of pick-and-place micromanipulation using a gripper constructed from two AFM tips. In order to validate the manipulation capability of the 3DMS, microspheres with a diameter of 3μm – 4μm were manipulated in air, and as a result, five 3-D micropyramids were constructed with the developed 3DMS. The 3DMS has made the automated 3-D micromanipulation and microassembly at scale of several micrometers in air feasible.

5.3 Atomic Force Microscopy Based Three-Dimensional Nanomanipulation

Pick-and-place nanomanipulation is a promising technique in 3D nanostructure fabrication since it is an indispensable step in the bottom-up building process. It can overcome limitations of bottom-up and top-down methods of nanomanufacturing and further combine advantages of these two methods to build complex 3D nanostructures. In literature, nanostructures have been manipulated, assembled and characterized by integrating nanomanipulators or nanogrippers into scanning electron microscopes (SEM) and transmission electron microscopes (TEM) [66, 67, 68, 69, 70]. Both the SEM and the TEM provide a vacuum environment where the van der Waals force is the main force to be overcome during the manipulation. 3D nanomanipulation could be also achieved with optical tweezers in liquid, where the adhesion forces are greatly reduced [71, 72, 73]. However, the pick-and-place nanomanipulation in air is still a great challenge due to the presence of strong adhesion forces, including van der Waals, electrostatic and capillary forces [74]. In this case, the main difficulties in achieving the 3D nanomanipulation are fabricating sharp end-effectors with enough grasping force, as well as capabilities of force sensing while controlling interactions between the nanoobject and the tool or the substrate.

Compared with the nanomanipulation carried out in the SEMs and the TEMs, the AFM based nanomanipulation has much more flexibility. It can be applied to various nanoobjects in different environments (air, liquid and vacuum). However, the conventional AFM based nanomanipulation is restricted to pushing and pulling nanoobjects on a single surface [75, 76, 77, 78, 79]. The pick-and-place manipulation of nanoobjects in air is still unresolved using the AFMs, although vertical pickup of atoms have be accomplished by electric field trapping [80], van der Waals forces [81, 82], tunneling current induced heating or by inelastic tunneling vibration [82]. Grasping submicron objects using a two-tip carbon nanotube (CNT) nanotweezer has been demonstrated [83], but the alignment between CNT arms remains a challenge and capabilities of the nanotube nanotweezer in overcoming large adhesion forces in air need further validation. Thus, a versatile 3D nanomanipulation system is crucial in achieving the pick-and-place manipulation in air to build complex 3D nanostructures. We have used the 3DMFM (presented in the previous section) to fabricate nanowire crosses by manipulating two types of cone-shaped silicon nanowires (SiNWs). During manipulation, interactive forces applied on the tips were recorded and analyzed.

5.3.1 System Configuration

Figure 5.19(a) shows the configuration of the 3DMFM, which is equipped with two sets of devices commonly used in conventional AFMs, including cantilevers, optical levers and nanopositioning devices. Two individually actuated cantilevers (ATEC-FM) with protrudent tips (as per right inset of Figure 5.19(a)), namely, Tip I and Tip II, are used as end-effectors to build a nanotweezer (as per left in-set of Figure 5.19(a)).

The full description is avalaible in the section 5.2.2.2 and can be summarized :

1. Tip I, utilized as an imaging tip before manipulation and then a manipulating end-effector, is fixed on a motorized X-Y-Z stage (not shown in Figure 1 (a)) for coarse positioning. On the other hand, Tip II, used as a manipulating end-effector supported by a compact manual X-Y-Z microstage for coarse positioning, is mounted on the X-Y-Z piezostage.
2. The X-Y-Z piezoscanner is fixed on the X-Y-Z piezostage to support and transport samples.

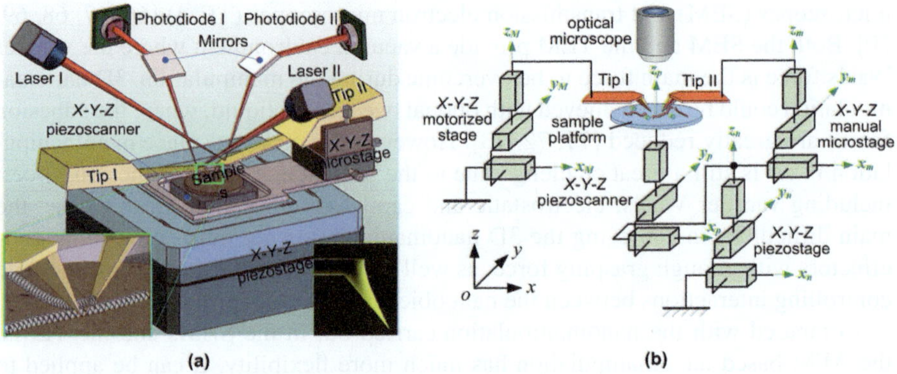

Fig. 5.19 Schematic structure of the 3DMFM. (a) System configuration of the 3DMFM, which consists of two sets of devices commonly used in the conventional AFM. A nanotweezer (left inset) is constructed from two protrudent tips of two cantilevers (right inset). (b) Mechanism illustration of the 3DMFM (see also figure 5.10)

5.3.2 *Manipulation Protocol*

5.3.2.1 Overview

As discussed above, the nanotweezer consists of two protrudent tips is used to achieve pick-and-place of nanoobjects deposited on the substrate due to it very sharp tip ends with a radius less than 10 nm and the capability of real-time force sensing. Once the tips and the nanoobjects are accurately located using the image scanning, the nanotweezer is formed to pick up and transport selected nanoobjects to target positions. The details of manipulation schemes and a protocol of the 3DMFM are discussed in following.

5.3.2.2 Pick-and-Place Protocol

Nanowires and nanotubes are being intensively investigated as promising nanomaterials for application to nanooptics, nanoelectronics and nanoelectromechanical systems (NEMS). Thus, a protocol of the 3DMFM is presented specifically for application to the nanowires or nanotubes deposited on a substrate. Applications of such a protocol can easily be extended to the pick-and-place of nanoparticles dispersed on a substrate.

Figure 5.20(a) shows a procedure of the pick-and-place nanowire manipulation, which involves following steps:

1. System initialization : once the manipulation area is selected, both the tips are aligned as a quasi-nanotweezer above the center of the manipulation area under the optical microscope. Each axis of the piezostage and the piezoscanner is initialized on a proper position. The initialization should provide enough motion range for the nanomanipulation within an image scanning area.

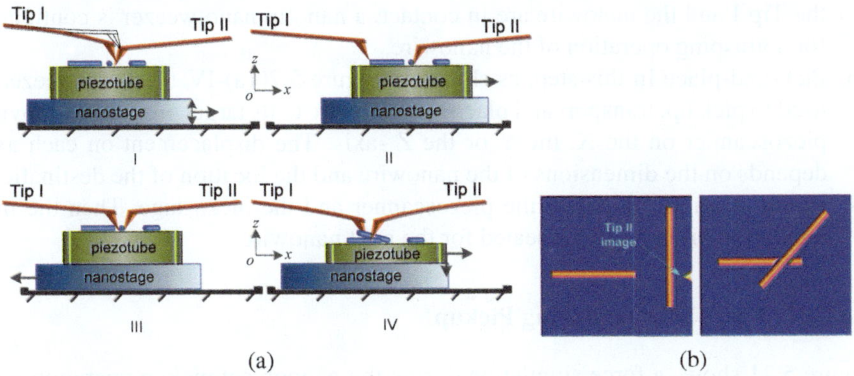

Fig. 5.20 Protocol of the pick-and-place of nanowires for building a nanowire cross. (a) Four main steps involved in a procedure of pick-and-place. Image scanning in step I; Tip II approaches nanowire by moving the X-axis of the piezoscanner in step II; In step III Tip I approaches the grasping location on the same nanowire to construct a nanotweezer by moving the X-axis of the piezostage; The pick-and-place manipulation is achieved in step IV. (b)Simulated image before the manipulation, by which locations of nanowires and Tip II are recorded. (c) Simulated image of a nanocross is built by the pick-and-place nanomanipulation.

2. Image scanning This step is started from locating Tip II by local scanning its end with Tip I. The image scanning length on Tip II is determined by its image height, which should have a roughly equal maximum image height as the nanoobjects. Once the local scanning on Tip II is completed, Tip I is used to fully scan the interested region, obtaining a topographic image that contains nanoobjects to be manipulated and of course, Tip II end. This step is depicted in Figure 5.20(a)-I, in which the piezostage is actuated for the image scan from left to right on the X-axis. Figure 5.20(b) shows a simulated image that contains topography of two nanowires and the end of Tip II,
3. Relocating Tip II After a long-playing image scanning, relocating Tip II is quite necessary to succeed in constructing the nanotweezer due to the system thermal drift, which will cause noticeable positioning errors if the nanomanipulation is performed in an uncontrolled environment. The relocation of Tip II is performed by the same local scanning operation as previously described.
4. Making Tip II-nanowire in contact Figure 5.20(a)-II shows that Tip II approaches the nanowire to make a contact by moving the X-axis of the piezoscanner. A gap (typically 5 20nm above the snap-in bound) between Tip II and the substrate should be kept during the approach because it enables a negative deflection response as a tiny force applied on Tip II, hence, a sensitive detection of the tip-nanowire contact.
5. Constructing the nanotweezer : similarly, as the step depicted in Figure 5.20(a)-III, Tip I approaches the nanowire by moving the X-axis of the piezostage. Once

the Tip I and the nanowire are in contact, a nanotip nanotweezer is configured for a grasping operation of the nanowire.

6. Pick-and-place In this step, as shown in Figure 5.20(a)-IV, the nanotweezer is used to pick up, transport and place the nanowire to its target position by moving piezoscanner on the X, the Y or the Z -axis. The displacement on each axis depends on the dimensions of the nanowire and the location of the destination.
7. Re-initialization Initialize the piezoscanner and the piezostage. Then the manipulation procedure is repeated for the next nanowire.

5.3.2.3 Force Sensing during Pickup

Figure 5.21 shows a force simulation during the nanoobject pickup operation with the nanotweezer.

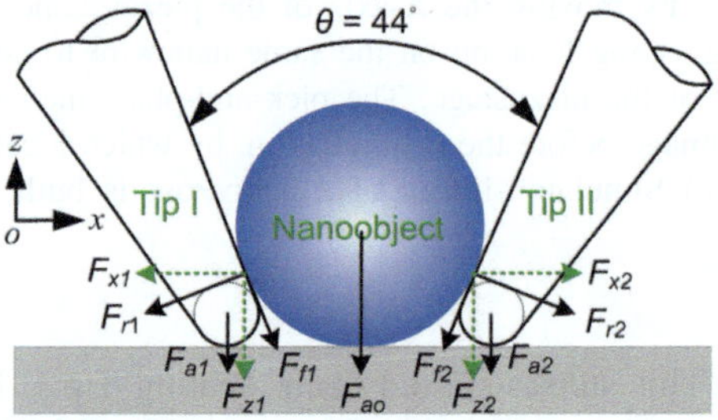

Fig. 5.21 Force simulation of pickup manipulation with a nanotweezer constructed from two cantilever's tips. The nanotweezer is with a clamping angle of 44°.

Interactive forces applied on the nanotweezer include repulsive forces, friction forces and adhesive forces. The interactive forces on the nanotweezer can be resolved into two components on the X-axis and the Z-axis in the defined frame, for example on Tip I, namely, F_{x1} and F_{z1}, respectively. The component force on the X-axis is the clamping force that holds the nanoobject. On the other hand, the component force on the Z-axis is the pickup force that balances the adhesion forces from the substrate. In order to measure the interactive forces, further analysis is carried out to estimate these two component forces, both of which result in bending deformation of the cantilevers.

Taking Tip I for instance, F_{x1} and F_{z1} can be calculated in the defined frame by:

$$F_{z1} = F_{r1}\left(\mu\cos\frac{\theta}{2} + \sin\frac{\theta}{2}\right) + F_{a1} \tag{5.38}$$

$$F_{x1} = F_{r1}\left(\mu\cos\frac{\theta}{2} - \sin\frac{\theta}{2}\right) \tag{5.39}$$

where F_{r1} is the repulsive force, $\mu F_{r1} = F_{f1}$ is the friction force, μ is the friction coefficient, θ is the clamping angle of the nanotweezer and F_{a1} is the adhesive force between Tip I and the substrate.

For the convenience of calculation, the angular deflection of the cantilever is adopted due to its linear relation with the voltage output of the optical lever. The bending angular deflection on the free end of the cantilever comprises two parts: ϕ_{z1} and ϕ_{x1}, which are caused by F_{z1} and F_{x1} respectively. These two parts can be calculated by:

$$\phi_{z1} = \frac{F_{z1}L^2}{2EI} \tag{5.40}$$

$$\phi_{x1} = \frac{F_{x1}Ll}{EI} - F_{f1}\sin\frac{\theta}{2} \tag{5.41}$$

where L is the effective length of the cantilever, l is the cantilever's tip height, E is Young's modulus of the silicon and I is moment of inertia on the cantilever's cross-section. L and l are measured as 250μm and 9.5μm, respectively.

Here we set $\mu = 0.33$, which is one of experimental values obtained from an AFM lateral force calibration using a silicon grating [87]. As seen in Figure 5.21, the nanotweezer has a clamping angle $\theta = 44°$. Thus, without consideration of F_{a1}, a simplified relation between ϕ_{z1} and ϕ_{x1} can be estimated as:

$$\phi_{x1} = 0.089\phi_{z1} \tag{5.42}$$

Actually, magnitude of the adhesion force F_{a1} is often comparable to that of the repulsive force F_{r1} until the nanotweezer pull off the substrate. In this case, the angular deflection produced by F_{x1} will be approximately reduced to less than half of the estimation from the Equation 5.42. Therefore, for simplifying the estimating process of the adhesion forces F_{ao}, the cantilever deflection caused by F_{x1} will be neglected in following calculations. Thus, the F_{z1} can be easily estimated from the normal voltage output of the optical lever of Tip I by [88]:

$$F_{z1} = \beta_1 \Delta V_1 \tag{5.43}$$

where β_1 and ΔV_1 are the calibrated normal force sensitivity [89] and voltage output of the optical lever of Tip I, respectively. A similar result can be obtained from Tip II. Once the F_{z1} and F_{z2} are known, for equilibrium, the adhesion force F_{ao} applied on the nanoobject can be estimated as:

$$F_{ao} = F_{z1} + F_2 = \beta_1 \Delta V_1 + \beta_2 \Delta V_2 \tag{5.44}$$

where β_2 and ΔV_2 are the normal force sensitivity and voltage output of the optical lever of Tip II, respectively.

5.3.2.4 Grasping Limit of the Nanotip Tweezer

However, every manipulation system has its own limitations on the size of the objects to be manipulated. In the proposed 3DMFM, the minimum size of the nanoobject that can be manipulated is limited by the tip size. For the nanowire grasping, nanoscale nanotweezer-nanowire contact area leads to a weak contact friction that is proportional to the contact area at nano scale [90].

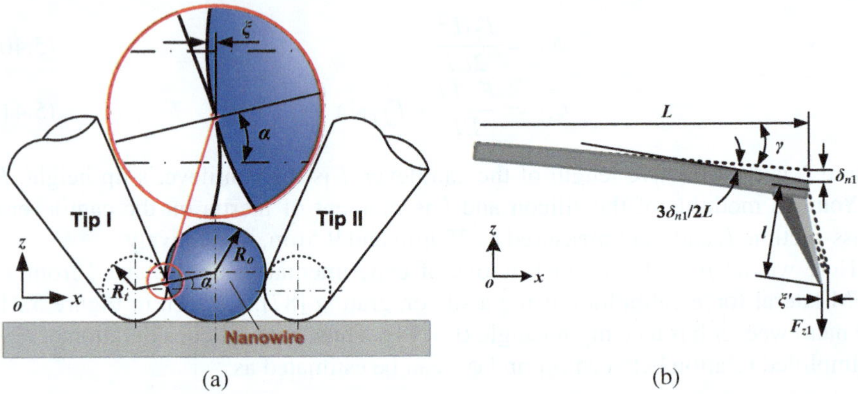

Fig. 5.22 Schematic for manipulation limits calculation. (a) In order to complete the grasping operation, the angle α and the dig-in distance ζ should be positive. The top inset shows details confined in the red circle. (b) Illustration for ζ' calculation.

These friction forces in general are not strong enough to overcome large adhesion forces of the nanowire-substrate contact. Fortunately, for the proposed nanotip tweezer, as seen in figure 5.22(a), a positive clamping angle is adopted that makes grasping stronger due to repulsive forces from the tip-nanowire contact which composites with the tip-nanowire friction forces to overcome strong adhesion forces from the contact with the substrate.

On the assumption that no preload is applied to the nanowire before the pickup operation, as shown in figure 5.22(a), in order to complete the grasping, the angle α and the dig-in distance ζ should be positive, which can be calculated via a simple geometric transform:

$$\alpha = \arcsin \frac{R_0 - R_t}{R_0 + R_t} \tag{5.45}$$

$$\zeta = R_0 (1 - \cos \alpha) \tag{5.46}$$

where R_o and R_t are the radiuses of the nanowire and the tip, respectively.

During the pickup operation, the tip end will retract a small distance ζ' because of the deflection on the cantilever beam. This retraction will reduce the effective dig-in distance. Figure 5.22(b) shows a illustration to calculate retracting distance

ζ' taking no account of the tip deformation based on the fact that the tip stiffness is typically two to three orders of the magnitude larger than the normal stiffness of the cantilever [91]. Therefore, with a very small deflection $\delta_{n1} = \frac{F_{z1}}{k_1}$ on the free end of the cantilever beam, ζ' can be estimated as:

$$\zeta' = -\frac{F_{z1}}{k_1}\left(\frac{3l}{2L}\cos\gamma + \tan\gamma\right) \quad (5.47)$$

where k_1 is the bending stiffness of the cantilever, γ is the cantilever's mounting angle, L is the effective length of the cantilever and l is the tip height. Here, deflection caused by the force F_{x1} is neglected as discussed. Therefore, in order to achieve the grasping operation with the nanotweezer, the effective dig-in distance should satisfy:

$$\zeta + \zeta' \geq 0 \quad (5.48)$$

Thus, from Equation 5.48, the minimum diameter of the nanowire R_{min} can be calculated by:

$$R_{\min}\left(1 - \sqrt{1 - \frac{(R_{\min} - R_t)^2}{(R_{\min} + R_t)^2}}\right) = \frac{F_{z1}}{k_1}\left(\frac{3l}{2L}\cos\gamma + \tan\gamma\right) \quad (5.49)$$

For example, during the pickup, if the maximum $F_{z1} = 50$nN, $R_t = 8$nm, $k_1 = 4$N/m, $\gamma = 5°, L = 250$μm and $l = 10$μm, the minimum diameter of the nanoobjects that can be picked up is $D_{min} = 2R_{min} = 39.4$nm. If a worn tip with $R_t = 15$nm is used, the minimum diameter of the nanoobjects will increase to $D_{min} = 61.1$nm.

It should be clear that this limit is calculated on the assumption that no grasping force exists prior to the pickup operation. If the nanotweezer holds the nanowire more tightly leading to a smaller gap between two tips before pickup, the minimum diameter that can be manipulated should be reduced a lot. However, the preload involves great risks in damaging the tips as well as the nanoobjects.

In addition, sharper AFM tips extended with single-walled carbon nanotubes (SWNT) or multi-walled carbon nanotubes (MWNT) may bring the 3DMFM a promising breakthrough in the grasping limit [92]. Another possibility on is to manipulate suspended nanoobjects, because in this case more space will be provided to obtain a closer grasping gap between the tweezer tips, bypassing the restriction from the dig-in distance.

5.3.3 Pick-and-Place Nanomanipulation

5.3.3.1 Sample Preparation

Uniformly well-dispersed deposition of the nanoparticles or the nanowires usually cannot be achieved by simply drying a colloid drop on a substrate. In our experiments, SiNW arrays were grown on Si wafers using chemical-vapor-deposition (CVD). In sample preparation, a freshly cleaned silicon wafer coated with 300nm silicon dioxide was placed in parallel with the sample surface of SiNWs. Then the SiNW array is lightly scratched, and SiNWs were uniformly attached to the silicon substrate due to the adhesive forces. AFM images show that the SiNWs have a taper shape. In pick-and-place experiments, two types of cone-shaped SiNWs diameters manipulated to build nanowire crosses. One type of the SiNWs has a diameter of 15nm (top) 70nm (root) and a length of 3μm 4μm. The other SiNWs have diameters of 25μm (top) 200μm (root) and lengths of about 7μm.

5.3.3.2 Contact Detection

In general, manipulating an object requires the ability to observe, position, and physically transport the object. Unlike the nanomanipulation performed in the SEM or the TEM, the AFM manipulation has no real-time vision feedback to monitor the manipulation process. In this case, a real-time force feedback is necessary to detect the interaction between the tips and the nanoobjects. In the 3DMFM pick-and-place manipulation, the contact detection is a significant step for a successful grasping of the nanoobjects. In order to detect the interaction between the nanoobject and the tip in the grasping operation, a tiny gap between the tip and the substrate (typically 5 nm 20nm) should be kept when the tip approaches the nanoobject. In the grasping operation, as the tip approach and further "dig into" the base of the nanoobject after contact, this gap enables cantilever's bending deformation, hence, precise normal force sensing.

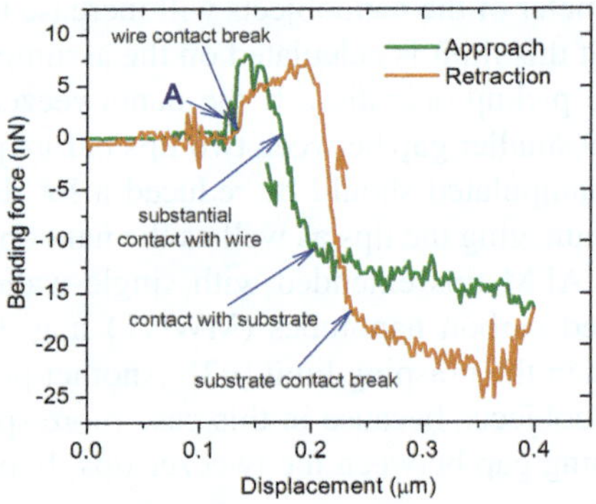

Fig. 5.23 Contact detection by the normal force sensing on Tip II. The force curve shows the bending behaviour of the cantilever during the steps of approach, contact and retraction.

Figure 5.23 shows an example of the contact detection with Tip II, which is obtained by grabbing a SiNW with diameters of 25nm (top) 200nm (root) and a length of 7μm. The grasping location is near the root of the nanowire. The curve starts from non-contact state with neither substrate nor the nanowire. As the tip further moves towards the nanowire, the cantilever is bent upward (about 3nm) leading to positive forces after the position "A" marked in the figure 5.23. Two factors could contribute to this positive response: chemical dirt on the border of the nanowire and adhesion forces from the nanowire.

The tip starts to dig into the base of the nanowire as its positive response reaches the first peak of 7.4nm and then contacts nanowire. In this part, before the tip contacts the substrate, the bending force decreases rapidly to -12nm with a short displacement about 35nm. The further move leads to the pushing of the nanowire, causing a slow decrease of the bending force due to the tip deformation. During the retraction, the adhesion forces between the tip and the substrate induces a sudden decrease of the bending force to -24nm. After the contact breaks with the substrate, the bending force sharply reaches a positive peak with a similar value than the former peak in approach path. Eventually, the tip returns to the position "A" with the same force response and reaches zero. This confirms that the positive peaks around the contact point in the approach and retraction paths are with the contribution of the unmoved chemical dirt.

The force response described in figure 5.23 is sufficient to detect not only the contact between the tip and the nanoobject, but the grasping state. In our experiments, as contact between Tip II and the nanowire is detected, Tip II retracts a little with a displacement of 5nm 20nm in order to keep a tiny gap between Tip II and the substrate, enabling the confirmation of grasping state as Tip I contact the nanowire.

5.3.3.3 Force Sensing in the Pick-and-Place Task

Figure 5.24 shows a curve of the peeling force spectroscopy on Tip II for the pick-and-place manipulation of the same nanowire on the same manipulating location used in the contact detection experiment.

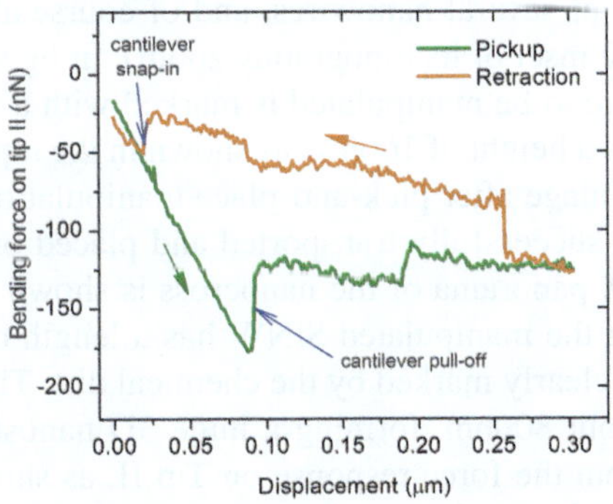

Fig. 5.24 Force detection on Tip II during the pickup operation. The force curve shows the bending behaviour of the cantilever during the processes of pickup and retraction for releasing.

The curve starts from contact state among the nanotweezer, the nanowire and the substrate. As the piezoscanner is moved down to pick up the nanowire, the cantilever is bent leading to negative forces. During the pickup, when the position reaches

88nm, the cantilever pulls off the substrate with a pull-off force of 52nN, and the force returns to −125nN. As the piezoscanner is moved down further, the force magnitude keeps slightly decreasing to a value of −136nN at 188nm. After this point, the magnitude of the force suddenly falls to −110nN, forming a tiny force peak of 5nN. Further pickup leads to another slow decrease of the peeling force to −125nN at 302nm. At the first part of the retraction, as the tips approach the substrate, the force slightly increases from −126nN to −118nN at 258nm. A sudden increase of 43 nN occurs at this point with a further retraction. Further retraction leads to a continuous increase except for a weak fluctuation between 135nm 81nm. Snap-in occurs at 21nm after a mild decrease of the force. The snap-in force reaches −21nN. Upon still further retraction, the magnitude of the normal force of Tip II approaches the previous grasping state.

The force responses on Tip I are similar in the shape of curve except for the force magnitude due to different force sensitivities on each tip and an uneven grasping due to asymmetric alignment of the nanowire related to the grasping direction.

Such a force spectroscopy curve during the pickup operation shows a stable grasping for further transport. A full force response of the nanowire pick-and-place is not shown here, but a simple description is shown in figure 5.25(d).

5.3.3.4 Manipulation Results

In order to validate the pick-and-place manipulation abilities of the developed 3DMFM, two types of cone-shaped SiNWs were manipulated and two nanocrosses were achieved. Figure 5.25 shows the manipulation processes and result with SiNWs with diameters of 25nm (top) 200nm (root).

A pre-scanned image (9μm × 9μm) is shown in figure 5.25(a), which includes the topographic image of several nanowires, and of course also involves the local image of Tip II (see the inset of its topography zoom). In figure 5.25(a), a grasping location of the nanowire to be manipulated is marked with a short green line A-A, where the nanowire has a height of 166nm, as shown in the top inset. Figure 5.25(b) shows the re-scanned image after pick-and-place manipulation. It can be seen that the nanowire has been successfully transported and placed onto another nanowire to build a nanocross. A panorama of the nanocross is shown in Figure 5.25(c), by which we can find that the manipulated SiNW has a length of 7.1μm and its trail before manipulation is clearly marked by the chemical dirt. The nanowire cross has a maximum height about 800nm, forming a huge 3D nanostructure. A full force curve was recorded from the force response on Tip II, as shown in figure 5.25(d). The force curve exhibits four main steps in the pick-and-place manipulation, including grasping, picking up, transporting and placing. The manipulation procedure is described as follows.

Once the SiNW was reliably grasped, piezoscanner moved down 650nm with a velocity of 65nm/s and the photodiode voltage output decreased to 0.24V before the third step. In this step, the SiNW was transported a distance of 4.6μm on the X-axis with a velocity of 120nm/s and 0.18μm on the Y-axis with a smaller velocity of 4.8nm/s. During the nanowire transport, a disorder (marked by a red circle)

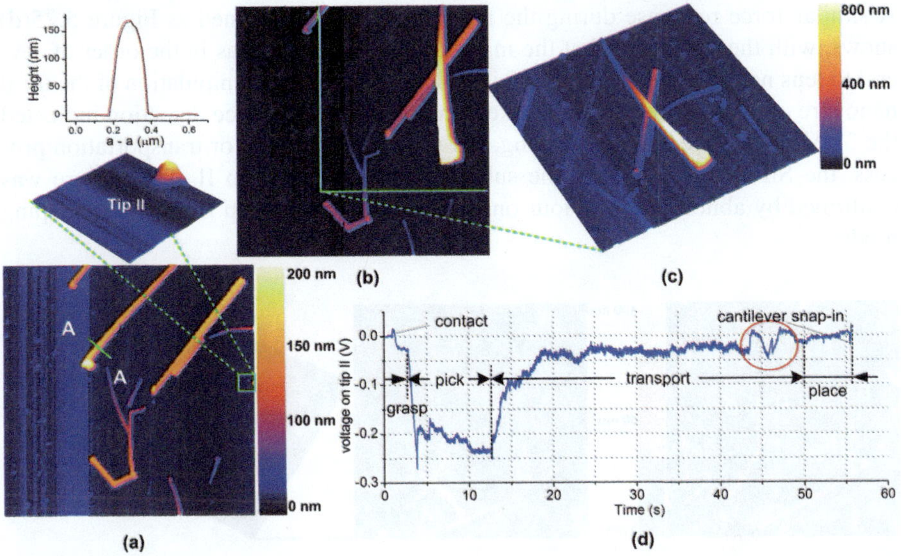

Fig. 5.25 Pick-and-place manipulation results of cone-shaped SiNWs with diameters of 25 nm (top) 200nm (root) and with a full length of about 7 m, in which the image scan size is 9 m 9 m. (a) An image scanned before the manipulation, which includes the topographic images of the nanowires and location information of Tip II. The top insets show the topography zoom of Tip II and provide height information at the grasping location A-A. (b) A new scanning image after pick-and-place manipulation verifies that a successful nanocross has been built. (c) Another new scanning image shows a panorama of the nanocross. (d) A full force curve on Tip II recorded during the whole procedure of pick-and-place manipulation.

occurred due to its contact with the supporting nanowire before releasing. In the releasing step, piezoscanner moved up with a velocity of 100nm/s. As Tip II was slightly bent upward leading a positive response of 0.015V, Tip I and Tip II were separated by moving the piezostage on the X-axis to open the nanotweezer.

Figure 5.26 shows the manipulation process and results of the other type of cone-shaped SiNWs with diameters of 15nm (top) 70nm (root). After image scanning, as shown in figure 5.26(a), three SiNWs were selected and the manipulation locations were placed near the top, near the root and on the middle part of the nanowires, marked by short green lines B-B, C-C and D-D, where heights of the SiNW are 46nm, 66nm and 58nm, respectively. During the first grasping on the location B-B, after successful contact detection on Tip II, as Tip I approach the nanowire to form a nanotweezer, the "dig-in" response as shown in Figure 5 did not happen. Consequently, the first SiNW could not be picked up but be pushed by Tip I, which is verified on figure 5.26(b). Two factors could contribute the failure of the grasping: some chemical dirt on the way of "dig into" and another significant factor is that the worn Tip I after image scanning reached its manipulation limit for pickup. On the second manipulation location C-C, the 3DMFM succeeded in pick-and-place manipulation and building a nanocross, as shown in the image zoom in Figure 5.26(c).

A similar force response during the manipulation was obtained as Figure 5.25(d) shows, with the difference that the magnitude of the force was in the order of several to tens nano-Newton. Another failure happened in the manipulation of the third nanowire on its middle part, as marked on location D-D. Force detection indicated the 3DMFM succeeded the grasping, but during the pick-up or transportation process, the SiNW jumped from the substrate and stuck to Tip II. This failure was confirmed by abnormal behaviour on local scanning with Tip II using the tapping mode.

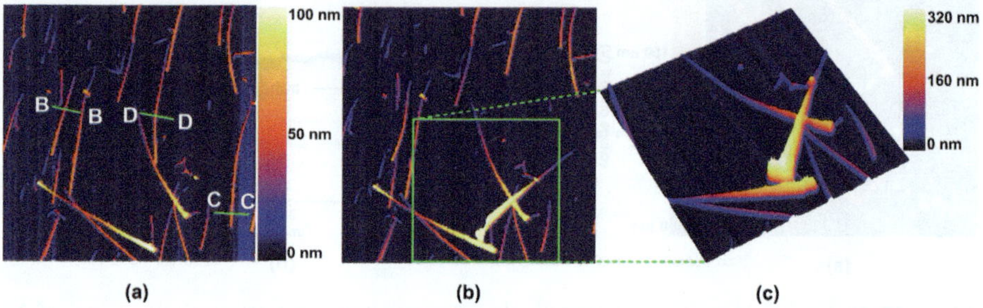

Fig. 5.26 Pick-and-place manipulation results of cone-shaped SiNWs with diameter of 15 nm (top) 70nm (root) and with lengths of 3 m 4 m. (a)An image scanned in size of 8 m 8 m before the manipulation, in which three nanowires were selected for the manipulation. Corresponding manipulation locations are marked as B-B, C-C and D-D where nanowire heights are 46 nm, 66nm and 58 nm, respectively. (b) A new image scanned after the manipulation, which shows that the first grasping on location B-B is unsuccessful due to the manipulation limit, the second manipulation on C-C build a perfect nanocross, the last manipulated nanowire stuck to Tip II due to the improper manipulating location D-D. (c) A topographic zoom of the nanocross.

The 3D manipulation force microscope (3DMFM) presented in this section has achieved this type of pick-and-place nanomanipulation using a nanotweezer formed by two AFM cantilevers. Each cantilever is equipped with a nanopositioning device and an optical lever, enabling the 3DMFM has capabilities of image scanning and real-time force sensing during pick-and-place manipulation. In order to validate the manipulation capabilities of the 3DMFM, two types of cone-shaped SiNWs were manipulated and two nanowire crosses were achieved. Moreover, force recording during the pick-and-place manipulation provides valuable data in the understanding of the nanomechanics and nanophysics. As a result, the 3DMFM makes the 3D nanomanipulation and nanoassembly, such as constructing 3D nanostructures, building 3D nanodevices, and revealing nanomechanics phenomena more practical.

5.4 Parallel Imaging/Manipulation Force Microscopy

Unlike nanomanipulation performed in a scanning electron microscope (SEM) and a transmission electron microscope (TEM), an inherent limitation of AFM-based nanomanipulation is that the AFM acts as an imaging sensor as well as a manipulation tool, and so can-not provide manipulation with real-time visual feedback, but rather an insufficient serial process of scanning-manipulation-scanning [104]. In order to facilitate nano-manipulation, haptic devices and virtual reality interfaces were introduced into AFM-based nanomanipulation systems [105, 106], thereby enabling an operator to directly interact with the real nano-world (see chapter 4). However, the time-consuming scanning-manipulation-scanning operation is still required, making mass production impossible, although virtual reality could provide users with a so-called "visualized" feedback using local scan to update a model-based virtual environment [106]. On the other hand, high-speed AFMs have succeeded in raising the scanning efficiency [107, 108, 109, 110]. In this case, the excellent imaging potential of high-speed AFM will be greatly reduced if it is used for nanomanipulation with such a serial scanning-manipulation-scanning process. The central problem in speeding up AFM-based nanomanipulation is in fact to develop sufficient harmony between image scan and manipulation processes. This section describes a parallel imaging/manipulation force microscope (PIMM) to improve the efficiency of AFM-based nanomanipulation.

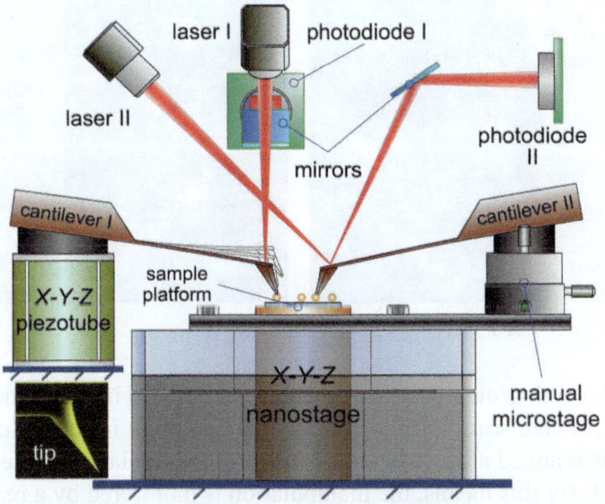

Fig. 5.27 Schematic diagram of the parallel imaging/manipulation force microscope (PIMM). The PIMM mainly consists of two individually actuated canti-levers with protrudent tips (left inset). Cantilever I is fixed on the X-Y-Z piezotube which acts as an imaging sensor to provide position information of nano-objects and the tip end of cantilever II. Cantilever II, fixed on the X-Y-Z nanostage, is used as a manipulation tool to push and pull nano-objects.

As shown in Fig. 5.27, the PIMM consists of two sets of similar devices commonly used in a conventional AFM, including two cantilevers with corresponding nanopositioning devices and laser measuring systems. Cantilever I (left), used as an imaging sensor, is fixed on an X-Y-Z piezotube (PI P-153.10H). Hysteresis of the piezotube is well compensated on each axis by the Prandtl-Ishlinskii operator [111].

In order to effectively avoid hysteresis and creep of the piezo-actuator, a closed-loop X-Y-Z piezo-actuated nanostage (MCL Nano-Bio2M) is used to actuate cantilever II (right), which is mounted on an X-Y-Z manual microstage for coarse posi-tioning, and mainly used as a manipulating end-effector. In order to achieve this parallel imaging/manipulation task, system control software with two parallel threads was developed to coordinate actions on these two cantilevers. A key component of the PIMM, as shown in the left-bottom inset of Fig. 1, is the special cantilever with a protrudent tip that makes interactions between cantilever I and cantilever II feasible. The protrudent tip enables image scan on the tip of cantilever II (tip II) using cantilever I. By this means, the relative positions between nano-objects and tip II end can be obtained from the image scan that is in-dispensable for AFM-based manipulation. The tilted angle of the cantilever tip is approximately $62^\circ - 64^\circ$ on the side view without a mounting angle. In the PIMM, the cantilevers are mounted with an angle of 8°. An excellent image quality can be achieved even though the scan length on tip II is more than 300nm, which is sufficient for tip positioning with a dynamic image display updated by real-time scanning data.

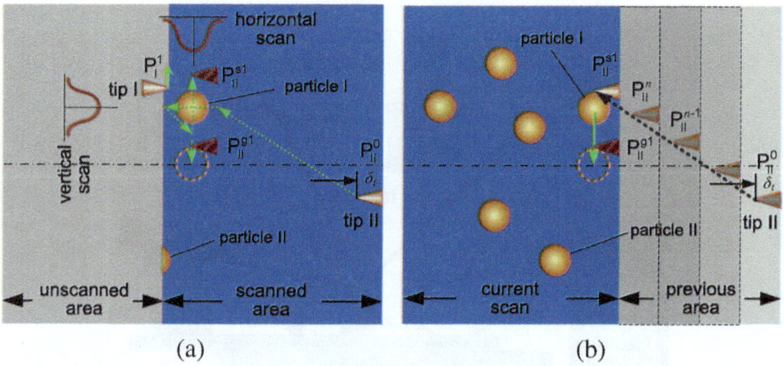

Fig. 5.28 Parallel imaging/manipulation schemes. (a) Parallel imaging/manipulation scheme with normal-speed image scan. In this case, the manipulation is controlled based on the image of the previous scanned area. (b) Parallel imaging/manipulation strategy with high-speed image scan on tip I. By this means, the manipulation is monitored by a real-time image feedback. The dynamic scan area is shifted by tracking the motion of tip II to keep a fixed scan length δ on tip II.

Fig. 5.28(a) is a diagram of the parallel im-age/manipulation with a normal-speed image scan using tip I. First, each axis of the nanostage and the piezotube is initialized in its proper position, which should provide an enough motion range for the nanomanipulation within an image scanning area. The next step is to locate tip II by local scan of tip II using tip I. Once tip II locating is ready, two parallel threads are started to perform a parallel imag-ing/manipulation task. One is used to control tip I to fulfill a full scan of the pertinent area selected under a optical microscope, resulting in a topographic image containing of nano-objects to be manipulated as well as tip II end (as a simulated image on position). The second thread is as-signed to dominate the performance of tip II, which is in idle process until a particle emerges on the dynamic image display. As particle I is entirely scanned while tip I reaches position , the first manipulation is started after the coarse position of particle I is estimated by processing the dynamic image.

In order to obtain an accurate position of particle I, tip II locally scans the nanoparticle horizontally and vertically along a line marked on Fig. 5.28(a) to locate the precise center of the nanoparticle and estimate the contact point between the tip and the nanoparticle. As tip II reaches P_{II}^{s1}, the manipulation is started until tip II reaches P_{II}^{g1} where particle I has been pushed to the target position. After the first manipulation, tip II is set to idle process again until the particle II emerges within the dynamic image.

Fig. 5.28(b) shows another strategy of the parallel imaging/manipulation with a high-speed image scan using tip I. As in the steps discussed above, once system initialization and local scanning on tip II are ready, the parallel image/manipulation task is started. Tip I is used for high-speed image scan of nano-objects and tip II end, pro-ducing a high-speed topographic image display which con-sists of tip II and nano-objects to be manipulated. Tip II is started to approach particle I for the first manipulation after the first frame of the dynamic image display. Simultaneously, the scan area is dynamically shifted by tracking the motion of tip II to keep a fixed scan length dt on tip II for a purpose of avoiding a block of the image scan due to excessive scan height on tip II. The scan area is moved from P_{II}^{0} to P_{II}^{*} until tip II reaches P_{II}^{s1}, at which point the manipulation is started to laterally push particle I until tip II reaches P_{II}^{g1}. After the first manipulation, tip II is moved on to the next nanoparticle.

Task time t_{task} of the parallel imaging/manipulation operation can be given :

$$t_{task} = \max(t_s, t_m) + t_m \tag{5.50}$$

where t_s is the scanning time of one image frame and t_m is the total manipulation time estimated from the sum of manipulation time of each single nano-object. For the normal-speed AFM, t_task is often equal to $2t_s$ except for a complex manipulation task that cannot be fulfilled within one frame period, and so has task time of $t_m + t_s$. This scheme can save much time, as compared with the serial imaging/manipulation operation that has task time of $t_m + 2t_s$. One disadvantage of this scheme is that environment-based motion planning is unavailable during the manipulation. However, the parallel imaging/manipulation can be perfectly performed by this scheme if a manipulation objective is defined before the operation. In contrast, for the second scheme, the task time with the high-speed AFM is approximately equal to the total manipulation time t_m due to very high frame rate of image scan. The manipulation process in this scheme is monitored by the high-speed visual feedback as the manipulation performed in the SEM. The second scheme is undoubtedly more efficient than the former one.

A parallel image/manipulation task was performed with a normal-speed image scan on the PIMM we developed. Manipulation results are shown in Fig. 5.29, in which four Ag nanoparticles with a diameter of 74nm – 82nm, emerged on dynamic

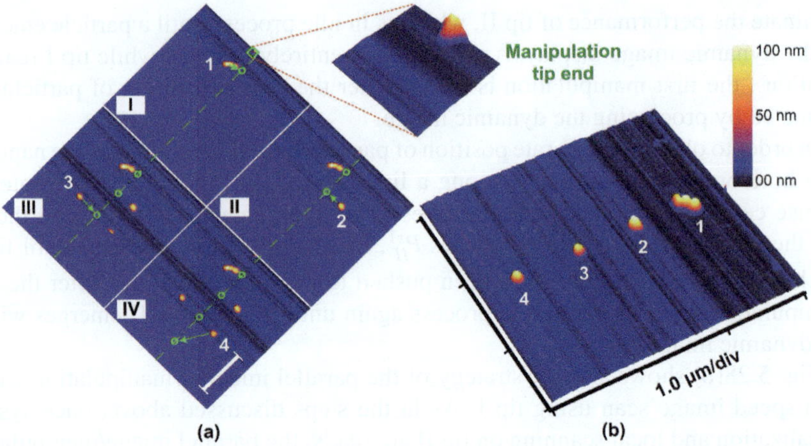

Fig. 5.29 A parallel imaging/manipulation result with a normal-speed image scan of tip I. (a) Emergences of four particles (with a diameter of 74nm - 82nm) on four different dynamic displays, namely, image I to IV. Once one particle has been fully scanned, a corresponding manipulation task will be activated to push it to its destination. (b) A manipulation result. Four nanoparticles were pushed along a line within the frame period of the image scan. The total manipulation time for these four particles was less than one minute, which is much less than the entire imaging time of ten minutes.

image I to IV in sequence, were pushed onto the image midline during the image scan. In the experiment, the frame period was about ten minutes. In contrast, the total manipulation time of these four nanoparticles was less than one minute with a pushing velocity of about 300nm/s. These results indicate that we can complete a more complex manipulation task during the image scanning thread to greatly increase the efficiency of the AFM-based nanomanipulation and making mass production feasible.

5.5 High-Efficiency Automated Nanomanipulation with Parallel Imaging/Manipulation Force Microscopy

The central problem in speeding up AFM-based nanomanipulation revolves around how to develop sufficient harmony between the image scan and manipulation processes of AFM-based nanomanipulation. As manipulation performed under optical microscopes or electron microscopes, the standard parallel imaging/manipulation method still offers promise for improving the efficiency of AFM-based nanomanipulation if the image scan and manipulation can be carried out in parallel.

This section, based on the previous description of PIMM (see 5.4), focuses on: (i) Automating parallel imaging/nanomanipulation by introducing effective manipulation strategies and control schemes, including image processing and feature recognition, tasking planning for both the normal- and high-speed image scan, and nanorobotic control for two tips coordination with coordinate transformations and detailed control flows; (ii) Introducing a dynamic pushing method for automated particle loss detection using amplitude feedback rather than the weak lateral force signal and the unstable normal force signal of the optical lever; (iii) Comparing efficiencies of automated serial and parallel nanomanipulation with experimental results; (iv) Performing a more complicated parallel nanomanipulation task with presented manipulation strategies and control schemes to validate the proposed parallel nanomanipulation method.

This section is organized as follows: subsection I briefly re-introduces the prototype and experimental setup of the PIMM. Two parallel manipulation schemes respectively with normal- and high-speed image scanning are presented in subsection II. Control flow and methods are introduced in subsection III. In section IV, we use one-tip and two-tip configurations of the PIMM to respectively complete serial and parallel nanomanipulation of nanoparticles to form nano-patterns with normal-speed image scanning.

5.5.1 System Set-Up of the PIMM

5.5.1.1 System Architecture

Figure 5.30 shows a schematic diagram of the developed PIMM. Cantilever I (tip I) and cantilever II (tip II), facing each other, are individually actuated by an open-loop *X*-*Y*-*Z* piezoscanner (PI P-153.10H) and an *X*-*Y*-*Z* closed-loop nanostage (MCL Nano-Bio2M on the *X*-*Y* axes, PI P-732.ZC on the *Z*-axis), respectively. An *X*-*Y*-*Z* motorized stage and an *X*-*Y*-*Z* manual stage are used for coarse positioning of tip I and tip II, respectively. The sample platform is immovable during manipulation that is fixed on the system base.

A data acquisition (DAQ) (NI 6289) card is used for high-speed capturing of photodiode voltage output from a lock-in to estimate deflections on both cantilevers induced by force load or resonant oscillation. The DAQ card is also used to actuate the piezoscanner by providing voltage signals to three independent amplifiers for each axis. A multi-thread planning and control system is developed to independently manage the AFM image scan and its dynamic display and processing, as well as two-tip coordination during parallel imaging/manipulation. Scanning data, a topographic image including nano-objects to be manipulated and the end of tip II, are shared by each of the threads for manipulation planning and system control.

5.5.1.2 Two-Tip Interactions

A key component of the PIMM is that the cantilever used has a protruding tip, which makes interactions between both cantilevers feasible. As shown in Fig. 5.31(a), the protruding tip enables image scan on the end of tip II using tip I. An AFM image of the end of tip II is shown in Fig. 5.31(b), in which the left inset shows a simulated scanning height when tip I is scanned tip II along the line a–a. By this means, the relative positions between the nano-objects and the end of tip II that is indispensable for AFM-based nanomanipulation can be obtained from the image scan. The tilted angle of the tip is approximately 63° on the side view. In the PIMM, the final tilted angle of each tip is about 68° to the substrate with a mounting angle of 5°, as seen in the right inset of Fig. 5.31(b). Thus, the scanning height of tip II can be estimated as:

$$H_t = L_t \tan 68^\circ \approx 2.48 L_t \tag{5.51}$$

where H_t and L_t are the scan height and the scan length on tip II, respectively. Excellent image quality can be achieved even though L_t is more than 300 nm, which is sufficient for tip positioning during the manipulation.

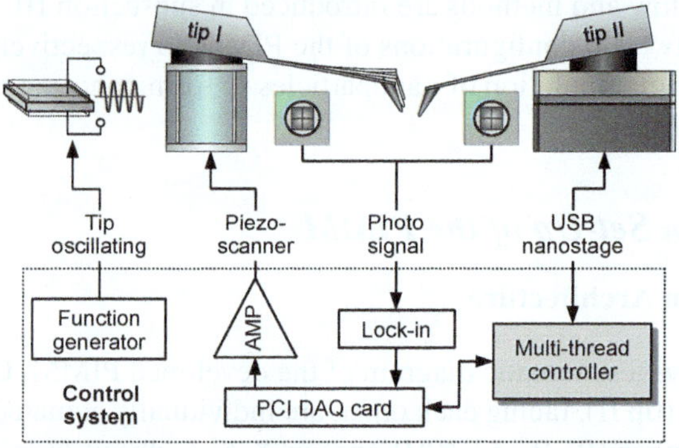

Fig. 5.30 System setup of the PIMM and a schematic diagram of the architecture of the control system.

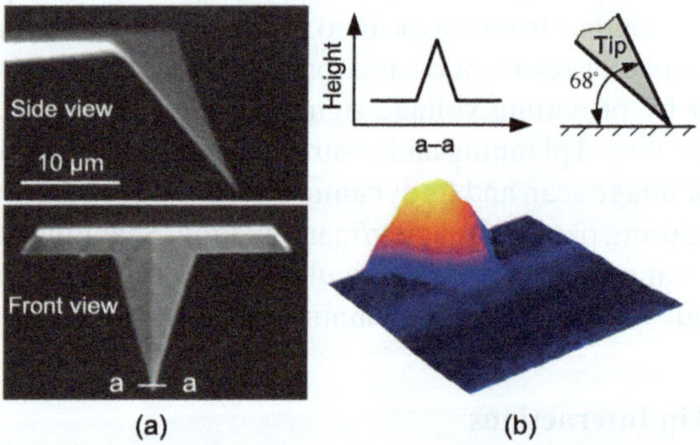

Fig. 5.31 Two tip interaction during the image scan. (a) An SEM image of tip's side and front view, in which a–a is a trace scanned at the end of tip II by tip I. (b) A topographic AFM image of the tip end. The left inset shows a simulated height of tip end on the scan line a–a, and the right inset shows the tip has a tilted angle of 68° with a mounting angle of 5°.

5.5.1.3 Coordinate Frames

Figure 5.32(a) shows coordinate frames of the PIMM that are summarized in Table II, including image scan frame s, manipulation frame m and image frame i. A point $P=(X,Y,Z)$ in the system coordinate X-Y-Z is mapped to a point $p=(u,v)$ in the image plane X_i-O_i-Y_i via a scaled orthographic projection:

$$\begin{bmatrix} p_x & 0 \\ 0 & p_y \end{bmatrix} \begin{bmatrix} u \\ v \end{bmatrix} = \begin{bmatrix} X \\ Y \end{bmatrix} \tag{5.52}$$

where p_x and p_y are scale factors, which are physical dimensions of pixels in the horizontal and vertical directions, respectively. Each scale factor is equal to the step length (s_x, s_y) of the image scan on the corresponding axis. The step lengths are determined by the resolution (r_x, r_y) and the length (l_x, l_y) of the scanned image:

$$\begin{bmatrix} s_x \\ s_y \end{bmatrix} = \begin{bmatrix} 1/r_x & 0 \\ 0 & 1/r_y \end{bmatrix} \begin{bmatrix} l_x \\ l_y \end{bmatrix}. \tag{5.53}$$

Therefore, the point $P=(X,Y,Z)$ in the system coordinate X-Y-Z can be determined by the point $p=(u,v)$ in the image plane X_i-O_i-Y_i via:

$$\begin{bmatrix} X \\ Y \end{bmatrix} = \begin{bmatrix} 1/r_x & 0 \\ 0 & 1/r_y \end{bmatrix} \begin{bmatrix} l_x & 0 \\ 0 & l_y \end{bmatrix} \begin{bmatrix} u \\ v \end{bmatrix}. \tag{5.54}$$

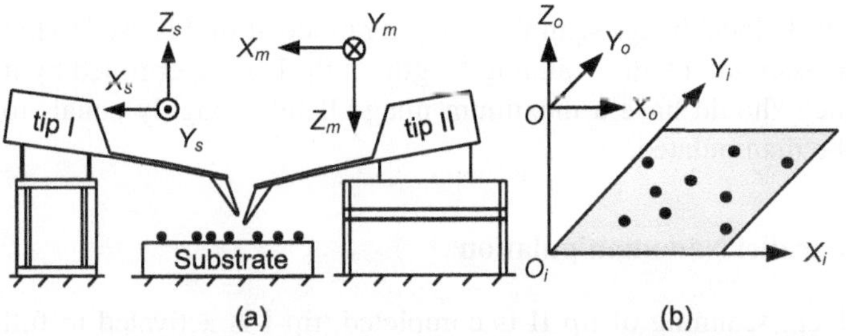

Fig. 5.32 (a) Coordinate frames of the system. (b) The coordinate frame of the image plane.

Table 5.2 Specifications of Coordinate Frames

Symbol	Coordinate frame definition
s	Imaging cantilever (tip I) coordinate frame X_s-Y_s-Z_s attached to the piezoscanner
m	Manipulating cantilever (tip II) coordinate frame X_m-Y_m-Z_m attached to the nanostage
o	Coordinate frame X_o-O_o-$Y_o$of the system
i	Coordinate frame X_i-O_i-Y_i of the scanned image plane

5.5.2 Strategies for Automated Parallel Manipulation

As mentioned above, image scanning with the cantilever's protruding tip makes it possible to locate the nano-objects as well as the tip end of the manipulating cantilever, providing positioning information for the further task planning. A manipulation protocol and two schemes are introduced respectively for parallel nanoparticle push with normal- and high-speed image scanning. Task planning based on the shortest path solution is proposed for linear trajectory planning on nano-pattern formations.

5.5.2.1 Parallel Imaging/Manipulation Protocol

A manipulation protocol is used to perform the image scan and the nanomanipulation in parallel. This protocol mainly includes the following steps.

5.5.2.2 System Initialization

In this step, both tips are aligned near the starting point of the selected scan area under the optical microscope. Each axis of the motion module is set on a proper position in the frames defined in Fig. 5.32. The initialization should provide enough motion travel for both the image scan and the manipulation.

5.5.2.3 Local Scanning on Tip II

Before the task, local image scan with tapping mode of tip I is used to locate the end of tip II. As seen in (1), the scanning length on tip II is determined by its scanning height, which should have a maximum image height roughly equals to the nano-objects to be manipulated.

5.5.2.4 Parallel Nanomanipulation

Once the local scanning of tip II is completed, tip I is activated to fully scan the region of interest, obtaining a topographic image that contains nano-objects to be manipulated and the end of tip II. Simultaneously, tip II is also managed by the manipulation control thread. After image processing and task planning, the nano-objects are manipulated with a controlled pushing operation. This step will be discussed in detail in the following sections.

5.5.2.5 Image Processing

Image processing is for task planning and the coarse positioning of each nanoparticle. Two steps are involved in the image processing: recognize nanoparticle distribution and calculate their central positions. Fig. 5.33(a) shows an AFM image, which contains the end of tip II and ten nanoparticles with a diameter of about 80 nm.

5.5.2.6 Recognition Nanoparticles and the End of Tip II

A feature in the scanned image may be a nanoparticle or the end of tip II. In order to recognize and classify these features, the template-matching method via correction coefficient $\gamma(x,y)$ is used [131]:

$$\gamma(x,y)=\frac{\sum_{s}\sum_{t}\left[(f-\bar{f})\right]\left[(w(x+s,y+t)-\bar{w})\right]}{\sqrt{\sum_{s}\sum_{t}\left[(f-\bar{f})\right]^{2}\sum_{s}\sum_{t}\left[(w(x+s,y+t)-\bar{w})\right]^{2}}} \tag{5.55}$$

where x and y are the width and height of the scanner image, $\bar{w}$ is the average gray value of the template window w, and $\bar{f}$ is the average gray value of the image f in the region coincident with the current location w. Using pre-defined templates, the nanoparticles and the end of tip II can be easily identified.

5.5.2.7 Positioning Nanoparticles and the End of Tip II

The AFM image is first convolved with a low-pass Gaussian filter for noise suppression, and then segmented to a binary image using the optimal global and adaptive thresholding method [131]. For further processing, the binary image is eroded to remove overly small or large areas that represent spurious or connected features with several nanoparticles stuck together.

Then the region-growing method [131] is used to label each feature and calculate its central position. After region-growing, each of the regions has certain gray values for feature identification. For each labeled feature, its center position $(\bar{x}_k,\bar{y}_k)$ can be calculated as:

$$\bar{x}_k=\frac{\sum_{j=1}^{m_k} x_j}{m_k}, \qquad \bar{y}_k=\frac{\sum_{j=1}^{m_k} y_j}{m_k} \tag{5.56}$$

where x_j is the x-coordinate and y_j is the y-coordinate of pixel j of the kth region that contains m_k pixels. Fig. 5.33(b) shows an image after region growing. Eleven features, including ten nanoparticles and the tip II end, are labeled from 1 to 10 by gray value 74 to 254 with an interval of 20 for the nanoparticles and gray value 255 for the end of tip II. If the number of nanoparticles is more than 255, feature labeling is performed within each sub-region of the image grid.

Once the features are characterized and their center positions are calculated, nanoparticles $p_k(k=1,2,\ldots,n)$, assuming k is a unique ID for each nanoparticle and its position in the image plane is denoted as $p_k=(\bar{x}_k,\bar{y}_k)$ for task planning of nano-pattern formation.

5.5.2.8 Manipulation Scheme with Normal-Speed Image Scan

The AFM tapping mode is required for nanoparticle scanning in ambient conditions, to avoid the tip from moving or removing the nanoparticle as it scans them. Several

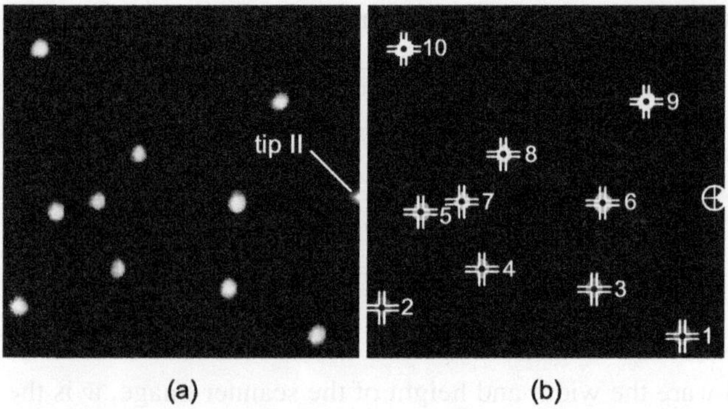

Fig. 5.33 An example of image processing. (a) An original gray-lever image with a scan area 5 μm $\times$ 5μm. (b) A binary image by global and adaptive thresholding; features are labeled and their centers are located.

minutes should then be taken for a full image scan using normal-speed scanning. In this case, the image display is updated pixel-by-pixel with the data sampling frequency, tracking the scan motion of tip I.

5.5.2.9 Manipulation Scheme

Figure 5.34 is a diagram of parallel imaging/manipulation with normal-speed image scanning. As for the parallel imaging/manipulation protocol mentioned above, once tip II locating is ready, two parallel threads are started to perform a parallel imaging/manipulation task. One is used to control tip I to complete a full scan of the pertinent area selected under the optical microscope, resulting in a topographic image containing nanoparticles to be manipulated as well as the end of tip II (as a simulated image on position P_{II}^{0}). The second thread is assigned to dominate the performance of tip II, which is in idle process until a particle emerges on the dynamic image display. As the predetermined target positions t_1 and t_2 as well as at least one nanoparticle are entirely scanned as tip I reaches position P_{I}^{1}, the first manipulation is started after the approximate position of particle I is estimated by processing the dynamic image.

In order to accurately position nanoparticle p_1, tip II locally scans the nanoparticle horizontally and vertically along a line marked to locate the precise center of the nanoparticle and estimate a pushing point. As tip II reaches P_{II}^{s1}, that is on the line connecting centers of p_1 and t_1, the manipulation is started until tip II reaches P_{II}^{g1} where particle p_1 has been pushed to the target position. After the first manipulation, tip II is started again to manipulate particle p_2 to its target position t_2. When the second manipulation is completed, tip II is set to idle process again until new target positions and nanoparticles for manipulation emerge.

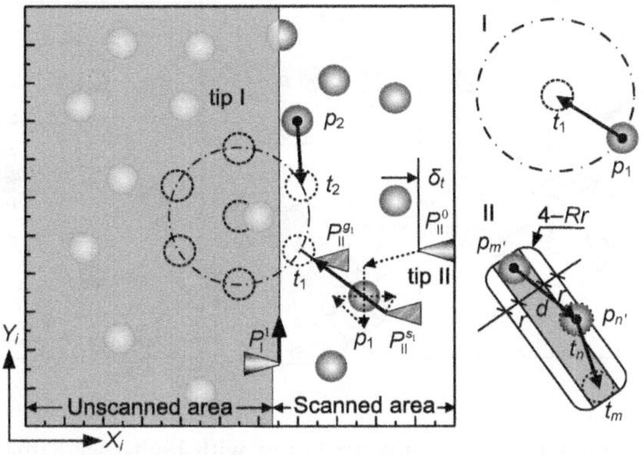

Fig. 5.34 Parallel imaging/nanomanipulation scheme with normal-speed image scan.

5.5.2.10 Task Planning

With normal-speed image scanning, global task planning is unavailable during manipulation due to the very low imaging rate. However, parallel imaging/nanomanipulation can certainly be performed if a manipulation objective is defined before the operation. As shown in Fig. 5.34, prior to the nanomanipulation, a regular hexagon pattern is predetermined within the image plane X_i–Y_i. As corners of the regular hexagon emerge on the dynamic image display and nanoparticles are present, the shortest path solution is used to plan the linear trajectory of the nanoparticle's motion.

As seen in inset I, the nanoparticle p_1, which has the shortest distance to t_1, is chosen as the first to be manipulated. The target position sequences are determined along the y-axis of the image coordinate frame when more than one target position emerges on the same effective image area, e.g. t_1 and t_2. During the whole process of parallel nanomanipulation, previously manipulated nanoparticles, which are in their target positions, are at times probably viewed as obstacles. As shown in inset II of Fig. 5.34, the manipulated nanoparticle $p_{n'}$, on its target position t_n, is within the obstacle zone of the nanoparticle $p_{m'}$ and its target position t_m. In this case, if no other nanoparticles exist within the currently scanned area, these two nanoparticles should exchange their target positions by removing $p_{n'}$ to t_m and $p_{m'}$ to t_n. The width of the obstacle zone is set at twice the diameter of the nanoparticle because of the possibility of the nanoparticle rotating around the AFM tip during the pushing operation.

5.5.2.11 Manipulation Scheme with High-Speed Image Scan

Unlike normal-speed image scanning, high-speed image scanning is expected to provide an image rate from several frames per second to video rate. In this case, image display can be updated rapidly. For task planning, image processing is performed for every frame of the image scan from the beginning.

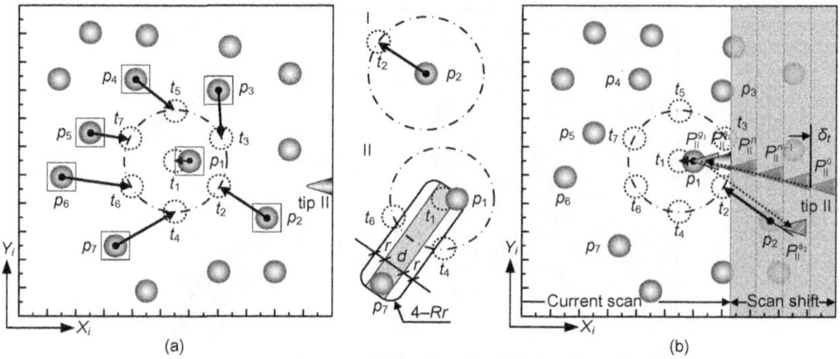

Fig. 5.35 Strategies for parallel nanomanipulation with high-speed image scan. (a) Simulated trajectory planning using the first frame of high-speed image scan. (b) Parallel imaging/nanomanipulation scheme with high-speed image scan.

5.5.2.12 Task Planning

With high-speed image scanning, global task planning is available that is performed and can be the first of scan frame. Fig. 5.35(a) shows the first frame of simulated topographic image that contains nanoparticles and the end of tip II. After image processing of this frame, trajectory planning is started to generate a manipulation sequence for each target position and the corresponding nanoparticle to be manipulated. The trajectory-planning algorithm is summarized as follow:

1. Share one frame of image from the imaging thread.
2. Detect the central position of each nanoparticle in the image space using the method mentioned above.
3. Determine an equal number of particles with the target positions.
4. Generate a manipulation sequence for the target positions.
5. **for** $k = 1$ to N (number of targets)
   a. Generate all possible linear trajectories between each particle and the target position $t_k(x_i, y_i)$ in the image space.
   b. Select the particle with a central position on $(x_i, y_i)(n = 1, 2, \ldots, N)$ that has the shortest path to the target position t_k.

   end for
6. Transform the position sequences $t_k(x_i, y_i)/p_n(x_i, y_i)$ from the image coordinate frame X_i-O_i-Y_i into manipulation coordinate frame X_m-Y_m-Z_m for actual motion planning of the nanomanipulation.

Since linear trajectory is more convenient than nonlinear for pushing nanomanipulation with a single nanotip, an efficient strategy was proposed for 2-D micro manipulation with a single probe [132]. In order to avoid losing the least problematic particles those have the shortest paths to the target positions, a strategy is introduced to combine distance planning with the method mentioned above. With this strategy, the shortest path method is used first to generate an equal number of particles with the reference target positions for trajectory planning only if there are more particles than target positions, as shown in inset I. Then by counting the number of the blockages of each target position to every previously selected particle, a manipulation sequence of the target positions is determined. Inset II shows an example for target position t_1, with three blockages counted with a possible trajectory to particle p_7. Finally, the shortest path method is used to find a particle for each target position with the generated sequence.

Once the trajectory planning is completed, for each of the target positions in sequence, tip II is used to push the nearest nanoparticle along the generated trajectories. Fig. 5.35(a) shows a predetermined nano-pattern formed by a regular hexagon with a central particle. Six nanoparticles labeled p_1 to p_6 are chosen to build this nano-pattern.

5.5.2.13 Manipulation Scheme

Figure 5.35(b) shows a diagram of the parallel manipulation with high-speed image scan. Tip I is used for high-speed image scanning of the nanoparticles and the end of tip II, producing a high-speed topographic image display, which consists of tip II and nanoparticles to be manipulated. After task planning, tip II is started to approach nanoparticle p_1 for the first manipulation after the first frame of the dynamic image display. Simultaneously, the scan area is dynamically shifted by tracking the motion of tip II to keep a fixed scan length δ_t on tip II for to avoid blocking the image scan due to excessive scan height on the end of tip II. The scan area is shifted from P_{II}^0 to P_{II}^n until tip II reaches $P_{\mathrm{II}}^{s_1}$, at which point the manipulation is started to push nanoparticle p_1 to its target t_1 while tip II travels to $P_{\mathrm{II}}^{g_1}$. After the first manipulation, tip II is moved on to the next nanoparticle.

5.5.2.14 Comparing Manipulation with Different Scan Rates

Obvious differences between the parallel nanomanipulation with normal- and high-speed scanning are summarized as:

5.5.2.15 Different Task Planning Methods

Task planning with normal-speed scanning is limited in local areas due to its low imaging rate. In contrast, with high-speed image scanning, the global task planning is available and, if necessary, dynamic task planning is feasible with high-speed visual feedback that provides more accurate and efficient trajectory planning for pattern formation.

5.5.2.16 Incomparable Manipulation Robustness

High-speed imaging rate make the parallel nanomanipulation more controllable due to its visual feedback, which can be used to track the nanoparticle's motion, monitor the tip-nanoparticle interactions, and compensate for system error induced by thermal drift. Combining visual feedback with force and amplitude feedback from the oscillating tip, the parallel manipulation control with high-speed image scan is undoubtedly more stable and robust.

5.5.2.17 Dissimilar Task Time

The comparison between the manipulation with normal- and high-speed image scanning centers around the manipulation efficiency. As shown in Fig. 5.36, task time t^p_{task} of the parallel imaging/manipulation can be given as:

$$t^p_{\text{task}} = \max(t_s, t_m) + t_s \tag{5.57}$$

where t_s is the scanning time of one image frame and t_m is the total manipulation time estimated from the sum of manipulation time of each single nano-object. For normal-speed AFM, the task time is often equal to $2t_s$ except for a complicated manipulation task that cannot be completed within one frame period, and so has task time of $t_m + t_s$.The task time with high-speed AFM is approximately equal to the total manipulation time t_m due to the very high frame rate of the image scan. The manipulation process in this scheme is monitored by the high-speed visual feedback like the manipulation performed in the SEM. The parallel nanomanipulation with high-speed image scan is undoubtedly more efficient than normal-speed image scanning.

In contrast, the task time t^s_{task} in the serial imaging/manipulation operation:

$$t^s_{\text{task}} = t_m + 2t_s. \tag{5.58}$$

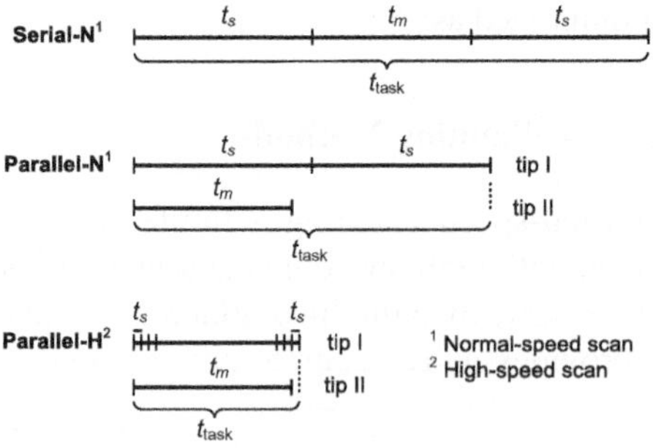

Fig. 5.36 The timing model of serial conventional AFM-based nanomanipulation, and parallel nanomanipulation with normal- and high-speed image scanning.

From Eq. (5.57) and Eq. (5.58), it is found that with the normal-speed image scan the parallel nanomanipulation can save a lot of time, e.g., when $t_m = t_s$, task time of the serial nanomanipulation will increase by 50% compared with the parallel nanomanipulation.

5.5.3 Automated Control of the Parallel Nanomanipulation

5.5.3.1 Nanorobotic Control on Both Tips

Once the manipulation thread is started, the motion control of tip II is based on the position information from the image scan, irrespective of normal- or high-speed image scanning. There is the difference, however, that high-speed image scanning provides dynamic image feedback to track the motion of tip II and the nanoparticle being manipulated, compared with normal-speed image scanning in which the positions of tip II and the nanoparticles to be manipulated are just generated once before the manipulation. As discussed in section III-B, feature recognition from image processing is performed to identify the nano-particle position T_i, in which case control is in the form of manipulation under the optical microscope using the "look-then-move" strategy. In order to determine the kinematic parameters of tip II from image recognition, coordinate transformations are performed between coordinate frames of the image plane and the tips.

Denote ${}^{o}P \in \Re^2$ the coordinate of point P with respect to coordinate frame o; Denote ${}^{m}_{o}R \in \Re^{2\times 2}$ the rotation matrix that represents the orientation of frame o with respect to frame m; Denote ${}^{m}_{o}T \in \Re^2$ the origin location of frame o with respect to frame m

$$ {}^{m}P = {}^{m}_{o}R\,{}^{o}P + {}^{m}_{o}T, $$

$$ {}^{s}P = {}^{s}_{o}R\,{}^{o}P + {}^{s}_{o}T. \tag{5.59} $$

Re-writing Eq. (5.54)

$$ {}^{i}_{s}P = {}^{o}P \tag{5.60} $$

where

$$ s = \begin{bmatrix} 1/r_x & 0 \\ 0 & 1/r_y \end{bmatrix} \begin{bmatrix} l_x & 0 \\ 0 & l_y \end{bmatrix}. $$

From Eq. (5.59) and Eq. (5.60)

$$ {}^{m}P = {}^{m}_{o}R\,{}^{i}_{s}P + {}^{m}_{o}T, $$

$$ {}^{s}P = {}^{s}_{o}R\,{}^{i}_{s}P + {}^{s}_{o}T. \tag{5.61} $$

According to the coordinate frames defined in Fig. 5.32.

$$ {}^{m}_{o}R = \begin{bmatrix} -1 & 0 \\ 0 & 1 \end{bmatrix}, \quad {}^{s}_{o}R = \begin{bmatrix} -1 & 0 \\ 0 & -1 \end{bmatrix}. \tag{5.62} $$

With a known s, the transformations between frame i, frame m and frame s are uniquely determined by ${}^{s}_{o}T$ and ${}^{m}_{o}T$. ${}^{m}_{o}T$ is solved from ${}^{m}P$ and ${}^{i}P$, as tip II position is identified after image processing. Similarly, ${}^{s}_{o}T$ is solved from ${}^{s}P$ and ${}^{i}P$ with known dimensions of the image scan.

5.5.3.2 Control Flow

The control system is developed upon the platform C++, in which two subsidiary threads are respectively established for image scanning and manipulation. The main thread of the task here is used to store the image data and provide the user with an interface with dynamic image display and other real-time data, e.g., force responses from both the cantilevers and position feedback from each of the motion modules.

Figure 5.37 shows the control flow of the parallel nanomanipulation with the normal-speed image scanning. In this flow, before the next manipulation, task planning on a newly updated scanned area is performed when new target(s) and new nanoparticle(s) emerge. The image scan thread is ended with the last scan point; meanwhile, *m_img* is set to zero. The manipulation thread is ended if all the target positions are filled or there are no nanoparticle(s) for the remaining target(s).

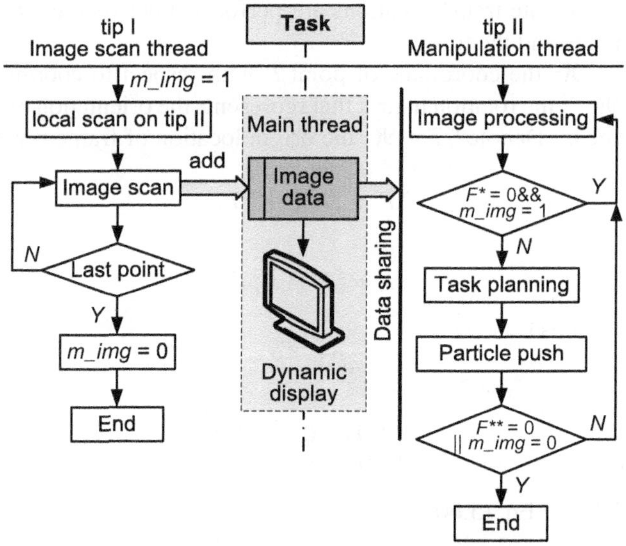

Fig. 5.37 Control flow of automated parallel nanomanipulation with normal-speed image scan. $F^* = 0$ presents a logic: no new target(s) and new nanoparticle(s) emerge; $F^{**} = 0$ presents a logic: no target or no nanoparticle remain. *m_img* is a global variable used as a switch for image scan.

Figure 5.38 shows the control flow with high-speed image scanning, in which image processing is performed for each frame of the scan and the pushing operation are control with visual feedback. The manipulation task is ended if there are no targets or any nanoparticles emerge. *m_map* here is used as a global switch to end the image scan thread, as it is set to zero when the manipulation is ended. The last frame is scanned as a post-manipulation image. Notice that the next frame of the image scan shifts backwards a distance of δ_{t-1} from the last start point, where δ_{t-1} is tip II's displacement on the x-axis of frame s during last frame period t_s.

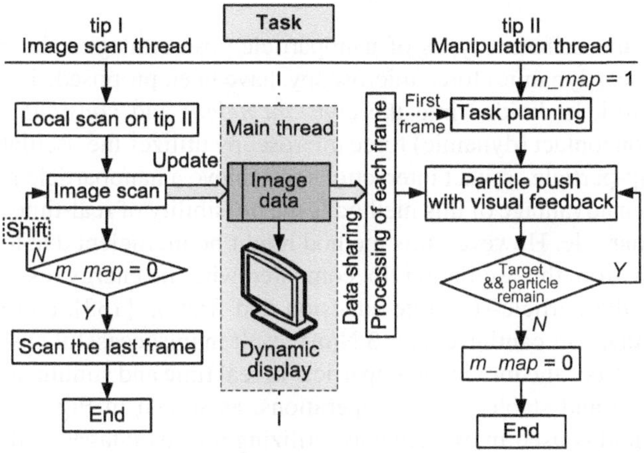

Fig. 5.38 Control flow of automated parallel nanomanipulation with high-speed image scan. *m_map* is a global variable used as a switch for parallel nanomanipulation.

5.5.3.3 Force Sensing during the Nanoparticle Push

In experiments, for a cantilever with a beam length L, beam width w, beam thickness t and tip height l, its normal stiffness k_n and lateral stiffness k_l are calculated by:

$$k_n = \frac{Ewt^3}{6L^2} \quad \text{(N/rad)} \tag{5.63}$$

$$k_l = \frac{Gwt^3}{3L(l+t/2)} \quad \text{(N/rad)} \tag{5.64}$$

where E and G are, respectively, Young's and shear modulus of the cantilever. Assuming the optical lever has the same angular sensitivities on lateral and normal directions, the ratio of the cantilever's normal and lateral force sensitivities S_{ratio} can be obtained from (13) and (14):

$$S_{\text{ratio}} = \frac{k_l}{k_n} = \frac{2LG}{E(l+t/2)}. \tag{5.65}$$

With commercially provided dimensions of the cantilevers used in our system: $L = 240\ \mu$m, $E = 160$ GPa, $G = 58$ GPa, $l = 10\ \mu$m and $t = 3\ \mu$m, a value of the ratio is calculated as $S_{\text{ratio}} = 15.1$. The result indicates that the lateral voltage output, compared with the normal voltage output, is too weak to monitor the nanoparticle pushing operation. That is the reason why the normal voltage rather than the lateral voltage of the optical lever has been commonly used to monitor the pushing/pulling nanomanipulation [114, 115, 116, 117, 121].

5.5.3.4 Controlled Push of Nanoparticles

In the literature, different types of nanoparticle push strategies, including contact and noncontact (dynamic) force microscopy, have been proposed. The former, with pushing control, is divided into the *constant height* and *constant force methods* [114]. The noncontact (dynamic) force microscopy utilizes the oscillating energy of the tip and tip-particle contact interactions to achieve a nanoparticle pushing operation. The main advantage of this method is the possibility of real-time monitoring of the moving particle. However, this method might be inefficient due to very the low power supply from the oscillating tip, compared with the energy dissipation needed to overcome the particle-substrate adhesion and friction [133], and investigations have shown that soft cantilevers (< 3 N/m) result in unsuccessful push [134].

In order to "see" the moving nanoparticle in real time and simultaneously achieve high-efficiency and stable pushing operations, as shown in Fig. 5.39, a dynamic pushing method is used in experiments, utilizing the advantages of the contact and noncontact pushing control methods. With this method, a constant distance d_0 between the tip and substrate is kept during the push. The tip-substrate separation d_0 is more or less equal to the radius R_p of the nanoparticle, which should provides enough maximum effective deflection D_0 to overcome strong static friction at the beginning of the push while the tip slides up the nanoparticle. Once contact with the nanoparticle is established, the amplitude of the oscillating tip reduces form A_0 to zero and the contact push is started. During pushing operations, the tip-nanoparticle contact may be lost, e.g. due to in-plane rotation/spinning of the nanoparticle as the result of a positioning error on the contact point or non-uniform interactions between the nanoparticle-substrate contact. In this case, the tip returns to oscillating with the same amplitude before contact; this, can be used to detect the contact loss automatically to restart the manipulation. This will be discussed in the following part.

During the pushing operation, possible nanoparticle motion modes are sliding, in-plane rotating and out-plane rotating (rolling). Fig. 5.39 shows an example of the interactive forces involved in the pushing-sliding motion of the nanoparticle. For other modes, detailed descriptions can be found in [121]. Let ${}^{t}F_f$ and ${}^{t}F_n$ be the friction force and the normal force applied on the nanoparticle due to the tip-nanoparticle interface, respectively. Therefore, to successfully push-slide the particle, a threshold force F_l^t along the push direction is needed:

$$F_l^t = {}^{s}F_f^{\max} = {}^{t}F_f \sin\theta + {}^{t}F_n \cos\theta, \tag{5.66}$$

$${}^{t}F_f^{\max} \geq F_l \sin\theta - F_n \cos\theta \tag{5.67}$$

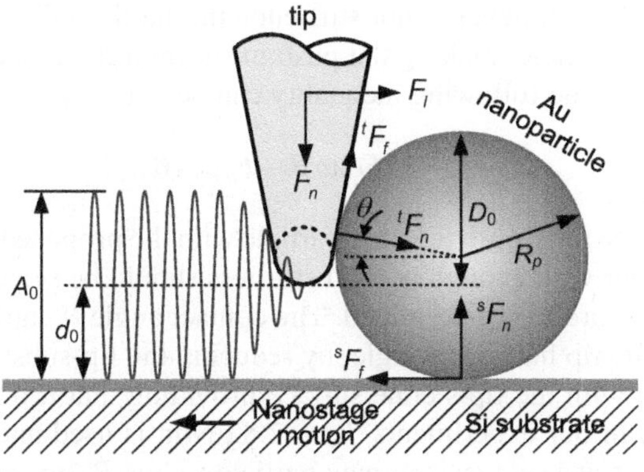

Fig. 5.39 Dynamic pushing schemes and interaction forces of the AFM tip apex and particle during pushing-sliding operation.

where ^sF_f and ^sF_n are the friction force and the repulsive force applied on the nanoparticle from the substrate, respectively; θ is the contact angle and its minimum value θ_{min} is equal to tip's half cone angle if tip-substrate separation d_0 is small enough; F_l and F_n are the lateral and the normal forces applied on the tip from the cantilever. For low speed and quasi-static pushing, it can be deduced that $F_l = {}^sF_f$ and $F_n = {}^sF_n$.

The friction force in nanoscale contact has been known to be proportional to the contact area A. Therefore, ^sF_f can be described by [135]:

$$^sF_f = \tau A \tag{5.68}$$

where the proportional constant τ is the so-called shear strength. Consequently, the force needed to push the nanoparticle is proportional to the particle-substrate contact area A. The contact area depends on the interaction forces of the nanoparticle-substrate contact. According to the JKR model, the contact A of the nanoparticle with a radius R_p can be estimated from [136]:

$$A = \pi \left[\frac{R_p}{K}(F_{ext} + 3\pi R_p \Delta\gamma + \sqrt{6\pi R_p \Delta\gamma F_{ext} + (3\pi R_p \Delta\gamma)^2}) \right]^{2/3} \tag{5.69}$$

where $K = (4/3)[(1-v_p^2)/E_p + (1-v_s^2)/E_s]^{-1}$ is the reduced elastic modulus for the particle-substrate interface, in which v_p and v_s are the Poisson's coefficients, and E_p and E_s are the Young moduli of the nanoparticle and substrate, respectively; $\Delta\gamma$ is the particle-substrate critical energy needed to separate the nanoparticle from the substrate. Measuring $F_{ext} = {}^sF_n = F_n$ and using the known values of R_p, K, $\Delta\gamma$ and shear strength τ, the friction force ^sF_f can be calculated from Eq. (5.68).

When starting the push operation, the tip will slide up a little along the surface of the nanoparticle to obtain a new equilibrium. However, if the nanoparticle is too sticky to push or the cantilever is not stiff enough, the tip will slide up a lot, even slide over the nanoparticle, making the push unsuccessful. As the tip slides on the nanoparticle surface, the following inequality can be obtained:

$$ {}^{t}F_f^{\max} < F_l \sin\theta - F_n \cos\theta. \tag{5.70}$$

In order to successfully push the nanoparticle with the proposed dynamic pushing method, a cantilever with enough normal stiffness should be selected, and a proper contact angle θ be carefully determined. The contact angle θ can be set at its minimum value, i.e., the tip half cone angle, by reducing the tip-substrate separation d_0 to make a contact with the tip's cone surface providing that it does not result in a sticking problem. In our experiments, $k_n = 2.82$ N/m is used and the tip-substrate separation d_0 is set at less than the nanoparticle radius R_p to get a small contact angle θ and enough effective deflection D_0.

5.5.3.5 Manage Abnormity Status

5.5.3.6 Sticking Problem

The commercial AFM tip is very sharp (typically around 10 nm in radius), and that minimizes the adhesion forces between the nanoparticle and the tip. Nonetheless, unlike the nanoparticle push performed in liquid or vacuum [115, 137], the problem of nanoparticle sticking to the tip has been observed during nanoparticle push in ambient conditions, even in image scan with the AFM tapping mode. These phenomena are are mainly induced by the capillary forces [138]. In order to greatly reduce the possibilities of the particle sticking, it is necessary to modify the substrate to increase adhesion to the nanoparticle or functionalize the AFM tip, e.g., with hydrophobic interfaces that will strongly reduce adhesion to the nanoparticle. Moreover, environment control is also an effective method for changing the adhesion forces. In our experiments, the environment is controlled with an ambient temperature of $20 \pm 1^\circ$C and relative humidity of $40 \pm 5\%$. In addition, nanoparticles with a diameter of about 80 nm are used and in contrast, the diameter of the AFM tip apex is much smaller. In this case, tip-particle adhesion forces are relatively lower than the particle-substrate interface. Therefore, the tip-particle sticking problem during the pushing operation can be minimized. The experimental results verified that sticking phenomena seldom occurred.

5.5.3.7 Particle Loss

A positioning error of the contact point and non-uniform interactions of the nanoparticle-substrate contact will result in in-plane rotation/spinning of the nanoparticle, especially over a long-distance pushing operation. Nanoparticle rotation/spinning leads to a push failure because of the contact loss. Second, strong static sticking of some nanoparticles to the substrate requires a stronger pushing force. Otherwise, the tip will slip over the nanoparticle, losing the push. In order to promptly detect the particle loss, as seen in Fig. 5.39, the tip II is kept in oscillation during the whole pushing procedure. By this means, not only can the tip-nanoparticle contact be detected by amplitude fading of the manipulating tip, but also the particle loss when the tip amplitude suddenly increases from zero.

5.5.4 Experimental Results

High-speed image scanning is not yet available on the current system: that forms the basis of our future work on substantially improving nanomanipulation efficiency. However, a parallel image/manipulation task was first performed in air with normal-speed image scanning on the developed PIMM.

5.5.4.1 Sample Preparation

The manipulation samples were prepared by depositing 80 nm gold colloidal particles (Ted Pella, Inc) from an aqueous solution on freshly cleaned silicon substrates using the boiling method [139]. The Si substrate was placed on a hot plate that was set at an appropriate temperature of $100 \sim 110°C$. Drops of the gold nanoparticle solution were released on the substrate and were left to boil. The experiments were carried out in air (an ambient temperature of 20°C and relative humility of 40%) with AFM probes with a spring constant of 2.82 N/m (ATEC-FM, Nanosensors).

5.5.4.2 Image Scanning Rate and Pushing Parameters

The imaging rate of the normal-speed scan and pushing parameters are set as follows:

- A scan frequency of 0.4 Hz on Tip II in tapping mode was used to fully scan the manipulation area of interest, e.g., for an image resolution of 200×200, the overall frame period of the image scan is about eight minutes and twenty seconds.
- The scan length on the end of tip II is set as about 40 nm, which results in a maximum scan height of 94 nm according to (1).
- Nanoparticle pushing velocity was set at about 200 nm/s and a small velocity of 80 nm/s was used for local nanoparticle scanning to achieve fine positioning.
- In the experiments, to improve the convergence rate of the particle repositioning and increase manipulation efficiency, a lower particle releasing accuracy is set at step length of the image scan, representing one pixel on the image plane.
- The tip-substrate separation d_0 during push was kept $10 \sim 20$ nm that depends on the nanoparticle diameter and snap-in boundary of the cantilever due to the adhesion forces.

5.5.4.3 Particle Loss Detection

Figure 5.40 how normal force and amplitude responses in a case of a successful gold nanoparticle (80 nm in diameter) push. When the tip snaps in the nanoparticle at point A, the oscillating amplitude disappears while starting the push. At point B, motion is started when static friction is broken, where the normal force on the cantilever is about $F_n = 24$ nN with a voltage output 18 mV (amplified) of the optical lever and force sensitivity 1.35 μN/V of the cantilever. From Eq. (5.68) and Eq. (5.69), the friction force is estimated as ${}^sF_f = 28$ nN. For calculation, $\Delta\gamma = 0.248$ J/m^2 is computed for SiO$_2$–Au contact in air with $\gamma_{\text{SiO}_2} = 0.16$ J/m^2, $\gamma_{Au} = 1.5$ J/m^2 and $\gamma_{\text{H2O}} = 0.073$ J/m^2; $K = 56$ GPa and $\tau \approx G^*/29 = 30$ MPa [140] with $G^* = [(2 - \nu_{\text{Au}})/G_{\text{Au}} + (2 - \nu_{\text{SiO2}})/G_{\text{SiO2}}]^{-1}$, where $G_{\text{SiO}_2} = 31.4$ GPa, $\nu_{\text{SiO}_2} = 0.17$, $G_{\text{Au}} = 30$ GPa, and $\nu_{\text{Au}} = 0.42$. With this friction force, the calculated (from Eq. (5.65)) and measured maximum lateral voltage output are both less than 3 mV, which is insufficient to monitor the pushing operation.

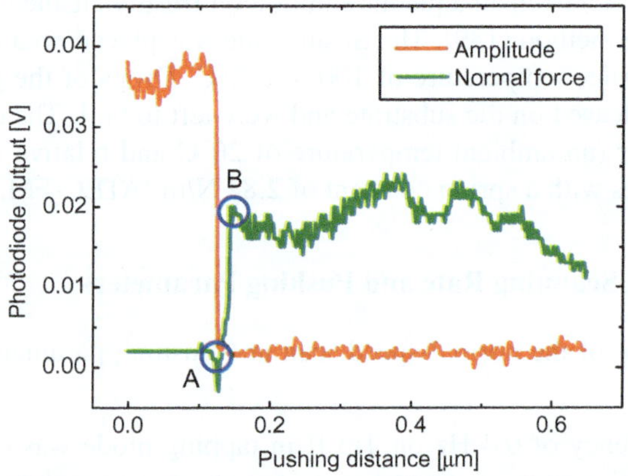

Fig. 5.40 A experimental example of tip II responses during a particle push: point A shows tip II snaps into contact with the nanoparticle, point B is where motion occurs when the static friction is broken.

Figure 5.41 shows a cases of a nanoparticle loss at C point. Once the particle loss is detected, repositioning the nanoparticle is necessary to restart the unfinished push. As at the beginning of the push, a local scan on vertical and horizontal directions is used to relocate the lost nanoparticle with a scan distance of two to three times the nanoparticle's diameter. In the case of strong sticking friction from the substrate, tip-substrate separation d_0 is reduced to increase the maximum re-pushing force, or d_0 is even reduced to zero with normal preloading on the tip II for rare strong sticking cases.

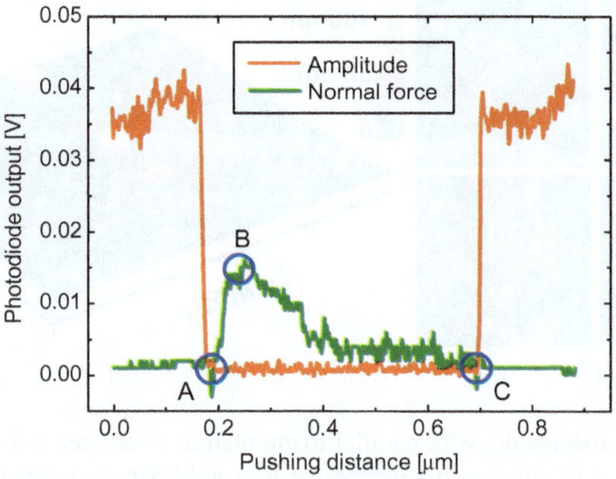

Fig. 5.41 A experimental example of tip II responses during a particle push then loss: point A shows tip II snaps into contact with the nanoparticle, point B is where motion occurs and point C is where the particle loss occurs.

5.5.4.4 Comparisons of Nanoparticle Parallel and Serial Push

With the settings described above, as seen in Fig. 5.42, the system was first tested by a lateral push of four gold nanoparticles; each particle was moved back and forth for four times between its original position and its target. The results show that no particle loss occurred; this might be due to a purely lateral push being used. In long-distance pushing, there is a greater tendency to rotation around the tip, resulting in larger pushing errors. The overall manipulation time, including nanoparticle positioning and pushing, is about $t_m = 320$ s, i.e, $t_m < t_s$, here $t_s = 500$ s. Thus, the task time of this work is $t^p_{\text{task}} = 2t_s = 1000$ s.

For comparison, a serial nanoparticle scan-manipulation-scan task was performed by Tip II. As seen in Fig. 5.43, four gold nanoparticles were also lateral pushed back and forth for four time as the same task in the above experiment. The overall manipulation time in this task is about $t_m = 280$ s. Thus, the task time $t^s_{\text{task}} = 2t_s + t_m = 1280$ s. By comparison, 28% more task time needed in this case than the former parallel nanomanipulation.

5.5.4.5 Parallel Manipulation Results

Figure 5.44 shows another parallel nanomanipulation result in which seven gold nanoparticles with a diameter of 80 nm, emerging in dynamic images I to IV in sequence, were pushed onto corners and the center of a regular hexagon during the period of the image scan. The frame period was about ten minutes. In contrast, the total manipulation time of these seven nanoparticles was less than seven minutes. The results indicate that more complex tasks can be completed during the image

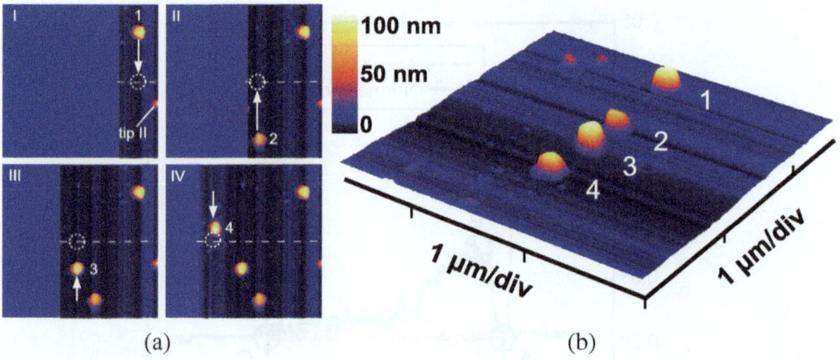

Fig. 5.42 Lateral push results with parallel manipulation (scan area 2.6 μm $\times$ 2.6 μm). (a) Dynamic image I to IV display emergences of four gold nanoparticles (with a diameter of 70 $\sim$ 80 nm). Image processing and task planning were performed every 16 scan lines that present the particle size on the image plane. (b) A post-manipulation image verifys the result. Each particle was pushed back and forth for four cycles between its original position and its target position before ultimately being placed on the target position on the middle line.

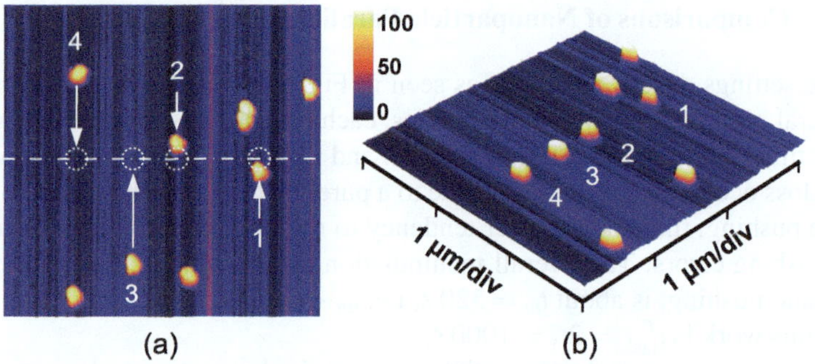

Fig. 5.43 Lateral push results with serial manipulation (scan area 3 μm $\times$ 3 μm). (a) Pre-scanned image of gold nanoparticles (with a diameter around 70 nm). (b) A post-manipulation image verifys that each particle was laterally pushed to its target position distributed on the middle line.

scanning thread, thereby greatly increasing the efficiency of AFM-based nanomanipulation.

In the experiments, image processing and task planning were first perform at the 95th (image I) and the 145th (image II) scan lines, which are, respectively, middle lines between the first and second, and second and third columns of the regular hexagon. After the 210th scan line, which is close to the left side of p_6 (image III), task planning had been arranged to perform once every 10 scan lines that present the particle size on the image plane. Fortunately, particles p_6 and p_7 were discovered in this frame. Thus, 3 times of image processing and task planning were performed during the manipulation.

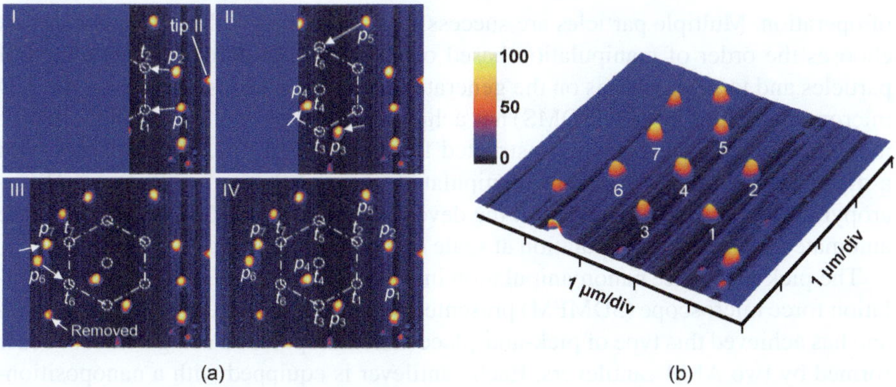

Fig. 5.44 A parallel imaging/manipulation result with normal-speed image scan with tip I (scan area 4.8 μm × 4 μm). (a) Emergences of three parts of target positions and gold nanoparticles (with a diameter of 80 nm) are on three different dynamic displays, namely, image I to III. A full scan is shown in IV. (b) A post-manipulation image shows a regular nano-hexagon with a central nanoparticle is built with seven nanoparticles.

From the experimental records, four particle losses occurred during the whole manipulation procedure, among which one for particle p_5, one for particle p_7 and two for particle p_6. It seems that the inward push along the long axis of the cantilever is more subject to pushing loss by sliding over the particle, which might to due to the large contact angle of about 52° from the side view of the titled tip. However, in this case, particle losses caused by in-plane rotation were seldom observed, which might be due to the contact with the backside of the cantilever's tetrahedral-shape tip. In contrast, the negative contact angle, which is about –22°, makes the forward push along the cantilever more vigorous. Additionally, apart from the final position check, a repositioning operation occurred five times to eliminate the final pushing error before releasing the particles on the target positions, once for particles p_1, p_2 and p_5 to p_7. A particle releasing accuracy is set at 20 nm, the step length of the image scan, representing one pixel on the image plane.

In this section, a high-efficiency automated parallel imaging/manipulation microscopy was presented. Experimental results validated the automated parallel manipulation scheme with normal-speed image scanning.

5.6 Conclusion

The mechanical pick-and-place manipulation at the scale of several microns to the submicron scale is still not well resolved, especially manipulation confined in air. In this chapter, autonomous manipulation of 4.5 μm PS particles with a single AFM probe based on visual information by an optical microscope is demonstrated with a positioning accuracy less than 0.64 μm in average. Smooth linear trajectories are generated for the motion of the probe relative to the substrate for speed and ease of operation. Multiple particles are successfully positioned with a task planner that chooses the order of manipulation based on a metric that defines the blockage of particles and target positions on the generated linear trajectories. Moreover, the 3-D micromanipulation system (3DMS) has achieved such type of pick-and-place micromanipulation using a gripper constructed from two AFM tips. Microspheres with a diameter of 3μm - 4μm were manipulated in air, and as a result, five 3-D micropyramids were constructed with the developed 3DMS. The 3DMS has made the automated 3-D micromanipulation at scale of several micrometers in air feasible.

The pick-and-place nanomanipulation in air is still a challenge. The 3D manipulation force microscope (3DMFM) presented here offer new interesting perspectives and has achieved this type of pick-and-place nanomanipulation using a nanotweezer formed by two AFM cantilevers. Each cantilever is equipped with a nanopositioning device and an optical lever, enabling the 3DMFM has capabilities of image scanning and real-time force sensing during pick-and-place manipulation. In order to validate the manipulation capabilities of the 3DMFM, two types of cone-shaped SiNWs were manipulated and two nanowire crosses were achieved. Moreover, force recording during the pick-and-place manipulation provides valuable data in the understanding of the nanomechanics and nanophysics. As a result, the 3DMFM makes the 3D nanomanipulation, such as constructing 3D nanostructures, building 3D nanodevices, and revealing nanomechanics phenomena more practical.

Finally, a high-efficiency automated parallel imaging/manipulation microscopy was presented. Two automated parallel manipulation schemes with normal- and the high-speed image scanning were introduced. Experimental results validated the automated parallel manipulation scheme with normal-speed image scanning.

References

1. Menciassi, A., Eisinberg, A., Izzo, I., Dario, P.: From "macro" to "micro" manipulation: models and experiments. IEEE Transactions on Mechatronics 9-2, 311–320 (2004) 237
2. Fearing, R.S.: Survey of sticking effects for micro parts handling. In: Proc. of the IEEE/RSJ Intelligent Robots and Systems Conference, pp. 212–217 (1995) 237
3. Sitti, M.: Micro- and Nano-Scale Robotics. In: Proc. of the American Control Conference, pp. 1–8 (2004) 237
4. Gorman, J.J., Dagalakis, N.G.: Probe-Based Micro-Scale Manipulation and Assembly Using Force Feedback. In: 1st Joint Emer. Prep. & Response/Robotic & Remote Sys. Top. Mtg., pp. 621–628 (2006) 238
5. Burkle, A., Schmoeckel, F., Worn, H., Amavasai, B.P., Caparrelli, F., Travis, J.R.: A versatile vision system for micromanipulation tasks. In: Proc. of the Int. Conference on Multisensor Fusion and Integration for Intelligent Systems, pp. 271–276 (2001)
6. Ahn, Y., Ono, T., Esashi, M.: Si multiprobes integrated with lateral actuators for independent scanning probe applications. J. Micromech. Microeng. 15, 1224–1229 (2005)
7. Dionnet, F., Haliyo, D.S., Regnier, S.: Autonomous Micromanipulation Using a New Strategy of Accurate Release by Rolling. In: Proc. of the IEEE/International Conference on Robotics and Automation (2004)
8. Kasaya, T., Miyazakia, H.T., Saito, S., Koyano, K., Yamaura, T., Sato, T.: Image-based autonomous micromanipulation system for arrangement of spheres in a scanning electron microscope. Review of Scientific Instruments 75-6, 2033–2042 (2004)

9. Saito, S., Motokado, T., Obata, K.J., Takahashi, K.: Capillary force with a concave probe-tip for micromanipulation. Applied Physics Letters 87–234103 (2005)
10. Castelino, K., Satyanarayana, S., Sitti, M.: Manufacturing of Two and Three-Dimensional Micro/Nanostructures by Integrating Optical Tweezers with Chemical Assembly. Robotica 23, 435–439 (2005)
11. Zheng, L., Li, S., Burke, P.J., Brody, J.P.: Towards Single Molecule Manipulation with Dielectrophoresis Using Nanoelectrodes. In: Proc. of the IEEE Conference on Nanotechnology, pp. 437–440 (2003)
12. Subramanian, A., Vikramaditya, B., Nelson, B.J., Bell, D., Dong, L.: Dielectrophoretic Micro/Nanoassembly with Microtweezers and Nanoelectrodes. In: Proc. of the International Conference on Advanced Robotics, pp. 208–215 (2005)
13. Bohringer, K.F., Goldberg, K., Cohn, M., Howe, R., Pisano, A.: Parallel Microassembly with Electrostatic Force Fields. In: Proc. of the IEEE International Conference on Robotics & Automation, pp. 1204–1211 (1998)
14. Requicha, A.A.G., Meltzer, S., Resch, R., Lewis, D., Koel, B.E., Thompson, M.E.: Layered Nanoassembly of Three-Dimensional Structures. In: Proc. of the IEEE International Conference on Robotics & Automation, pp. 3408–3411 (2001)
15. Utkin, V.: Variable structure systems with sliding modes. IEEE Transactions on Automatic Control 22-2, 212–222 (1977) 238, 243
16. Onal, C.D., Sabanovic, A.: Plant Behaviour Dictation using a Sliding Mode Model Reference Controller. In: International Workshop on Advanced Motion Control (AMC), pp. 243–248 (March 2006) 238, 242
17. Yilmaz, B., Unel, M., Sabanovic, A.: Rigid and Affine Motion Estimation in Vision Using Sliding Mode Observers. In: IEEE International Symposium on Industrial Electronics (ISIE), pp. 31–36 (June 2005) 242
18. Sabanovic, A., Fridman, L.M., Spurgeon, S.: Variable Structure Systems: from Principles to Implementation. IEE Control Series, vol. 66. The Institute of Electrical Engineers, London (2004) 238
19. Lee, E., Hong, H.S.: An Integrated System of Microcantilever Arrays with Carbon Nanotube Tips for Bio/Nano Analysis: Design and Control. In: Proc. of the IEEE Conference on Automation Science and Engineering, pp. 113–117 (2005)
20. Pozidis, H., Haberle, W., Wiesmann, D., Drechsler, U., Despont, M., Albrecht, T.R., Eleftheriou, E.: Demonstration of Thermomechanical Recording at 641 Gbit/in^2. IEEE Transactions on Magnetics 40-4, 2531–2536 (2004)
21. Sitti, M.: Atomic Force Microscope Probe Based Controlled Pushing for Nanotribological Characterization. IEEE/ASME Transactions on Mechatronics 9-2, 343–349 (2004) 245, 246, 254
22. Venture, G., Haliyo, D.S., Regnier, S., Micaelli, A.: Force-feedback micromanipulation with unconditionally stable coupling. In: Proc. of the IEEE/International Conference on Intelligent Robots and Systems, pp. 784–789 (2005)
23. Ferreira, A., Cassier, C., Hirai, S.: Automatic Microassembly System Assisted by Vision Servoing and Virtual Reality. IEEE/ASME Transactions on Mechatronics 9-2, 321–333 (2004)
24. Shen, Y., Xi, N., Li, W.J., Tan, J.: A High Sensitivity Force Sensor for Microassembly: Design and Experiments. In: Proc. of the IEEE/ASME International Conference on Advanced Intelligent Mechatronics, pp. 703–708 (2003)
25. Tafazzoli, A., Pawashe, C., Sitti, M.: Force-Controlled Microcontact Printing using Microassembled Particle Templates. In: Proc. of the IEEE Robotics and Automation Conf., pp. 263–268 (May 2006) 254

26. Duda, R., Hart, P.: Use of the Hough Transform to Detect Lines and Curves in Pictures. Communications of the ACM 15, 11–15 (1972) 240
27. Ballard, D.H.: Generalizing the Hough transform to detect arbitrary shapes. Pattern Recognition 13, 111–122 (1981) 241
28. Xu, L., Oja, E., Kultanan, P.: A new curve detection method: randomized Hough transform (RHT). Pattern Recognition Letters 11-5, 331–338 (1990) 241
29. Rad, A.A., Faez, K., Qaragozlou, N.: Fast Circle Detection Using Gradient Pair Vectors. In: Sun, C., Talbot, H., Ourselin, S., Adriaansen, T. (eds.) Proc. of the VIIth Digital Image Computing: Techniques and Applications, December 10-12 (2003) 241
30. Bhushan, B.: Handbook of Micro/Nano Tribology, 2nd edn. CRC Press, Boca Raton (1999) 246
31. Saito, S., Miyazaki, H.T., Sato, T., Takahashi, K.: Kinematics of mechanical and adhesional micromanipulation under a scanning electron microscope. Journal of Applied Physics 92-9, 5140–5149 (2002) 246
32. Dedkov, G.V.: Friction on the nanoscale: new physical mechanisms. Materials Letters 38, 360–366 (1999) 246
33. `http://www.andrew.cmu.edu/user/cdonal/uassembly.avi` 253
34. Dechev, N., Cleghorn, W.L., Mills, J.K.: Microassembly of 3-D microstructures using a compliant, passive microgripper. J. Microelectromech. Syst. 13(2), 176–189 (2004)
35. Xie, H., Rong, W.B., Sun, L.N.: A flexible experimental system for complex microassembly under microscale force and vision-based control. Int. J. Optome-chatro 1, 80–102 (2007)
36. Yang, G., Gaines, J.A., Nelson, B.J.: A supervisory wafer-level 3D microassembly system for hybrid MEMS fabrication. J. Intell. Robot. Syst. 37, 43–68 (2003)
37. Aokil, K., Miyazakl, H.T., Hirayama, H., Inoshita, K., Baba, T., Sakoda, K., Shinya, N., Aoyagl, Y.: Microassembly of semiconductor three dimensional photonic crystals. Nature Material 2, 117–121 (2003)
38. Wang, W.H., Liu, X.Y., Gelinas, D., Brian Ciruna, B., Sun, Y.: A fully automated robotic system for microinjection of zebrafish embryos. Plos ONE 9, e862 (2007)
39. Sun, Y., Nelson, B.J.: Journal of Robotics Research 21, 861–868 (2002)
40. Lu, Z., Chen, P.C.Y., Nam, J., Ge, R.W., Lin, W.: A micromanipulation system with dynamic force-feedback for automatic batch microinjection. J. Micromech. Microeng. 17, 314–321 (2007)
41. Gauthier, M., Piat, E.: Control of a particular micro-macro positioning system applied to cell micromanipulation. IEEE T. Autom. Sci. Eng. 3(3), 264–271 (2006)
42. Marcy, Y., Prost, J., Carlier, M.F., Sykes, C.: Forces generated during actin-based propulsion: A direct measurement by micromanipulation. P. Natl. Acda. Sci. USA 101, 5992–5997 (2004)
43. Onal, C.D., Sitti, M.: Visual servoing-based autono-mous 2-D manipulation of microparticles using a nano-probe. IEEE T. Contr. Syst. Tech. 15(5), 842–852 (2007) 259
44. Nguyen, N.T., Ho, S.S., Low, C.L.N.: A polymeric microgripper with integrated thermal actuators. J. Micromech. Microeng. 14, 969–974 (2004)
45. Millet, O., Bernardoni, P., Régnier, S., Bidaud, P., Tsitsiris, E., Collard, D., Buchaillot, L.: Electrostatic actuated micro gripper using an amplification mechanism. Sensor Actuat A: Phys. 114, 371–378 (2004)
46. Clévy, C., Hubert, A., Agnus, J., Chaillet, N.: A micro-manipulation cell including a tool changer. J. Micromech. Microeng. 15, S292-S301 (2005) 256
47. Pérez, R., Agnus, J., Clévy, C., Hubert, A., Chaillet, N.: Modeling, fabrication, and validation of a high-performance 2-DoF piezoactuator for microma-nipulation. IEEE-ASME T. Mechatro. 10(2), 161–171 (2005)

48. Carrozza, M.C., Eisinberg, A., Menciassi, A., Campolo, D., Micera, S., Dario, P.: Towards a force-controlled microgripper for assembling biomedical microdevices. J. Micromech. Microeng. 10, 271–276 (2000)
49. Neild, A., Oberti, S., Beyeler, F., Dual, J., Nelson, B.J.: A micro-particle positioning technique combining an ultrasonic manipulator and a microgripper. J. Micromech. Microeng. 16, 1562–1571 (2006)
50. Walle, B.L., Gauthier, M., Chaillet, N.: Principle of a submerged freeze gripper for microassembly. IEEE T. Robot. 24(4), 897–902 (2008)
51. Kim, K.Y., Liu, X.Y., Zhang, Y., Sun, Y.: Nanonewton force-controlled manipulation of biological cells using a monolithic MEMS microgripper with two-axis force feedback. J. Micromech. Microeng. 18, 055013 (2008)
52. Haliyo, D.S., Dionnet, F., Régnier, S.: Controlled rolling of microobjects for autonomous manipulation. J. Micromechatro 3(2), 75–101 (2006)
53. Driesen, W., Varidel, T., Régnier, S., Breguet, J.M.: Micromanipulation by adhesion with two collaborating mobile micro robots. J. Micromech. Microeng. 15, S259–S267 (2005)
54. Park, I.-Y., Sung, S.-Y., Lee, J.-H., Lee, Y.-G.: Manufacturing micro-scale structures by an optical tweezers system controlled by five finger tips. J. Micromech. Micro-eng. 17, 82–89 (2007)
55. Menciassi, A., Eisinberg, A., Izzo, I., Dario, P.: From "macro" to "micro" manipulation: models and experiments. IEEE-ASME T. Mechatro 9(2), 311–320 (2004) 254, 257
56. Lambert, P., Régnier, S.: Surface and contact forces models within the framework of microassembly. J. Micromechatro 3(2), 123–157 (2006) 254
57. Sitti, M.: Microscale and nanoscale robotics systems-characteristics, state of the art, and grand challenges. IEEE Robot Autom. Mag. 14(1), 53–60 (2007) 254, 257
58. Xie, H., Rong, W.B., Sun, L.N.: Construction and evaluation of a wavelet-based Focus measure for microscopy. Imaging Microsc. Res. Techniq. 70, 987–995 (2007) 254
59. Liu, X.Y., Wang, W.H., Sun, Y.: Dynamic evaluation of autofocusing for automated microscopic analysis of blood smear and pap smear. J. Microsc. 227, 15–223 (2007) 254
60. Xie, H., Rong, W.B., Sun, L.N., Chen, W.: Image fusion and 3-D surface reconstruction of microparts using complex valued wavelet transforms. In: IEEE Intl. Conf. Image Process, Atlanta GA, pp. 2137–2140 (2006)
61. Xie, H., Vitard, J., Haliyo, S., Régnier, S.: Optical lever calibration in atomic force microscope with a mechanical lever. Rev. Sci. Instrum. 79, 096101 (2008) 254, 256
62. Xie, H., Rakotondrabe, M., Régnier, S.: Characterizing piezoscanner hysteresis and creep using optical levers and a reference nanopositioning stage. Rev. Sci. Instrum. 80, 46102 (2009) 256
63. Vögeli, B., von Känel, H.: AFM-study of sticking effects for microparts handling. Wear 238, 20–24 (2000) 257
64. Varenberg, M., Etsion, I., Halperin, G.: An improved wedge calibration method for lateral force in atomic force microscopy. Rev. Sci. Instrum. 74, 3362–3367 (2003) 262
65. Xie, H., Vitard, J., Haliyo, S., Régnier, S.: Enhanced accuracy of force application for AFM nanomanipulation using nonlinear calibration of optical Levers. IEEE Sensors 8(8), 1478–1485 (2008) 263, 266
66. Fukuda, T., Arai, F., Dong, L.X.: Assembly of nanode-vices with carbon nanotubes through nanorobotic manipulations. Proc. IEEE 91, 1803–1818 (2003) 267
67. Dong, L.X., Arai, F., Fukuda, T.: Electron-beam-induced deposition with carbon nanotube emitters. Appl. Phys. Lett. 81, 1919–1921 (2002) 267

68. Dong, L.X., Arai, F., Fukuda, T.: Destructive construc-tions of nanostructures with carbon nanotubes through nano-robotic manipulation. IEEE/ASME Trans. Mechatron 9, 350–357 (2004) 267
69. Molhave, K., Wich, T., Kortschack, A., Boggild, P.: Pick-and-place nanomanipulation using microfabricated grippers. Nanotechnology 17, 2434–2441 (2006) 267
70. Dong, L.X., Tao, X.Y., Zhang, L., Nelson, B.J., Zhang, X.B.: Nanorobotic spot welding: controlled metal deposition with attogram precision from copper-filled carbon nanotubes. Nano. Lett. 7, 58–63 (2007) 267
71. Leach, J., Sinclair, G., Jordan, P., Courtial, J., Padgett, M.J., Cooper, J., Laczik, Z.J.: 3D manipulation of particles into crystal structures using holographic optical tweezers. Optical Express 12, 220–226 (2004) 267
72. Yu, T., Cheong, F.C., Sow, C.H.: The manipulation and assembly of CuO nanorods with line optical tweezers. Nanotechnology 15, 1732–1736 (2004) 267
73. Bosanac, L., Aabo, T., Bendix, P.M., Oddershede, L.B.: Efficient optical trapping and visualization of silver nanoparticles. Nano Letters 8, 1486–1491 (2008) 267
74. Rollot, Y., Régnier, S., Guinot, J.C.: Simulation of microma-nipulations: Adhesion forces and specific dynamic models Inte. J. Adhes. Adhes. 19, 35–48 (1999) 267
75. Falvo, M.R., Taylor, R.M.I., Helser, A., Chi, V., Brooks, F.P.J., Washburn, S., Superfine, R.: Nanometre-scale rolling and sliding of carbon nanotubes. Nature 397, 236–238 (1999) 267
76. Baur, C., Bugacov, A., Koel, B.E., Madhukar, A., Montoya, N., Ramachandran, T.R., Requicha, A.A.G., Resch, R., Will, P.: Nanoparticle manipulation by mechanical pushing: underlying phenomena and real-time monitoring. Nanotechnology 9, 360–364 (1998) 267
77. Resch, R., Lewis, D., Meltzer, S., Montoya, N., Koel, B.E., Madhukar, A., Requicha, A.A.G., Will, P.: Manipulation of gold nanoparticles in liquid environments using scanning force microscopy. Ultramicroscopy 82, 135–139 (2000) 267
78. Sitti, M., Horiguchi, S., Hashimoto, H.: Controlled pushing of nanoparticles: modeling and experiments. IEEE/ASME Trans. Mechatron. 5, 199–211 (2000) 267
79. Yun, Y.J., Ah, C.S., Kim, S., Yun, W.S., Park, B.C., Ha, D.H.: Manipulation of freestanding Au nanogears using an atomic force microscope. Nanotechnology 18, 505304 (2007) 267
80. Whitman, L.J., Stroscio, J.A., Dragoset, R.A., Cellota, R.J.: Manipulation of adsorbed atoms and creation of new structures on room-temperature surfaces with a scanning tunneling microscope. Science 251, 1206–1210 (1991) 267
81. Eigler, D.M., Schweizer, E.K.: Positioning single atoms with a scanning tunneling microscope. Nature 344, 524–526 (1990) 267
82. Avouris, P.: Manipulation of matter at the atomic and molecular-levels Accounts. Chem. Res. 28, 95–102 (1995) 267
83. Kim, P., Lieber, C.M.: Nanotube nanotweezers. Science 286, 2148–2150 (1999) 267
84. Xie, H., Vitard, J., Haliyo, S., Régnier, S.: Optical lever calibration in atomic force microscope with a mechanical lever. Rev. Sci. Instrum. 79, 96101 (2008)
85. Krejci, P., Kuhnen, K.: Inverse control of systems with hysteresis and creep. IEE Proc.-Control Theory Appl. 148, 185–192 (2001)
86. Jung, H., Gweon, D.: Creep characteristics of piezo-electric actuators. Rev. Sci. Instru. 71, 1896 (2000)
87. Varenberg, M., Etsion, I., Halperin, G.: An improved wedge calibration method for lateral force in atomic force microscopy. Rev. Sci. Instrum. 74, 3362–3367 (2003) 271
88. Xie, H., Vitard, J., Haliyo, S., Régnier, S.: Enhanced accuracy of force application for AFM nanomanipulation using nonlinear calibration of optical Levers. IEEE Sensors J. 8, 1478–1485 (2008) 271

89. Xie, H., Vitard, J., Haliyo, S., Régnier, S., Boukallel, M.: Calibration of lateral force measurements in atomic force microscopy with a piezoresistive force sensor. Rev. Sci. Instrum. 79, 33708 (2008) 271
90. Carpick, R.W., Ogletree, D.F., Salmeron, M.: Lateral stiffness: A new nanomechanical measurement for the determination of shear strengths with friction force micros-copy. Appl. Phy.s Lett. 70, 1548–1550 (1997) 272
91. Lantz, M.A., O' shea, S.J., Hoole, A.C.F., Welland, M.E.: Lateral stiffness of the tip and tip-sample contact in frictional force microscopy. Appl. Phys. Lett. 70, 970 (1997) 273
92. Lee, H.W., Kim, S.H., Kwaka, Y.K., Han, C.S.: Nanoscale fabrication of a single multiwalled carbon nanotube attached atomic force microscope tip using an electric field. Rev. Sci. Instrum. 76, 046108 (2005) 273
93. Eigler, D.M., Schneizer, E.K.: Nature 344, 524 (1990)
94. Junno, T., Carlsson, S.B., Xu, H.Q., Montelius, L., Samuelson, L.: Appl. Phys. Lett. 72, 548 (1998)
95. Martin, M., Roschier, L., Hakonen, P., Parts, U., Paalanen, M., Schleicher, B., Kauppinen, E.I.: Appl. Phys. Lett. 73, 1505 (1998)
96. Resch, R., Lewis, D., Meltzer, S., Montoya, N., Koel, B.E., Madhukar, A., Requicha, A.A.G., Will, P.: Ultramicroscopy 82 135 (2000)
97. Dietzel, D., Mönninghoff, T., Jansen, L., Fuchs, H., Ritter, C., Schwarz, U.D., Schirmeisen, A.: J. App. Phys. 102, 084306 (2007)
98. Tong, L.M., Zhu, T., Liu, Z.F.: Appl. Phys. Lett. 92, 023109 (2008)
99. Li, X.D., Gao, H.S., Murphy, C.J., Caswell, K.K.: Nano Lett. 3, 1945 (2003)
100. Samitsu, S., Shimomura, T., Ito, K., Fujimori, M., Heike, S., Hashi-zume, T.: Appl. Phys. Lett. 86, 233103 (2005)
101. Falvo, M.R., Taylor, R.M., Helser, A., Chi, V., Brooks Jr. F.P., Washburn, S., Superne, R.: Nature 397, 236 (1999)
102. Prior, M., Makarovski, A., Finkelstein, G.: Appl. Phys. Lett. 91, 053112 (2007)
103. Albrecht, P.M., Lyding, J.W.: Small 3, 146 (2007)
104. Dong, L.X., Nelson, B.J.: IEEE Robot. Automat. Mag. 14, 111 (2007) 279
105. Kim, S.G., Sitti, M.: IEEE Trans. Autom. Sci. Eng. 3, 240 (2006) 279
106. Li, G.Y., Xi, N., Chen, H.P., Pomeroy, C., Prokos, M.: IEEE T. Nanotechnology 4, 605 (2005) 279
107. Fantner, G.E., Hegarty, P., Kindt, J.H., Schitter, G., Cidade, G.A.G., Hansma, P.K.: Rev. Sci. Instrum. 76, 026118 (2005) 279
108. Hansma, P.K., Schitter, G., Fantner, G.E., Prater, C.: Science, 601 (2006) 279
109. Seo, Y., Choi, C.S., Han, S.H., Han, S.-J.: Rev. Sci. Instrum. 79, 103703 (2008) 279
110. Howard-Knight, J.P., Hobbs, J.K.: Appl. Phys. Lett. 93, 104101, (2008) 279
111. Krejci, P., Kuhnen, K.: IEE Proc.-Control Theory Appl. 148, 185 (2001) 279
112. Strus, M.C., Zalamea, L., Raman, A., Pipes, R.B., Nguyen, C.V., Stach, E.A.: Nano Lett. 8, 544 (2008)
113. Eigler, D.M., Schweizer, E.K.: Positioning single atoms with a scanning tunneling microscope. Nature 344(6266), 524–526 (1990)
114. Sitti, M., Hashimoto, H.: Controlled pushing of nanoparticles: modeling and experiments. IEEE/ASME Trans. Mechatron. 5(2), 199–211 (2000) 296
115. Resch, R., Lewis, D., Meltzer, S., Montoya, N., Koel, B.E., Madhukar, A., Requicha, A.A.G., Will, P.: Manipulation of gold nanoparticles in liquid environments using scanning force microscopy. Ultramicroscopy 82(1-4), 135–139 (2000) 296, 298
116. Albrecht, P.M., Lyding, J.W.: Lateral manipulation of single-walled carbon nanotubes on H-passivated Si(100) surfaces with an ultra-high-vacuum scanning tunneling microscope. Small 3(1), 146–152 (2007) 296

117. Tranvouez, E., Orieux, A., Boer-Duchemin, E., Devillers, C.H., Huc, V., Comtet, G., Dujardin, G.: Manipulation of cadmium selenide nanorods with an atomic force microscope. Nanotechnology 20(16), 165304 (2009) 296
118. Li, X., Gao, H., Murphy, C.J., Caswell, K.K.: Nanoindentation of silver sanowires. Nano Lett. 3(11), 1495–1498 (2003)
119. Xu, B., Tao, N.J.: Measurement of single-molecule resistance by repeated formation of molecular junctions. Science 301(5637), 1221–1223 (2003)
120. Grill, L., Rieder, K., Moresco, F., Stojkovic, S., Gourdon, A., Joachim, C.: Exploring the interatomic forces between tip and single molecules during STM manipulation. Nano Lett. 6(12), 2685–2689 (2006)
121. Sitti, M.: Atomic force microscope probe based controlled pushing for nanotribological characterization. IEEE/ASME Trans. Mechatron. 9(2), 343–349 (2004) 296
122. Palacio, M., Bhushan, B.: A nanoscale friction investigation during the manipulation of nanoparticles in controlled environments. Nanotechnology 19(31), 315710 (2008)
123. Bordag, M., Ribayrol, A., Conache, G., Fröberg, L.E., Gray, S., Samuelson, L., Montelius, L., Pettersson, H.: Shear Stress measurements on InAs nanowires by AFM manipulation. Small 3(8), 1398–1401 (2007)
124. Dong, L.X., Arai, F., Fukuda, T.: Destructive constructions of nanostructures with carbon nanotubes through nanorobotic manipulation. IEEE/ASME Trans. Mechatron. 9(2), 350–357 (2004)
125. Li, G.Y., Xi, N., Chen, H.P., Pomeroy, C., Prokos, M.: 'Videolized' atomic force microscopy for interactive nanomanipulation and nanoassembly. IEEE Trans. Nanotechnol. 4(5), 605–615 (2005)
126. Vogl, W., Ma, B.K.-L., Sitti, M.: Augmented reality user interface for an atomic force microscope based nanorobotic system. IEEE Trans. Nanotechnol. 5(4), 397–406 (2006)
127. Paloczi, G.T., Smith, B.L., Hansma, P.K., Walters, D.A., Wendman, M.A.: Rapid imaging of calcite crystal growth using atomic force microscopy with small cantilevers. Appl. Phys. Lett. 73(12), 1658–1660 (1998)
128. Kodera, N., Yamashita, H., Ando, T.: Active damping of the scanner for high-speed atomic force microscopy. Rev. Sci. Instrum. 76(5), 053708 (2005)
129. Schitter, G., Åström, K.J., DeMartini, B.E., Thurner, P.J., Turner, K.L., Hansma, P.K.: Design and modeling of a high-speed AFM-Scanner. IEEE Trans. Cont. Syst. Technol. 15(5), 906–915 (2007)
130. Xie, H., Haliyo, D.S., Régnier, S.: Parallel imaging/manipulation force microscopy. Appl. Phys. Lett. 94(15), 153106 (2009)
131. Russ, J.C.: The image processing handbook, 5th edn. CRC Press, Boca Raton (2006) 287
132. Onal, C.D., Sitti, M.: Visual servoing-based autonomous 2-D manipulation of microparticles using a nanoprobe. IEEE Trans. Cont. Syst. Tech. 15(5), 842–852 (2007) 290
133. Martin, M., Roschier, L., Hakonen, P., Parts, Ü., Paalanen, M., Schleicher, B., Kauppinen, E.I.: Manipulation of Ag nanoparticles utilizing noncontact atomic force microscopy. Appl. Phys. Lett. 73(11), 1505–1507 (1998) 296
134. Resch, R., Bugacov, A., Baur, C., Koel, B.E., Madhukar, A., Requicha, A.A.G., Will, P.: Manipulation of nanoparticles using dynamic force microscopy: simulation and experiments. Appl. Phys A: Mater. Sci. Proces. 67(3), 265–271 (1998) 296
135. Bhushan, B.: Introduction to tribology. Wiley, New York (2002) 297
136. Johnson, K.L., Kendall, K., Roberts, A.D.: Surface energy and the contact of elastic solids. Proc. R. Soc. Lond. A 324(1558), 301–313 (1971) 297

137. Dietzel, D., Monninghoff, T., Jansen, L., Fuchs, H., Ritter, C., Schwarz, U.D., Schirmeisen, A.: Interfacial friction obtained by lateral manipulation of nanoparticles using atomic force microscopy techniques. J. Appl. Phys. 102(8), 084306 (2007) 298
138. Mougin, K., Gnecco, E., Rao, A., Cuberes, M.T., Jayaraman, S., McFarland, E.W., Haidara, H., Meyer, E.: Manipulation of gold nanopar-ticles: influence of surface chemistry, temperature, and environment (vacuum versus ambient atmosphere). Langmuir 24(4), 1577–1581 (2008) 298
139. Lee, K., Duchamp, M., Kulik, G., Magrez, A., Seo, J.W., Jeney, S., Kulik, A.J., Forro, L.: Uniformly dispersed deposition of colloidal nanoparticles and nanowires by boiling. Appl. Phys. Lett. 91(17), 173112 (2007) 299
140. Dedkov, G.: Friction on the nanoscale: New physical mechanisms. Materials Letters 38(5), 360–366 (1999) 300
141. Mokaberi, B., Requicha, A.A.G.: Drift compensation for automatic nanomanipulation with scanning probe microscopes. IEEE Trans. Autom. Sci. Eng. 3(3), 199–207 (2006)
142. Xie, H., Haliyo, D.S., Régnier, S.: A versatile atomic force microscope for three-dimensional nanomanipulation and nanoassembly. Nanotechnology 20(21), 215301 (2009)

Chapter 6
Applications of AFM Based Nanorobotic Systems

This chapter deals with potential applications of atomic force microscopy based robotic system. In a first part, a flexible robotic system developed for multi-scale manipulation and assembly from nanoscale to microscale is presented. This system is based on the principle of atomic force microscopy and comprises two individually functionalized cantilevers. After reconfiguration, the robotic system could be used for pick-and-place manipulation from nanoscale to the scale of several micrometers, as well as parallel imaging/nanomanipulation. Flexibilities and manipulation capabilities of the developed system are validated by pick-and-place manipulation of microspheres and silicon nanowires to build three-dimensional micro/nano structures in ambient conditions. Moreover, the capability of parallel nanomanipulation is certified by high-efficiency fabrication of a two-dimensional pattern with nanoparticles. Complicated micro/nano manipulation and assembly can be reliably and efficiently performed using the proposed flexible robotic system.

The second part of this chapter describes a method for in situ peeling force measurement of one-dimensional nanostructures using a dual-probe nanotweezer, which is developed on the principle of the robotized atomic force microscope. Benefiting from capabilities of image scanning and accurate force sensing, the nanotweezer is capable of positioning one-dimensional nanostructures deposited on a surface, and then performing in situ peeling tests with pick-and-place operations at different peeling locations of interest along a selected nanostructure. In experiments, nanoscale peeling of silicon nanowires (SiNW) on a silicon substrate are studied. Peeling locations at the end and middle of the SiNW are tested and results indicate that approximate peeling energies are needed.

6.1 Flexible Robotic System for Multi-scale Applications of Micro/NanoscaleManipulation and Assembly

We present in this part a concept to make a prototype of an AFM-based flexible robotic system (AFM-FRS) which is equipped with two collaborative cantilevers. By reconfiguring the modular hardware and software of the AFM-FRS, three

H. Xie et al.: Atomic Force Microscopy Based Nanorobotics, STAR 71, pp. 313–342.
springerlink.com

different configurations can be obtained. With different manipulation strategies and protocols, two of the configurations can be respectively used to complete pick-and-place manipulations from the microscale of several microns to the nanoscale that are still challenges. In addition, the third system configuration of the AFM-FRS can be easily used for high-efficiency nanomanipulation by assigning imaging and 2-D pushing/pulling tasks in parallel to the individually actuated cantilevers. Automation on different applications could be also implemented benefiting from accurate force sensing and positioning of the system.

The three kinematic configurations and necessities to reconfigure the AFM-FRS for multi-scale applications are detailed. For automated parallel nanomanipulation, a dynamic pushing scheme with amplitude feedback, rather than the weak lateral force signal and the unstable normal force signal of the cantilever, is introduced for stable particle loss detection, and the shortest path solution was used for trajectory planning. Moreover, flexibilities and manipulation capabilities of the AFM-FRS are validated by performing three new experiments, including 3-D microsphere manipulation, nanowire pick-and-place and nanoparticle parallel nanomanipulation on each corresponding system configuration. All the experiments, including tasks of manipulation and system reconfiguring, were carried out in a regular workday.

This section is organized as follows. Subsection I introduces the prototype and experimental setup of the AFM-FRS. Various system configurations of the AFM-FRS are presented in subsection II to subsection IV. In subsection V, pick-and-place manipulation of microspheres and nanowires for building 3-D micro/nano structures are performed using the AFM-FRS. Performance of the AFM-FRS was also validated by parallel imaging/manipulation of nanoparticles to form a nanopattern with a commonly used pushing manipulation.

6.1.1 AFM-Based Flexible Robotic System

6.1.2 AFM-FRS Setup

Figure 6.1(a) shows the AFM-FRS system setup, which is in the configuration for nanoscale pick-and-place. The system is equipped with an optical microscope and two sets of modules commonly used in a conventional AFM, mainly including two AFM cantilevers (namely, tip I and tip II, ATEC–FM Nanosensors, see Fig. 6.1(b)), two sets of nanopositioning devices and optical levers. The motion modules include an open-loop X-Y-Z piezoscanner (PI P-153.10H), an X-Y-Z closed-loop nanostage (MCL Nano-Bio2M on the X- and Y-axis, PI P–732. ZC on the Z-axis), an X-Y-Z motorized stage and an X-Y-Z manual stage. Detailed specifications of the motion modules are summarized in Table 6.1. Figure 6.1(c) shows an optical microscope image of the collaborating tips in microsphere grasping mode.

A data acquisition (NI 6289) card is used for high-speed (500$\sim$800 Hz of sampling frequency for force and 600 kHz for amplitude) capture of the photodiode voltage output to estimate deflections on both tips induced by force loading or resonant oscillation. A multi-thread planning and control system based on the C++

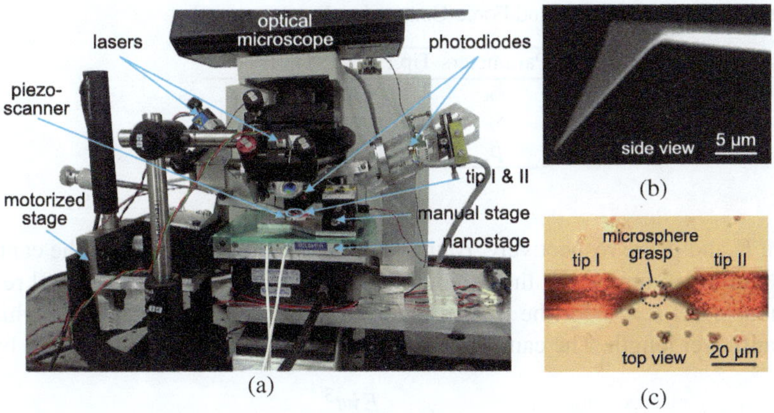

Fig. 6.1 (a) A photo of the AFM-FRS on the configuration for nanoscale pick-and-place manipulation. (b) A SEM image of the cantilever fabricated with a protruding tip. (c) An optical microscopic image of the collaborating tips in microsphere grasping operation.

Table 6.1 Specifications of Each Motion Module

Actuator	Travel range	Resolution
X-Y-Z piezoscanner	$10 \times 10 \times 10(\mu\text{m})$	sub-nm
X-Y-Z nanostage	$50 \times 50 \times 10(\mu\text{m})$	0.1 nm
X-Y-Z motorized stage	$25 \times 25 \times 25$(mm)	50 nm
X-Y-Z manual stage	$5 \times 5 \times 5$(mm)	0.5 μm

is developed for AFM image scan and two-tip coordination control during manipulation. This control system enables programming of complex tasks on the highly distributed reconfigurable system.

6.1.2.1 Forces Calibration

Accurate interactive force detection between the AFM tips and the micro/nanoobjects or the micro/nano-environment is a prerequisite for achieving precise and stable manipulation. In our experiments, normal forces from both the tips are used to monitor the whole manipulation process, e.g., precise contact detection, grasping force sensing and controlled nanoobject pushing.

The method of forced oscillation was first used to determine the thickness t of the cantilever from its natural frequency. As an Euler-Bernoulli beam, if resonant frequencies of the cantilever are known, its thickness t can be determined by:

$$t = \frac{\omega_n}{K_n^2}\sqrt{\frac{12\rho}{E}} \tag{6.1}$$

Table 6.2 Normal Stiffness and Force Conversion Factor

Parameters	Tip I	Tip II
k_n	2.72 N/m	2.78 N/m
S_n	0.51 μm/V	0.49 μm/V
β	1.39 μN/V	1.36 μN/V

where E and ρ are respectively Young's modulus and the density of the cantilever, K_n is the wave number on the oscillating cantilever, ω_n is the nth flexural resonant frequency. Set $n = 1$ for the first resonant mode, then $K_n l = 1.8751$, in which l is the cantilever length. The cantilever's normal stiffness k_n can be calculated by:

$$k_n = \frac{Ewt^3}{4l^3} \tag{6.2}$$

where w is the width of the cantilever.

In order to convert signals from the optical levers into normal forces, it is necessary to calibrate the normal sensitivity of the optical lever S_n and then the normal force conversion factor $\beta = S_n k_n$, for which the corresponding bending force F_n is given by:

$$F_n = \beta \times \Delta V_n \tag{6.3}$$

where ΔV_n represents the photodiode voltage output attributable to an applied force in the normal direction.

The normal stiffness of cantilevers and sensitivities of optical levers used in the experiments are shown in Table 6.2.

6.1.2.2 Force Sensing during Pick-and-Place

Figure 6.2 shows a schematic diagram of the nanotip gripper for the micro/nanoscale pick-and-place operation that has a clamping angle $\theta \approx 44°$ micro/nanoscale pickup manipulation, for instance, interactive forces applied on tip I include repulsive forces F_{r1}, friction forces F_{f1}, and adhesive forces F_{a1}. The forces applied on tip I can be resolved into two components on the X-axis and the Z-axis in the defined frame, namely, F_{x1} and F_{z1}, respectively. F_{x1} is the clamping force that holds the micro/nanoobject. F_{z1} is the pickup force that balances adhesion forces from the substrate. To sense the pickup force, it is necessary to know the normal deflection on both cantilevers. The normal deflection ζ_{n1} associated with the normal voltage output of the optical lever on tip I is given by:

$$\zeta_{n1} = \frac{F_{z1}\cos\gamma + F_{x1}\sin\gamma}{k_n} + \frac{F_{z1}\sin\gamma + F_{x1}\cos\gamma}{k_{xz}} \tag{6.4}$$

where γ is the mounting angle of the cantilever, $k_{xz} = 2lk_n/3h$ is the bending stiffness due to the moment applied on the tip end, where h is the tip height. Assuming the magnitude of F_{z1} and F_{x1} are of the same order, contributions from the F_{x1} to

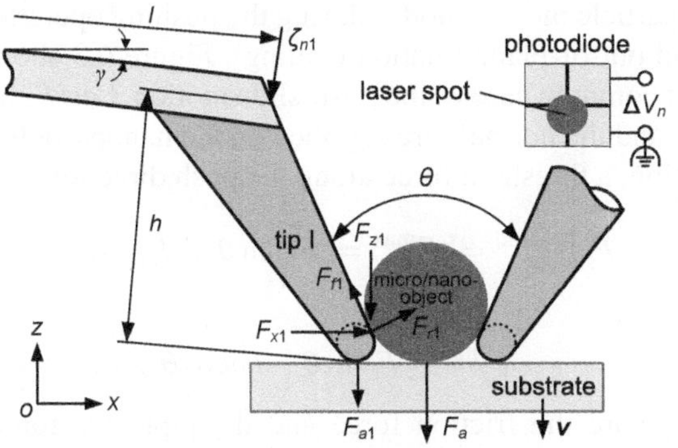

Fig. 6.2 Schematic diagram of the nanotip gripper and force simulation during the pickup manipulation.

the normal deflection of tip I are relatively very small since $k_{xz} \gg k_n$ and $\gamma = 5°$. Therefore, the normal deflection induced from F_{z1} is only considered in the following calculations of the adhesion force F_a applied on the micro/nanoobject. Thus, F_{z1} can be simplified by estimating the normal voltage output ΔV_{n1} from the tip I:

$$F_{z1} = \beta_1 \times \Delta V_{n1} \tag{6.5}$$

where β_1 is the normal force sensitivity the optical lever. A similar pickup force F_{z2} can be also obtained on Tip II. Before the gripper pulls off the substrate, the adhesion force F_a can be estimated as:

$$F_a = F_{z1} + F_{z2} = \beta_1 \Delta V_{n1} + \beta_2 \Delta V_{n2} \quad (F_{a1} + F_{a2}) \tag{6.6}$$

where β_2, ΔV_{n2} and F_{a2} are respectively the normal force sensitivity, normal voltage output and adhesive force on tip II. Once the gripper pulls off the substrate, e.g., in the case of nanowire/tube pick-and-place, the adhesion force F_a is estimated as:

$$F_a = F_{z1} + F_{z2} = \beta_1 \Delta V_{n1} + \beta_2 \Delta V_{n2}. \tag{6.7}$$

6.1.2.3 Controlled Nanoparticle Pushing

As shown in Fig. 6.3, a constant distance d_0 between the tip and substrate is maintained during the nanoparticle push. In order to promptly detect the tip-nanoparticle contact as well as nanoparticle loss during particle push, the tip should be kept oscillating during the whole push procedure. Once contact with the nanoparticle is established, the amplitude of the oscillating tip fades from A_0 to zero. Contrariwise, the oscillating tip returns to the amplitude A_0 when the tip-nanoparticle contact is lost; this can be used to detect nanoparticle loss automatically to relocate the nanoparticle and restart the pushing operation.

Possible nanoparticle motion modes during the pushing operation are sliding, in-plane rotation and out-of-plane rotation (rolling). Figure 6.3 shows the interactive forces of the most common case of the push-slide motion. Let ${}^tF_f$ and ${}^tF_n$ be respectively the friction and the normal force applied on the nanoparticle. For a successful push-slide operation, a threshold force along the push direction:

$$F_{\text{sliding}}^{\text{threshold}} = {}^sF_f^{\max} = {}^tF_f \sin\theta + {}^tF_n \cos\theta \tag{6.8}$$

$${}^tF_f^{\max} \geq F_l \sin\theta - F_n \cos\theta \tag{6.9}$$

where ${}^sF_f$ and ${}^sF_n$ are the friction force and the repulsive force applied on the nanoparticle from the substrate, respectively; θ is the contact angle between the tip side and the nanoparticle. The minimum value of θ is equal to tip's half cone angle if the tip-substrate separation d_0 is small enough to make a contact with the tip's cone surface rather than tip's apex; F_l and F_n are the lateral and the normal forces applied on the tip from the cantilever. For low-speed and quasi-static pushing, it can be deduced that $F_l = {}^sF_f$ and $F_n = {}^sF_n$.

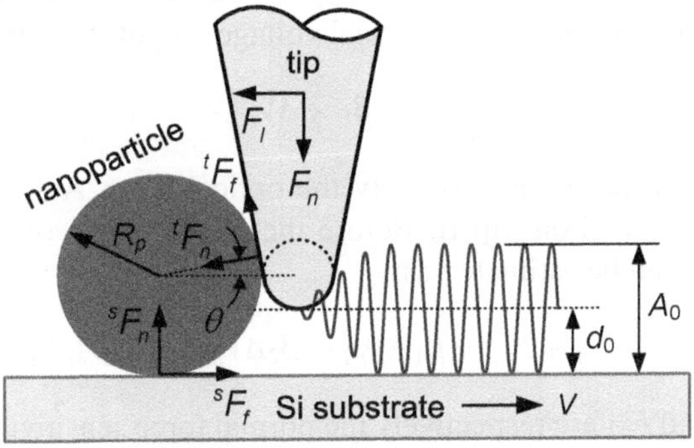

Fig. 6.3 Simulation of the pushing scheme and interaction forces during a nanoparticle push-slide operation.

6.1.2.4 Piezoscanner Hysteresis and Creep Compensation

Motion coordination between two cantilevers is achieved by the closed-loop nanostage and the open-loop piezoscanner. Thus, piezoscanner hysteresis and creep should be well compensated, and both the stages should have the same displacement scale to ensure prompt and accurate positioning between both the tips and samples.

To characterize hysteresis and creep, the piezoscanner is fixed on the nanostage with the same axis arrangement in all the three spatial directions. The nanostage here is used as a reference stage, and the optical lever on tip I is used as a displacement sensor to measure the relative movement between the nanostage and the piezoscanner by lateral tracking well-defined slopes (AFM calibration grating

TGF11) with the AFM tapping mode. During characterization, the piezoscanner is used to track the motion of the nanostage that provides a reference displacement thanks to the high resolution of the nanostage (0.1 nm) and the excellent dynamic performance of the piezoscanner (about 1 kHz on 1 μm motion amplitude). As the voltage-displacement response is characterized, the piezoscanner hysteresis and creep are predicated by Prandtl-Ishlinskii (PI) and logarithmic models, respectively [33]. After compensation, maximum errors on the full range of the X-axis, Y-axis and the Z-axis are compressed from 28.14%, 27.82% and 15.8% to 1.18%, 1.16% and 0.97%, respectively.

6.1.3 The Configuration of 3-D Micromanipulation

6.1.3.1 Overview

As the size of microobjects is reduced to several microns or submicrons, problems will arise with these conventional grippers: (i) Sticking phenomena becomes more severe due to the relatively larger contact area between the gripper and the microobject [2]. (ii) The tip diameters of the micro-fabricated clamping jaws are comparable in size to the microobjects to be grasped. Conventional grippers are not geometrically sharp enough to pick up microobjects of several micrometers deposited on the substrate. Fortunately, the AFM tip has a very tiny apex (typically ~10 nm in radius) with respect to the size of the microobject to be manipulated. Thus, the nanotip gripper can be used to achieve pick-and-place at the scale of several microns since the contact area of the gripper-microobject is much smaller than the microobject-substrate contact. Moreover, real-time force sensing makes the manipulation more controllable.

6.1.3.2 System Configuration for 3-D Micromanipulation

As shown in Fig. 6.4, the system configuration for 3-D micromanipulation is reconfigured as follows:

1. For a large manipulation travel range, the nanostage here is used to support the sample platform and transport the microobject during the manipulation.
2. Tip I, immovable during the pick-and-place micromanipulation, is fixed on the motorized stage for coarse positioning.
3. Tip II is actuated by the piezoscanner for gripper opening and closing operations. The piezoscanner is supported by the manual stage for coarse positioning.

6.1.3.3 Microscale Grasping Strategies

Benefiting from AFM-based accurate and stable amplitude feedback of a dithering cantilever, the grasping state can be successfully achieved by the amplitude feedback, with very weak interaction at the nano-Newton scale, protecting the fragile tips and the microobjects from damage during manipulation.

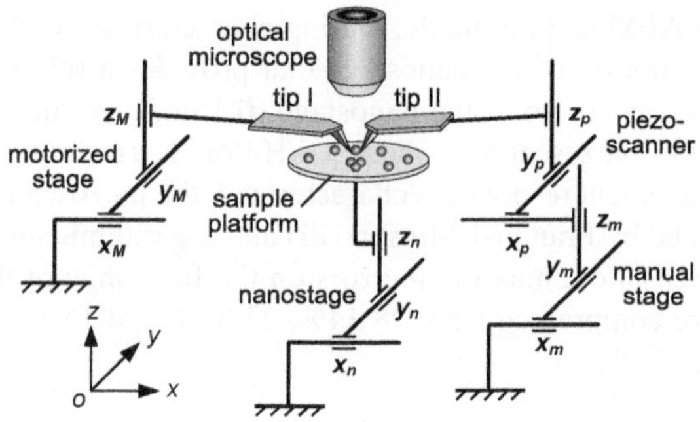

Fig. 6.4 A kinematic configuration of the AFM-RFS for 3-D micromanipulation at scale of several micrometers.

6.1.3.4 Grasping Points Search

As shown inset I of Fig. 6.5, the dithering cantilever with its first resonant mode is used to locate grasping points and detect contact. When approaching the microsphere with a separation between the tip and the substrate (typically 500 nm), the tip laterally sweeps the microsphere over the lower part of the microsphere. By this means, the grasping point can be accurately found by locating the minimum amplitude response of each single scan.

6.1.3.5 Contact Detection

From the scheme depicted in inset II of Fig. 6.5, the amplitude feedback is also used for contact detection. Tip-microsphere contact is detected as the amplitude reduces to a steady value close to zero.

6.1.3.6 3-D Micromanipulation Protocol

As shown in Fig. 6.6, a protocol for pick-and-place microspheres mainly consists of four steps:

6.1.3.7 System Initialization and Task Planning

Each axis of the nanostage and the piezoscanner are set in a proper position, supplying the manipulation with enough travel range on each axis. Then the task is planned in Fig. 6.6(a) with a global view of the manipulation area that provides coarse positions of the microspheres and tips.

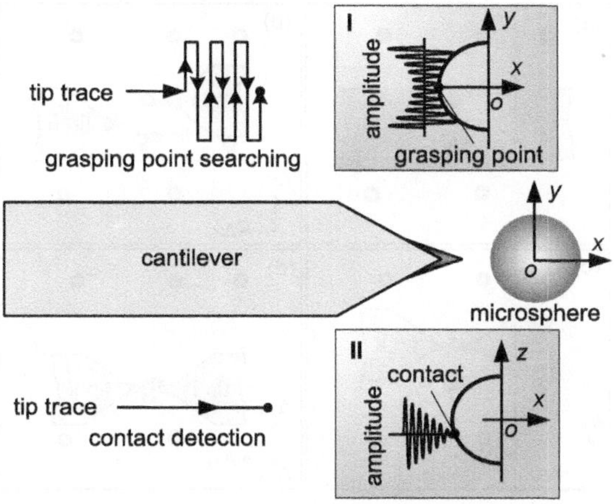

Fig. 6.5 Schematic diagrams for grasping point searching (inset I) and contact detection (inset II) with amplitude feedback of the dithering cantilever.

6.1.3.8 Making Tip I-Microsphere in Contact

In Fig. 6.6(b), tip I is started to approach the microsphere by moving the nanostage with amplitude feedback to search for the grasping point and detect contact.

6.1.3.9 Forming the Gripper

Similarly, tip II approaches the microsphere by moving the piezoscanner. Once tip II and the microsphere are in contact, a nanotip gripper is configured in Fig. 6.6(c) for a manipulation.

6.1.3.10 Pick-and-Place Micromanipulation

In Fig. 6.6(d), the microsphere is picked up, transported and released by moving the nanostage with a proper displacement on each axis that depends on the diameter of the microsphere and its destination. The whole process of 3-D micromanipulation is monitored by real-time force sensing.

6.1.4 The Configuration of 3-D Nanomanipulation

6.1.4.1 Overview

In addition to the microscale pick-and place, the proposed nanotip gripper can also be used to achieve nanoscale pick-and-place in ambient conditions, since the tiny tip is geometrically sharp enough to pick up nanoobjects deposited on the substrate, and simultaneously has capabilities of sensing nanoscale force between the

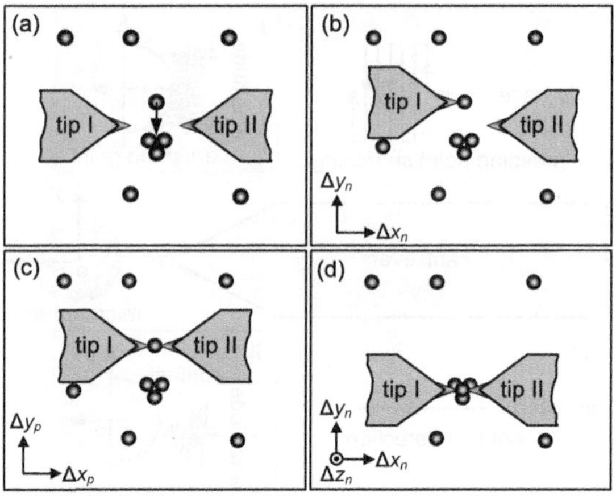

Fig. 6.6 Microsphere manipulation protocol. (a) Task planning. (b) Tip I and the microsphere are in contact. (c) The nanotip gripper is formed. (d) Pick up and release the microsphere to its target position.

nanoobject and the tool. Moreover, the nanotip gripper itself acts as an image sensor for nanoobject positioning.

6.1.4.2 System Configuration for 3-D Nanomanipulation

Compared with the 3-D micromanipulation, tip alignment precision is the key factor in succeeding the nanoscale pick-and-place. Therefore, the closed-loop nanostage is considered in this configuration for accurate tip alignment. As shown in Fig. 6.7, the system configuration for 3-D nanomanipulation is reconfigured as follows:

1. The nanostage is used for image scan with tip I.
2. Nanoobjects are supported and transported by the piezoscanner.

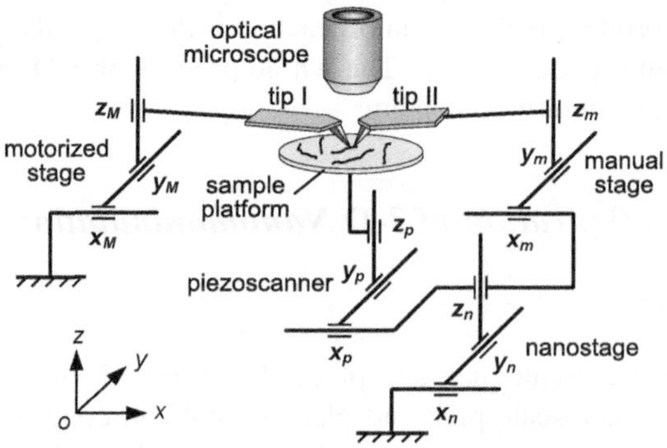

Fig. 6.7 A kinematic configuration of the AFM-RFS for nanoscale pick-and-place.

3. Tip I, fixed on the motorized stage for coarse positioning, is immovable during the pick-and-place micromanipulation. Before manipulation, cantilever I acts as an image sensor for nanoobject positioning.
4. For accurate gripper alignment between Tip I and Tip II, Tip II is fixed on the nanostage rather than the piezoscanner. Tip II is supported by the manual stage for coarse positioning.

6.1.4.3 3-D Nanomanipulation Protocol

Nanowires and nanotubes are being intensively investigated. Thus, a protocol is developed here for nanowire or nanotube pick-and-place. However, applications can easily be extended to, for example, pick-and-place of nano-rods or nanoparticles dispersed on a substrate.

6.1.4.4 System Initialization

Once the manipulation area is selected under the optical microscope, both the tips are aligned as a quasi-gripper above the center of the manipulation area. Each axis of the nanostage and the piezoscanner is initialized at an appropriate position to allow for enough manipulation motion travel.

6.1.4.5 Image Scan for Task Planning

In this step, tip I is used to fully scan the relevant area obtaining a topographic image that contains nanoobjects to be manipulated and the end of tip II. Figure 6.8(a) shows a simulated image that contains the topography of two nanowires and the end of tip II. The image provides the following pick-and-place with relative positions between tip I, tip II and the nanoobjects to be manipulated. However, after a long period image scan, relocating tip II is recommenced to eliminate system's thermal drift.

6.1.4.6 Making Tip II-Nanowire in Contact

As shown in Fig. 6.8(b), tip II approaches the nanowire to make contact by moving the X-axis of the piezoscanner. A gap (typically $\sim$20 nm above the snap-in boundary) between Tip II and the substrate should be maintained during the approach to enable a negative deflection response in the form of a tiny force applied on tip II, and hence, sensitive detection of the tip-nanowire contact.

6.1.4.7 Forming the Gripper

Similarly, in Fig. 6.8(c), once tip I is in contact with the nanowire, a nanotip gripper is configured for pick-and-place manipulation of the nanowire.

6.1.4.8 Pick-and-Place Nanomanipulation

The nanotip gripper in this step is used to pick up, transport and release the nanowire to its target position by moving the piezoscanner on the X, Y or Z-axis. The displacement on each axis depends on the dimensions of the nanowire and the location of the destination. Figure 6.8(d) shows a simulated post-manipulation image, in which a nanowire crossbar is built. The complete pick-and-place procedure is monitored by force sensing.

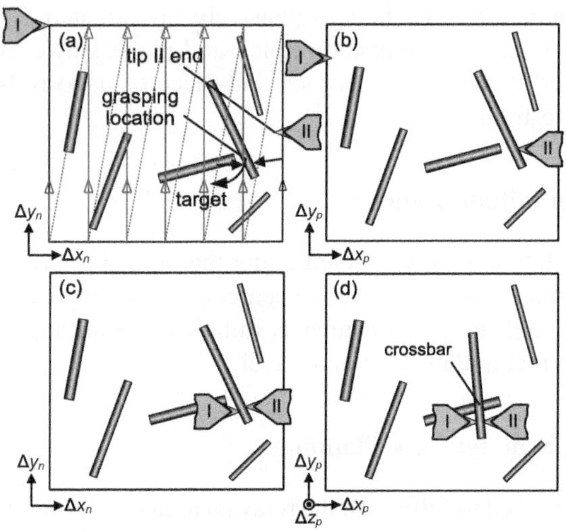

Fig. 6.8 A protocol for nanowire pick-and-place. (a) Image scan for task planning. (b) Tip II is in contact with the nanowire. (c) Tip I is in contact with the nanowire, forming a nanotip gripper. (d) Pick up and release a nanowire to its target position.

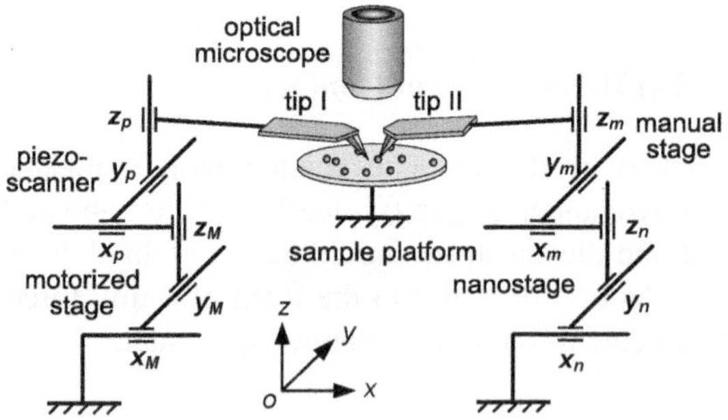

Fig. 6.9 A kinematic configuration of the AFM-RFS for parallel imaging/nanomanipulation.

6.1.5 *The Configuration for Parallel Nanomanipulation*

6.1.5.1 Overview

Conventional AFM-based manipulation has an inefficient scan-manipulation-scan process, which make mass production impossible. Fortunately, the AFM-FRS has two individually actuated cantilevers that can be used simultaneously as an image sensor and manipulation tool, respectively. By this means, image scan and nanomanipulation can be performed in parallel.

6.1.5.2 System Configuration for Parallel Nanomanipulation

Position precision is one of the critical issues in the AFM-based 2-D nanomanipulation. Among several factors that cause positioning errors, the most significant are piezoscanner hysteresis and creep. Thus, the closed-loop nanostage in this configuration is used to actuate the manipulating tip. As shown in Fig. 6.9, the system configuration for parallel nanomanipulation is reconfigured as follows:

1. Tip I, used as image sensor, actuated by the piezoscanner that is fixed on the motorized stage for coarse positioning. Excellent image quality can be obtained even though the scan length on tip II is more than 300 nm, which is sufficient for tip positioning with a dynamic image display updated by real-time scanning data.
2. To minimize positioning errors for a efficient manipulation, Tip II, used as an end-effector for nanoobjects manipulation, is actuated by the nanostage rather than the piezoscanner. The manual stage is used here for coarse positioning.
3. The sample platform is immovable during the manipulation and fixed on the system base.
4. The action of each cantilever is controlled by the independent thread of the control system and the image scanning data are shared by these two threads.

6.1.5.3 Parallel Nanomanipulation Scheme

Figure 6.10 is a diagram of the parallel image/manipulation with image scan using tip I. After system initialization, local scan is started to locate tip II with tip I. Once tip II locating is completed, two parallel threads are activated to control a parallel imaging/manipulation task. One is used for imaging with tip I to perform a full scan of the pertinent area, resulting in a topographic image that contains nanoobjects to be manipulated as well as the end of tip II (as seen in the simulated image at position P_{II}^{0}). The second thread is assigned to manage the performance of tip II, which is in idle process until first target position t_1 is entirely covered by the scanned area when tip I reaches position P_{I}^{1}.

The shortest path resolution with linear trajectories is used to choose a corresponding particle to target t_1. As seen the Fig. 6.10, particle p_1, which is with the shortest distance to t_1 compared with others scanned particles, is firstly pushed. The

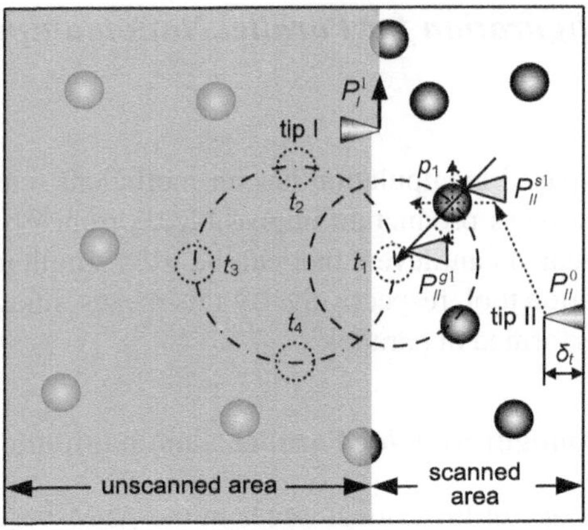

Fig. 6.10 A task planning scheme for the parallel imaging/nanomanipulation with the shortest path solution.

manipulation task is activated after the coarse positioning of particle p_1 by processing the scanned image. Accurate center position of particle p_1 are further obtained by locally scanning the nanoparticle horizontally and vertically with tip II along a defined line. When tip II reaches point P_{II}^{s1} on the pushing line, the push is started until tip II reaches P_{II}^{g1} where particle p_1 has been pushed to its target position t_1. After the first manipulation, tip II is set to the idle process again until target position t_2 emerges within the dynamic image.

6.1.5.4 Parallel Manipulation Efficiency

Task time of the parallel imaging/manipulation operation can be given as:

$$t_{task} = \max(t_s, t_m) + t_s \tag{6.10}$$

where t_s is the scanning time of one image frame and t_m is the total manipulation time estimated from the sum of manipulation time of each single nanoobject. For the normal-speed AFM, t_{task} is often equal to $2t_s$ except for a complex manipulation task that cannot be fulfilled within one frame period, and so has task time of $t_m + t_s$. This scheme can save a lot of time compared with the serial imaging/manipulation operation that has task time of $t_m + 2t_s$. One disadvantage of this scheme is that environment-based motion planning is unavailable during the manipulation. However, parallel imaging/manipulation can be perfectly performed by this scheme if a manipulation objective is defined before the operation.

6.1.6 Experimental Results

6.1.6.1 3-D Microsphere Assembly

Nylon microspheres with diameter of about 3∼4 μm were manipulated to build 3-D microstructures in experiments. The microspheres were deposited on a freshly cleaned glass slide and then an area of interest was selected under the optical microscope.

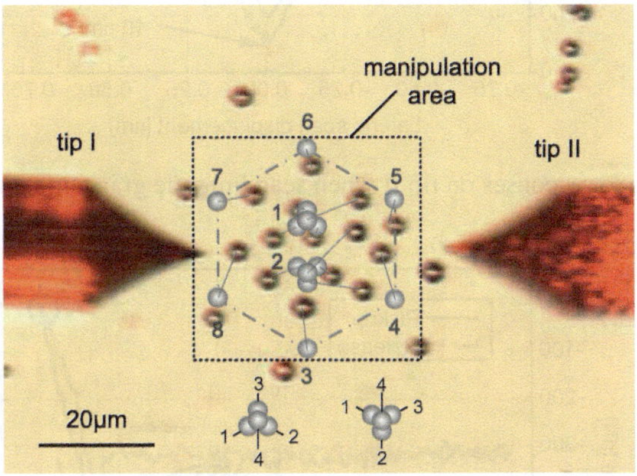

Fig. 6.11 Task descriptions of the microsphere assembly: a micro-pattern formed by a regular hexagon and two micropyramids are planned. Assembly sequences are described by numbers.

Figure 6.11 shows a plan view of the selected area, in which 14 microspheres separated in a 50 μm square frame are going to be manipulated to build two 3-D micropyramids and a regular 2-D hexagon labeled by assembly sequences from 1 to 8. Each pyramid is constructed from four microspheres with two layers and the assembly sequences are shown in the bottom insets for two different arrangements of the pyramids.

Figure 6.12 shows a result of grasping point searching, in which the dithering tip II laterally sweeps the microsphere within a range of 1.75 μm on the y-axis and with a free oscillating amplitude of about 285 nm. Ten different distances to the microsphere were tested from 100 nm to 10 nm with an interval of 10 nm and, consequently, the grasping point is well located with an accuracy of $\pm$10 nm.

Figure 6.13 shows a full force spectroscopy curve during the pick-and-place of a microsphere deposited on a glass slide with an ambient temperature of 20 °C and relative humidity of 40%. In this curve, point A represents the start of the pick-and-place, point B the pull-off location of the microsphere-substrate contact, point C nonlinear force restitution due to the tip-microsphere frictions and point D the snap-in point between the microsphere and the substrate. The force spectroscopy curve is synthesized from force responses on tip I and tip II.

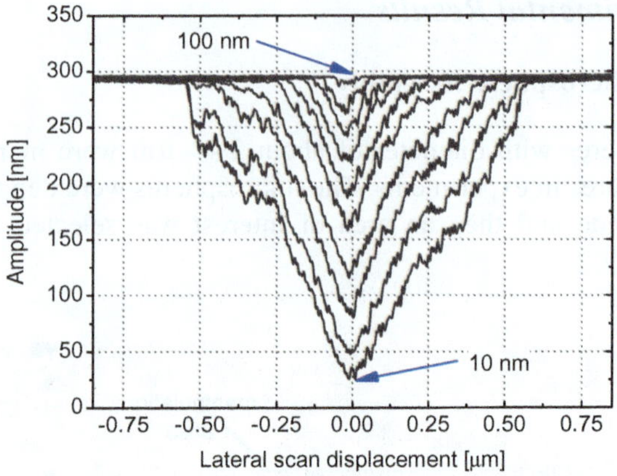

Fig. 6.12 Amplitude responses on tip II when searching the grasping points.

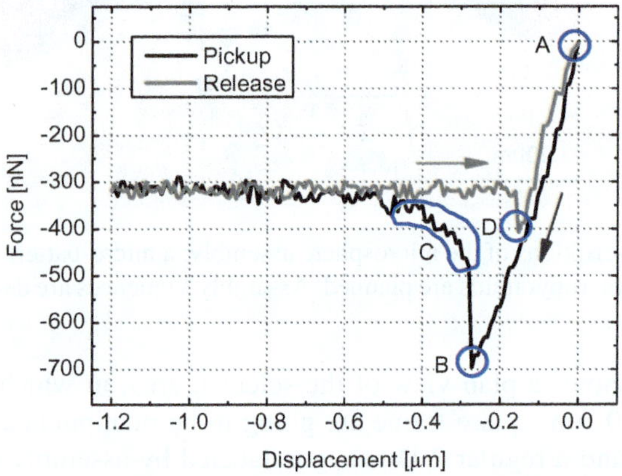

Fig. 6.13 Synthesized normal force responses from both the tips in the pick-and-place micromanipulation.

Figure 6.14 shows an automated microassembly result consisting of two 3-D micropyramids and a 2-D pattern of a regular hexagon. The whole manipulation process was completed in 11 minutes, so the average manipulation time for each microsphere is about 47 seconds, which mainly breaks down into about 20 seconds for microsphere grasping including the grasping point search and contact detection processes using amplitude feedback, 10∼35 s for microsphere release and, the remaining time for transport.

Assembly of the 4th is the key to success in building a micropyramid. During pick-and-place of the 4th microsphere, microscopic vision was firstly used for coarse positioning it target, then the normal force feedback of the gripper was used to detect the vertical contact between the 4th microsphere and other three microspheres on the

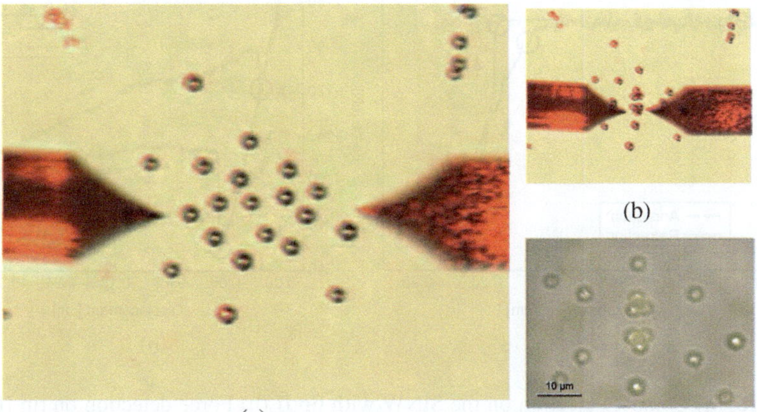

(a) (b) (c)

Fig. 6.14 A microassembly result shows that two micropyramids and a regular hexagon have been built. (a) Before the microassembly. (b) Micropyramids were built. (c) is an enlarged image of the microassembly result under a magnification of 100×.

base. When the contact is established, a small vertical force was applied on the 4th microsphere by moving the nanostage upward and it will adjust to contact with all the base microspheres.

6.1.6.2 3-D Silicon Nanowire Assembly

In experiments, silicon nanowires (SiNWs) were deposited on a freshly cleaned silicon wafer coated with 300 nm silicon dioxide. AFM images show that the SiNWs have a taper shape and have diameters of 25 nm (top) ~200 nm (root) and lengths of about 4~7 μm. Figure 6.15 shows an example of the contact detection with tip II: point A and point C are where the tip contacts with the SiNW and the Si substrate, respectively; Point B and point D are where the tip breaks the contact with the Si substrate and the SiNW, respectively. Figure 6.15 shows a curve of the peeling force spectroscopy on tip II for the pick-and-place manipulation of the SiNW: point A and point B are where the tip snaps in and pulls off the Si substrate, respectively. The shape of curve of the force responses on tip I are similar except for the force magnitude due to different force sensitivities on each tip and uneven grasping due to asymmetric alignment of the SiNW relative to the grasping direction. This force spectroscopy during the pickup operation shows stable grasping for further SiNW transport.

Figure 6.16 shows an experimental result of 3-D SiNW manipulation. A pre-scanned image (9 μm × 9 μm) is shown in Fig. 6.16(a), which includes the topographic image of SiNWs and the local image of tip II. A grasping location of the nanowire to be manipulated is marked with a–a, where the SiNW has a height of 160 nm. Figure 6.16(b) is a post-manipulation image. It can be seen that the

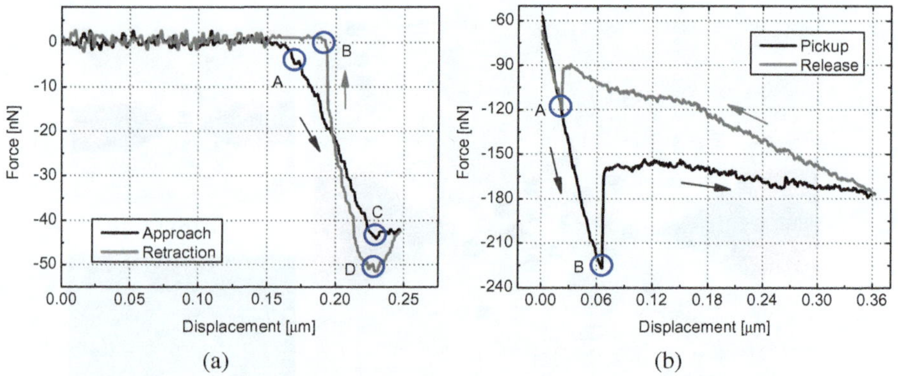

Fig. 6.15 (a) Contact detection on the SiNW with tip II (b) Force detection on tip II during the pick-and-place nanomanipulation of the SiNW

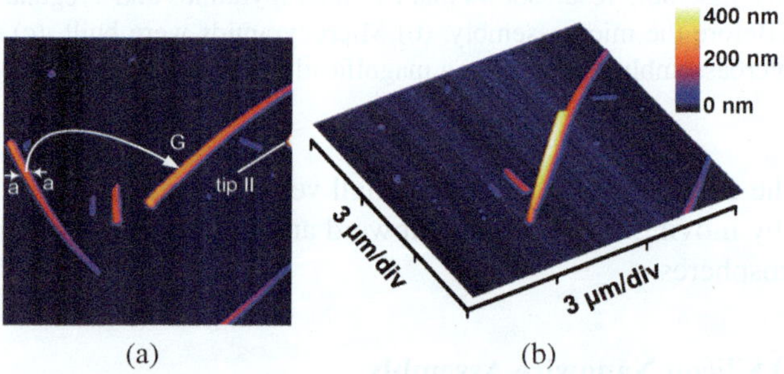

Fig. 6.16 Pick-and-place results of the SiNWs. (a) A pre-scanned image. In which a–a and G are the grasping location and the target position, respectively. (b) A post-manipulation image verifies that the manipulated SiNW is piled upon another nanowire deposited on the Si substrate.

SiNW has been successfully transported and piled onto another SiNW. The manipulation procedure is described as follows. Once the SiNW was reliably grasped, the piezoscanner moved down 560 nm at a velocity of 80 nm/s. In this step, the SiNW was transported a distance of 4.4 μm along the X-axis at a velocity of 120 nm/s and 0.12 μm along the Y-axis at a lower velocity of 3.3 nm/s. In the releasing step, the piezoscanner moved up at a velocity of 100 nm/s. As tip II was slightly bent upward leading to a positive response of 0.015 V, tip I and tip II were separated by moving both the nanostage and the piezoscanner on the X-axis to release the SiNW from the nanotip gripper.

6.1.6.3 Parallel Imaging/Nanomanipulation of Nanoparticles

The manipulation samples were prepared by depositing 50 nm gold colloidal particles (Ted Pella, Inc) from aqueous solution on a freshly cleaned silicon substrate using the boiling method [34]. The silicon substrate was placed on a hot plate that was set at an appropriate temperature of about 100∼110 °C. Drops of the gold nanoparticle solution were released on the hot substrate and were left to boil.

First, a contact point positioning error leads to contact loss during the push manipulation due to nanoparticle in-plane rotation, especially over a long-distance push. In order to detect particle loss promptly, tip II is kept in oscillating mode during the whole push procedure. Figure 6.17 (a) shows an example of the normal force and amplitude responses in the cases of successful push: point A shows the tip snaps into contact with a nanoparticle, and point B indicates motion occurs when the static friction is broken. Figure 6.17(b) shows a cases of nanoparticle loss: point A shows the tip and nanoparticle snap into contact. Motion occurs at point B and the particle loss occurs at point C. By this means, not only can the tip-nanoparticle contact be detected by amplitude fading, but also the particle loss when the amplitude suddenly increases from zero.

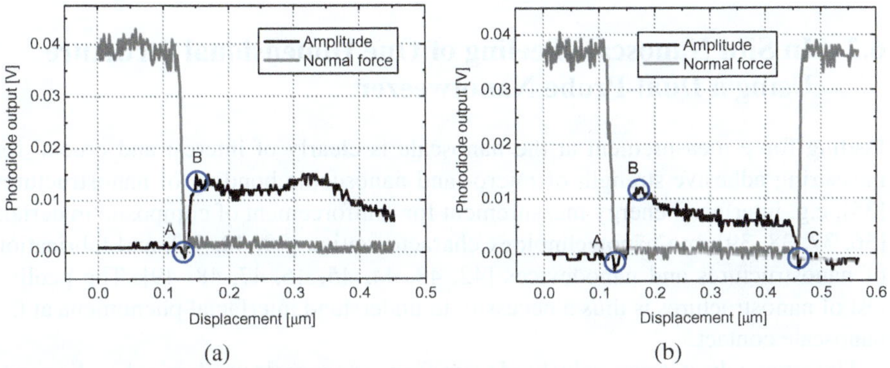

Fig. 6.17 (a) An example of the normal force and amplitude responses during a particle push (b) An example of the normal force and amplitude responses during a particle push followed by particle loss

A parallel image/manipulation task was performed with normal-speed image scan. In Fig. 6.18(a), five gold nanoparticles with a diameter of 50 nm and the corresponding target positions emerged on the dynamic image I to III in sequence. In Fig. 6.18(b), selected particles were pushed onto the corners and the middle of a square within the image scan period. The imaging frame period t_s was about eight minutes. In contrast, the total manipulation time t_m for these five nanoparticles was less than three and a half minutes with a pushing velocity of 200 nm/s. The result indicates that a more complicated task can be performed within an imaging period

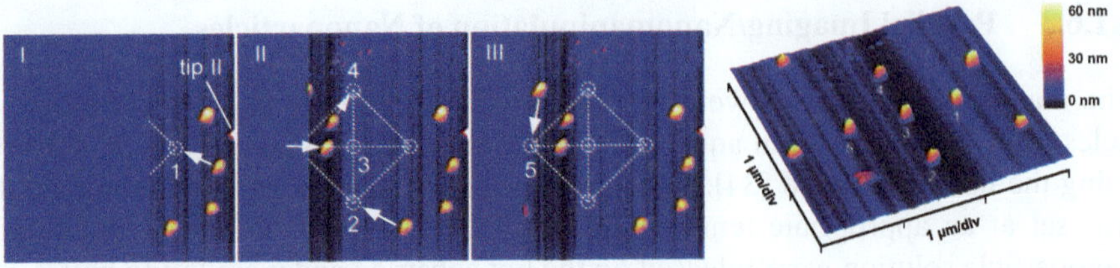

(a) (b)

Fig. 6.18 A parallel imaging/manipulation result with a normal-speed image scan of tip I. (a) Emergences of five particles (50 nm in diameter) and targets on three different dynamic displays I to III. (b) Five nanoparticles were pushed to form a square pattern with a central particle.

in order to increase AFM-based nanomanipulation efficiency and make the mass production feasible.

As a result, the AFM-FRS makes applications from the microscale to the nanoscale on the same flexible robotic system practical, such as building micro/nano 3-D structures and devices, and high-throughput 2-D nanomanipulation.

6.2 In Situ Nanoscale Peeling of One Dimensional Structure Using a Dual-Probe Nanotweezer

Peeling force measurement at the nanoscale is clearly of interest and crucial for measuring adhesive strength of micro- and nanoscopic bonding of nanostructures [35], e.g. interfacial energy measurement for reinforcement of composite materials [36, 37, 38, 39], bio-nanotechnology characterization [39, 40, 41], and fabrication of nanostructures and nanodevices [42, 43, 44, 45, 46, 47, 48, 49]. The peeling test of nanostructures is thus a necessity to understand interfacial phenomena at the nanoscale contact.

Unfortunately, it is generally hard to perform a nanoscale mechanical peeling test, which requires facilities with capabilities to locate, manipulate and displace nanostructures with nanoscale precision (~nm) while sensing extremely small peeling forces (~nN). As a uniquely suitable facility capable of nanoscale surface imaging and ultra-small force sensing, the conventional atomic force microscope (AFM) in the past two decades has succeeded in manipulating nanoparticles [50, 51, 52], single molecules [53], nanorods[54] and nanotubes [55] by typical used in-plane pushing or pulling manipulation. However, the nanoscale peeling force measurement, which typically requires out-of-plane manipulation, is still a challenge. Up to now, a few nanoscale peeling researches have been carried out. For instance, nanoscale peeling methods were developed to successfully peel multi-walled carbon nanotubes (MWCNT) attached to the end of a tipless microcantilever[35] in an AFM or the tip of a self-detective microcantilever[56] integrated in a scanning electron microscope that enables a so-called visualized nanoscale peeling process. The current methods might require strenuous labors of sample preparations by sorting straight MWCNTs out and then attaching them on the microcantilevers. An effective peeling tool is thus necessarily to be developed to facilitate the nanoscale peeling tests.

In this work, we present a new method for in situ nanoscale peeling of one-dimensional (1-D) nanostructures using a dual-probe nanotweezer, which is modified from our previous work of a parallel imaging/manipulation force microscope (PIMM) [57]. The system is equipped with two independently actuated and sensed microcantilevers with protruding tips that are in opposition to each other, forming a nanotweezer through coordination control. Due to the capability of nanoscale pick-and-place manipulation with accurate force sensing, the nanotweezer is suitable for in situ peeling of 1-D nanostructures without complicated sample preparation, e.g., single molecules, nanowires, and nanotubes.

6.2.1 Methods

The nanotweezer is formed by two force modulation microcantilevers (NANOSENSORS ATEC-FM, with a nominal stiffness of 2.8 N/m). Forces on each cantilever are independently detected by its own optical lever which is typically composed of a laser and a quartered photodiode. The calibrated [58] open-loop X-Y-Z piezotube (PI P-153.10H) in this configuration is removed from the system base and then fixed on the closed-loop X-Y-Z nanostage (MCL Nano-Bio2M on X- and Y-axis, PI P-732.ZC on Z-axis) to form a dual-driven nanopositioning stage. This stage can be used for image scanning, nanotweezer aligning and nano samples handling with a specially developed control system. The left microcantilever (probe I) is fixed to the base while the right microcantilever (probe II) is actuated by the nanostage. The proposed dual-probe nanotweezer is capable of the nanoscale peeling thanks to the followings:

i) In comparison with the normal diameter of 1-D nanostructures to be peeled, the tip apex of the probe is very tiny (typically with an apex radius of 10 nm or less), which leads to a geometrical condition for grasping at the nanoscale as well as smaller adhesive forces between the nanotweezer and the 1-D nanostructures that favor the release process.
ii) More importantly, the nanotweezer can be used as a normal AFM to image the 1-D nanostructures, as well as locate the probes for tweezer alignment. This function makes it possible to perform nanoscale grasping without visual feedback (normally in scanning and transmission electron microscopes) in ambient conditions.

Figure 6.19 shows a schematic diagram of the peeling test of 1-D nanostructures with the proposed nanotweezer. Figure 6.19(a) and 6.19(b) respectively show the peeling occurring at the end and at the middle of a 1-D nanostructure. For stiff 1-D nanostructures, peeling locations can be at either the end or the middle, whereas for soft 1-D nanostructures, notably with a small diameter or a long length, or made up of soft materials, the peeling location is recommended at the middle to avoid sliding off from the nanotweezer during the peeling process. A peeling protocol using the dual-probe nanotweezer can be summarized as follows:

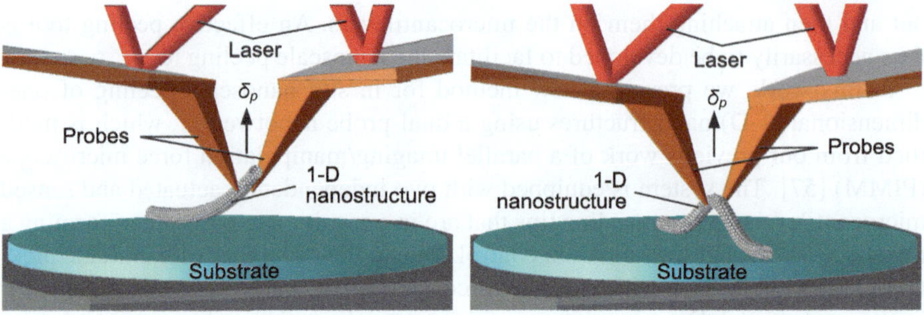

Fig. 6.19 A schematic diagram of a 1-D nanostructure peeling force measurement using the dual-probe nanotweezer. (a) Peeling occurs at the end of the 1-D nanostructure. (b) Peeling occurs at the middle of the 1-D nanostructure. δ_p is the gap between the substrate and the peeling location on the 1-D nanostructure. The peeling process is performed by moving the piezotube on the Z-axis while the nanotweezer is fixed. The peeling force is synthesized from bending forces on both cantilevers that are individually detected by their own optical levers.

i) First, testing 1-D nanostructures and the end of probe II are accurately located by image scan with probe I by moving the nanostage.
ii) Once the target 1-D nanostructure and peeling locations are determined from the scanned image, the nanotweezer is formed and then approaches to the target peeling location by moving the piezotube and the nanostage alternately.
iii) Peeling test is started by moving the piezotube on the Z-axis when a reliable grasping on the nanostructure is detected with force monitoring. For peeling energy computation, peeling forces are recorded during the whole procedure of the test.

Figure 6.20(a) and 6.20(b) respectively show elastic deformations of a 1-D nanostructure when the peeling occurs at the end and the middle. The gap between the peeling location and the substrate is defined as δ_p, while F_p is the peeling force applied by the nanotweezer at the peeling location. The parameter $z(x)$ describes the gap along the 1-D nanostructure from the substrate. As a peeling force computation model shown in Fig. 6.20(c), F_p can be calculated from the microcantilever's stiffness k_n and deflection δ_n by $F_p = 2k_n\delta_n$. In the models, δ_t is the peeling distance on the nanotweezer that can be described as a sum of δ_p and microcantilever's deflection δ_n. In the actual use, δ_p cannot be directly measured. Instead, $F_p - \delta_t$ curve during the whole peeling procedure with pickup and release processes is used to estimate the peeling energies.

Figure 6.21 shows a mechanics analysis of a nanotweezer probe during a peeling operation. Forces applied on the probe tip can be resolved into two components on the X-axis and the Z-axis in the defined frame, i.e, F_x and F_z, respectively. F_x is the clamping force that holds the 1-D nanostructure and F_z is the peeling force that balances adhesion forces from the substrate. To sense the peeling force, it is

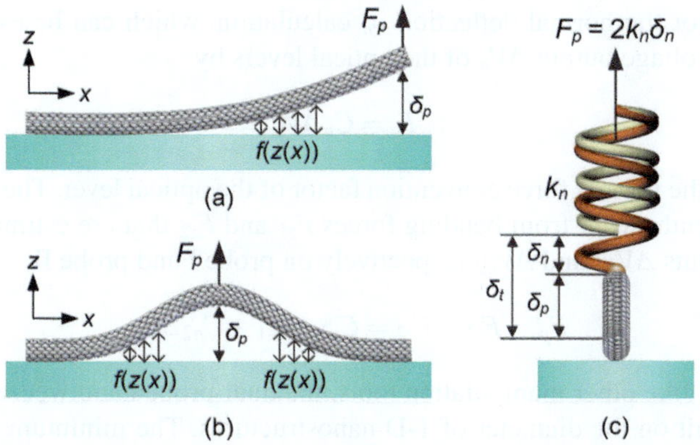

Fig. 6.20 (a) and (b) are respectively deflection simulations for peeling locations at the end and at the middle. (c) Peeling force computation model of the peeling test. F_p is the peeling force applied on the peeling location, $f(z(x))$ is the separating gap along the 1-D nanostructure, k_n is the stiffness of the cantilever, δ_n, δ_p and δ_t are respectively the cantilever's normal deflection, the gap on the peeling location and the displacement on the nanotweezer.

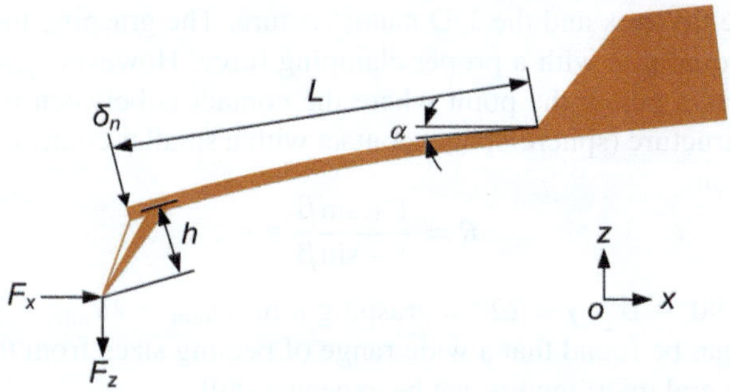

Fig. 6.21 Analysis of mechanics of the nanotweezer probe (a cantilever with beam length L, mounting angle α and tip height h) during a peeling test. A normal deflection δ_n on a microcantilever is caused by both the clamping force F_x and the peeling force F_z applied on the end of the tip.

necessary to detect the normal deflection δ_n on each cantilever. δ_n can be calculated by:

$$\delta_n = \frac{F_z \cos\alpha + F_x \sin\alpha}{k_n} + \frac{F_z \sin\alpha + F_x \cos\alpha}{k_{xz}} \tag{6.11}$$

Where $\alpha = 5°$ is the mounting angle of the cantilever, $k_{xz} = 2lk_n/3h$ is the bending stiffness due to the moment applied on the tip end, here, h is the tip height and L is the beam length of the cantilever. Assuming magnitudes of F_z and F_x are of the same order, contributions from F_x to δ_n can be ignored since it is very small compared with that of F_z due to $k_{xz} \gg k_n$ with $L = 250\ \mu$m and $h = 15\ \mu$m. Thus, only F_z is considered for the normal deflection δ_n calculation, which can be estimated from the normal voltage output ΔV_n of the optical levers by:

$$F_z = C_n \Delta V_n \tag{6.12}$$

where C_n is the normal force convention factor of the optical lever. The peeling force F_p can be synthesized from bending forces F_{z1} and F_{z2} that are estimated from the voltage outputs ΔV_{n1} and ΔV_{n2} respectively on probe I and probe II:

$$F_p = F_{z1} + F_{z2} = C_{n1} \Delta V_{n1} + C_{n2} \Delta V_{n2} \tag{6.13}$$

However, like other manipulation tools, the dual-probe nanotweezer has its own grasping limit on the diameter of 1-D nanostructures. The minimum diameter that can be peeled is generally determined by the size of the tip apex and tips' deformations during peeling. Effects from the deformations can be counteracted by preloading a clamping force F_x. Thus, the size of the tip apex is the determinative factor. A geometric simulation of the grasping limit is seen in the bottom inset of Fig. 6.22, in which $\phi = 60°$ is the tip's tilted angle through its rotation axis relative to the substrate; $\gamma = 8°$ is the half-cone angle of the tip; $r = 8$ nm and R are respectively the radii of the tip apex and the 1-D nanostructure. The grasping limit $R_{\min}$ can be theoretically equal to r with a proper clamping force. However, grasping will lose when R decreases below the point where the contact is between the tip apex and the 1-D nanostructure (sphere-sphere contact with a smaller contact angle β). Thus, from the relation:

$$R = \frac{1+\sin\beta}{1-\sin\beta} \times r \tag{6.14}$$

where $\beta_{\max} = 90° - \phi - \gamma = 22°$, a grasping limit $D_{\min} = 2R_{\min} = 35.2$ nm can be calculated. It can be found that a wide range of peeling sizes from the nanoscale to the scale of several micrometers can be expected [59].

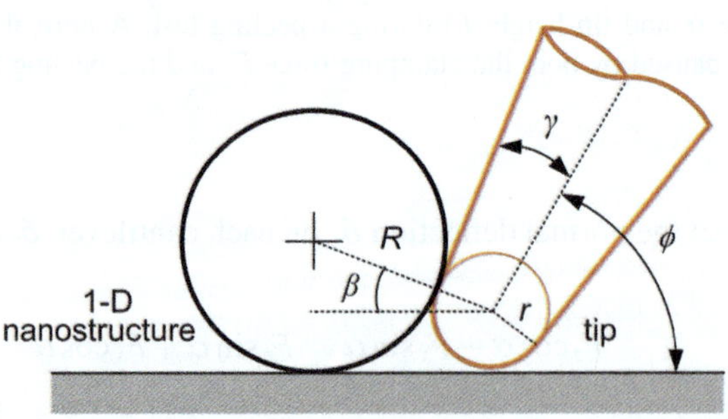

Fig. 6.22 A schematic diagram of a geometry simulation of grasping limit on the size of the 1-D nanostructure. ϕ and γ are respectively the tip's titled angle relative to the substrate and half-cone angle, R and r are respectively radii of 1-D nanostructure and the tip apex, β is defined as a contact angle between the tip apex and the 1-D nanostructure.

6.2.2 Results and Discussion

Peeling tests of a silicon nanowire (SiNW) have been processed with the proposed nanotweezer in an ambient temperature of 20°C and a relative humidity of 40%. As seen in Fig. 6.23(a), the cone-shape SiNWs, with a diameter of 25 nm (top)∼200 nm (root) and a length of about 7 μm, were deposited on a silicon substrate (with 300 nm silicon dioxide) for peeling tests. Once the SiNW was clamped, the peeling was started by moving the piezotube down while nanotweezer kept immovable, followed by a retraction for releasing as the SiNW was completely peeled from the substrate. The tests were repeated dozens of times. Fig. 6.23(b) shows a post-peeling image after 25 tests at the end. Then 20 tests were performed at the middle of the SiNW. As seen in the post-peeling image Fig. 6.23(c), the peeled SiNW was displaced due to complicated interactions among the SiNW, nanotweezer and the substrate. 6.23(d) shows heights at peeling locations: 196 nm at the end and 145 nm at the middle of the SiNW.

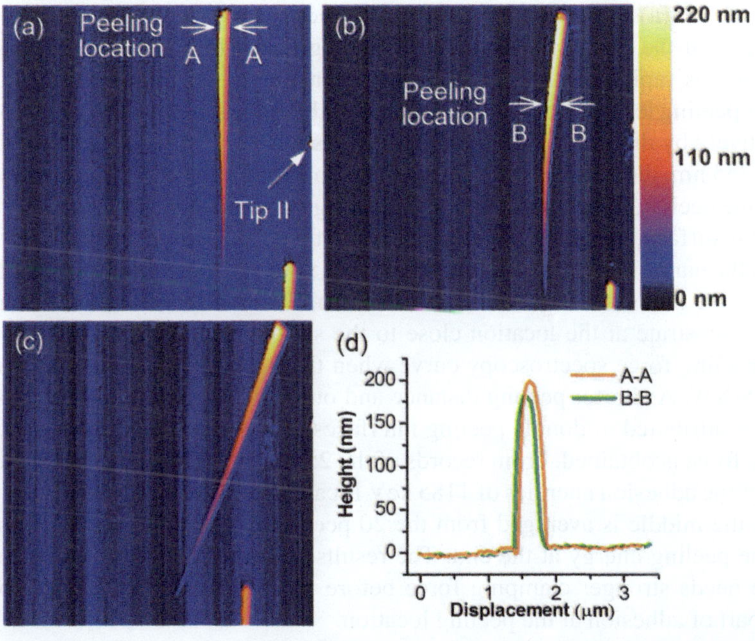

Fig. 6.23 (a) Pre-scanned image of SiNWs and the end of probe II. A peeling location is firstly defined at the end (root) of a selected SiNW. (b) A post-peeling image after 25 peeling tests at the end. Another peeling location is then defined at the middle of the SiNW. (c) A post-peeling image after 20 peeling tests at the middle, in which the peeled SiNW was displaced due to complicated interactions among the nanotweezer, the SiNW and the substrate. (d) Line scans show heights of the peeling locations at the end (196 nm) and the middle of the SiNW (145 nm).

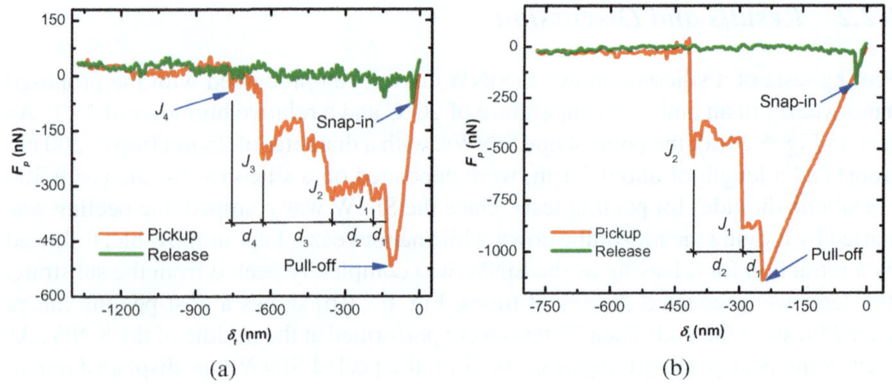

Fig. 6.24 A full peeling force spectroscopy curve recorded by peeling the end of the SiNW. (b) A full peeling force spectroscopy curve recorded by peeling the middle of the SiNW.

Figure 6.24(a) shows a full peeling force spectroscopy curve for the peeling point at the end of the SiNW. When the peeling begins (yellow line), the peeling force F_p decreases rapidly. At $\delta_t = -115$ nm, the nanotweezer pulls off the substrate. Further peeling leads to four local increases of the force magnitude that are followed respectively by discontinuous jumps $J_1 - J_4$ respectively with distances $d_1 - d_4$. At $\delta_t = -755$ nm, the peeling force jumped to zero. Upon still further moving substrate down, the peeling force remains zero, indicating that the SiNW has been fully peeled from the surface. As seen in the green part of Fig. 6.24(a), during the retraction, neither the nanotweezer nor the SiNW interacts with the approaching surface and the force on the nanotweezer remains around zero. Eventually, the nanotweezer snaps into the substrate at the location close to the starting point. Figure 6.24(b) shows a full peeling force spectroscopy curve when the peeling location is at the middle of the SiNW. A shorter peeling distance and only two distinct jumps are observed, which is attributed to double peeling interfaces in this case, and reasonably, larger peeling force is obtained. From records of the 25 peeling tests at the end, an average value of the adhesion energies of 1185 keV is calculated. An adhesion energy of 906 keV at the middle is averaged from the 20 peeling tests, which is slightly smaller than the peeling energy at the end. The results are reasonable since the peeling at middle needs stronger clamping force before peeling that might already break a small part of adhesion at the peeling location.

In summary, we have presented a method for in situ peeling test of 1-D nanostructures. This method uses a dual-probe nanotweezer that has capabilities of pick-and-place the 1-D structures as well as accurate force sensing during the manipulation. With this method, a 1-D nanostructure can be easily peeled from the surface by grasping it at various locations without complicated sample preparation. Experimental results of SiNW peeling tests on a silicon surface validate the proposed method. Moreover, peeling applications of the dual-probe nanotweezer could be used to peel other types of nanostructures, e.g. 0-D nanoparticles and 3-D nanostructures. In addition, peeling applications of the dual-probe nanotweezer can be extended from the nanoscale to the scale of several micrometers.

6.3 Conclusion

The first part of this chapter focuses on an robotized AFM for multi-scale micro/nano manipulation and assembly applications. Using AFM-based accurate force sensing and the tiny apex of the AFM tip, a nanotip gripper able to pick-and-place very tiny samples was constructed from two tips fitted to the AFM-FRS. Using this nanotip gripper, 3-D manipulation from the nanoscale to the scale of several-microns was achieved. Moreover, the two-tip configuration of the AFM-FRS also makes high-efficiency parallel imaging/nanomanipulation feasible by performing imaging and manipulation in parallel. Three different configurations and corresponding experiments respectively for 3-D microsphere manipulation, 3-D nanowire manipulation and parallel nanomanipulation of nanoparticles, were presented in detail.

The second part describes a method for in situ peeling test of 1-D nanostructures. This method uses a dual-probe nanotweezer that has capabilities of pick-and-place the 1-D structures as well as accurate force sensing during the manipulation. With this method, a 1-D nanostructure can be easily peeled from the surface by grasping it at various locations without complicated sample preparation. Experimental results of SiNW peeling tests on a silicon surface validate the proposed method. Moreover, peeling applications of the dual-probe nanotweezer could be used to peel other types of nanostructures, e.g. 0-dimensional nanoparticles and 3-dimensional nanostructures. In addition, peeling applications of the dual-probe nanotweezer can be extended from the nanoscale to the scale of several micrometers.

References

1. Rollot, Y., Régnier, S., Guinot, J.C.: Simulation of micro-manipulations: adhesion forces and specific dynamic models. Int. J. Adhes. Adhes. 19(1), 35–48 (1999)
2. Menciassi, A., Eisinberg, A., Izzo, I., Dario, P.: From 'macro' to 'micro' manipulation: models and experiments. IEEE/ASME Trans. Mechatro. 9(2), 311–320 (2004) 319
3. Dong, L.X., Nelson, B.J.: Robotics in the Small. IEEE Robot. Autom. Mag. 14(3), 111–121 (2007)
4. Savia, M., Koivo, H.N.: Contact Micromanipulation–Survey of Strategies. IEEE/ASME Trans. Mechatron. 14(4), 504–514 (2009)
5. Millet, O., Bernardoni, P., Régnier, S., Bidaud, P., Tsitsiris, E., Collard, D., Buchaillot, L.: Electrostatic actuated micro gripper using an amplification mechanism. Sensor Actuat A: Phys. 114(2-3), 371–378 (2004)
6. Pérez, R., Agnus, J., Clévy, C., Hubert, A., Chaillet, N.: Modeling, fabrication, and validation of a high-performance 2-DoF piezoactuator for micromanipulation. IEEE/ASME Trans. Mechatron. 10(2), 161–171 (2005)
7. Driesen, W., Varidel, T., Régnier, S., Breguet, J.M.: Micromanipulation by adhesion with two collaborating mobile micro robots. J. Micromech. Microeng. 15(10), S259–S267 (2005)

8. Neild, A., Oberti, S., Beyeler, F., Dual, J., Nelson, B.J.: A micro-particle positioning technique combining an ultrasonic manipulator and a microgripper. J. Micromech. Microeng. 16(8), 1562–1571 (2006)
9. Onal, C.D., Sitti, M.: Visual servoing-based autonomous 2-D manipulation of microparticles using a nanoprobe. IEEE Trans. Cont. Syst. Tech. 15(5), 842–852 (2007)
10. Kim, K.Y., Liu, X.Y., Zhang, Y., Sun, Y.: Nanonewton force-controlled manipulation of biological cells using a monolithic MEMS microgripper with two-axis force feedback. J. Micromech. Microeng. 18(5), 055013 (2008)
11. Walle, B.L., Gauthier, M., Chaillet, N.: Principle of a submerged freeze gripper for microassembly. IEEE Trans. Robot. 24(4), 897–902 (2008)
12. Elbuken, C., Khamesee, M.B., Yavuz, M.: Design and Implementation of a Micromanipulation System Using a Magnetically Levitated MEMS Robot. IEEE/ASME Trans. Mechatron. 14(4), 434–445 (2009)
13. Liaw, H.C., Shirinzadeh, B.: Neural Network Motion Tracking Control of Piezo-Actuated Flexure-Based Mechanisms for Micro-/Nanomanipulation. IEEE/ASME Trans. Mechatron. 14(5), 517–527 (2009)
14. Fukuda, T., Arai, F., Dong, L.X.: Assembly of nanodevices with carbon nanotubes through nanorobotic manipulations. Proc. IEEE 91(11), 1803–1818 (2003)
15. Molhave, K., Wich, T., Kortschack, A., Boggild, P.: Pick-and-place nanomanipulation using microfabricated grippers. Nanotechnology 17(10), 2434–2441 (2006)
16. Dong, L.X., Arai, F., Fukuda, T.: Destructive constructions of nanostructures with carbon nanotubes through nanorobotic manipulation. IEEE/ASME Trans. Mechatron. 9(2), 58–63, 350–357 (2004)
17. Fatikow, S., Wich, T., Hulsen, H., Sievers, T., Jahnisch, M.: Microrobot system for automatic nanohandling inside a scanning electron microscope. IEEE/ASME Trans. Mechatron. 12(3), 244–252 (2007)
18. Bosanac, L., Aabo, T., Bendix, P.M., Oddershede, L.B.: Efficient optical trapping and visualization of silver nanoparticles. Nano Lett. 8(5), 1486–1491 (2008)
19. Pacoret, C., Bowman, R., Gibson, G., Haliyo, S., Carberry, D., Bergander, A., Régnier, S., Padgett, M.: Touching the microworld with force-feedback optical tweezers. Optics Express 17(12), 10260–10264 (2009)
20. Sitti, M., Hashimoto, H.: Controlled pushing of nanoparticles: modeling and experiments. IEEE/ASME Trans. Mechatron. 5(2), 199–211 (2000)
21. Guthold, M., Falvo, M.R., Matthews, W.G., Paulson, S., Washburn, S., Erie, D.A., Superfine, R., Brooks, F.P., Taylor, R.M.: Controlled manipulation of molecular samples with the nanomanipulator. IEEE/ASME Trans. Mechatron. 5(2), 189–198 (2000)
22. Resch, R., Lewis, D., Meltzer, S., Montoya, N., Koel, B.E., Madhukar, A., Requicha, A.A.G., Will, P.: Manipulation of gold nanoparticles in liquid environments using scanning force microscopy. Ultramicroscopy 82(1-4), 135–139 (2000)
23. Sitti, M.: Atomic force microscope probe based controlled pushing for nanotribological characterization. IEEE/ASME Trans. Mechatron. 9(2), 343–349 (2004)
24. Yun, Y.J., Ah, C.S., Kim, S., Yun, W.S., Park, B.C., Ha, D.H.: Manipulation of freestanding Au nanogears using an atomic force microscope. Nanotechnology 18(50), 505304 (2007)
25. Sitti, M., Aruk, B., Shintani, K., Hashimoto, H.: Scaled teleoperation system for nanoscale interaction and manipulation. Advanced Robotics 17(3), 275–291 (2003)

26. Sitti, M.: Teleoperated 2-D Micro/Nanomanipulation using an Atomic Force Microscope. PhD thesis, University of Tokyo, Japan (September 1999)
27. Sitti, M., Hashimoto, H.: Teleoperated touch feedback of surfaces at the nanoscale: Modeling and experiments. IEEE/ASME Trans. Mechatron. 8(2), 287–298 (2003)
28. Li, G.Y., Xi, N., Yu, M.M., Fung, W.K.: Development of augmented reality system for AFM-based nanomanipulation. IEEE/ASME Trans. Mechatron. 9(2), 358–365 (2004)
29. Liu, L.Q., Luo, Y.L., Xi, N., Wang, Y.C., Zhang, J.B., Li, G.Y.: Sensor referenced real-time videolization of atomic force microscopy for nanomanipulations. IEEE/ASME Trans. Mechatron. 13(1), 76–85 (2008)
30. Xie, H., Régnier, S.: Three-dimensional automated micromanipulation using a nanotip gripper with multi-feedback. J. Micromech. Microeng. 19(7), 075009 (2009)
31. Xie, H., Haliyo, S., Régnier, S.: A versatile atomic force microscope for three-dimensional nanomanipulation and nanoassembly. Nanotechnology 20(21), 215301 (2009)
32. Xie, H., Haliyo, S., Régnier, S.: Parallel imaging/manipulation force microscopy. App. Phy. Lett. 94(15), 0153106 (2009)
33. Xie, H., Rakotondrabe, M., Régnier, S.: Characterization of the piezoscanner with an optical lever and a reference nanopositioning stage. Rev. Sci. Instrum. 80(4), 046102 (2009) 319
34. Lee, K., Duchamp, M., Kulik, G., Magrez, A., Seo, J.W., Jeney, S., Kulik, A.J., Forro, L.: Uniformly dispersed deposition of colloidal nanoparticles and nanowires by boiling. Appl. Phys. Lett. 91(17), 173112 (2007) 331
35. Barber, A.H., Cohen, S.R., Wagner, H.D.: Appl. Phys. Lett. 82, 4140 (2003) 332
36. Breuer, O., Sungararaj, U.: Polym. Compos. 25, 630 (2004) 332
37. Strus, M.C., Cano, C.I., Pipes, R.B., Nguyen, C.V., Raman, A.: Compos. Sci. Technol. 69, 1580 (2009) 332
38. Sever, M.J., Weisser, J.T., Monahan, J., Srinivasan, S., Wilker, J.J.: Angew. Chem. Int. Ed. 43, 448 (2004) 332
39. Shi, X., Kong, Y., Zhao, Y., Gao, H.: Acta Mech. Sin. 21, 249 (2005) 332
40. Ge, L., Sethi, S., Ci, L., Ajayan, P.M., Dhinojwala, A.: PNAS 104, 10792 (2007) 332
41. Jang, J.E., Cha, S.N., Choi, Y., Amaratunga, G.A.J., Kang, D.J., Hasko, D.G., Jung, J.E., Kim, J.M.: Appl. Phys. Let. 87, 163114 (2005) 332
42. Dujardin, E., Derycke, V., Goffman, M.F., Lefèvre, R., Bourgoin, J.P.: Appl. Phys. Let. 87, 193107 (2005) 332
43. Cha, S.N., Jang, J.E., Choi, Y., Amaratungab, G.A.J., Kang, D.J., Hasko, D.G., Jung, J.E., Kim, J.M.: Appl. Phys. Let. 86, 083105 (2005) 332
44. Ekinci, K.L., Roukes, M.L.: Rev. Sci. Instrum. 76, 061101 (2005) 332
45. Mahar, B., Laslau, C., Yip, R., Sun, Y.: IEEE Sens. 7, 266 (2007) 332
46. Dragoman, M., Takacs, A., Muller, A.A., Hartnagel, H., Plana, R., Grenier, K., Dubuc, D.: Appl. Phys. Let. 90, 113102 (2007) 332
47. Li, Q., Koo, S.M., Richter, C.A., Edelstein, M.D., Bonevich, J.E., Kopanski, J.J., Suehle, J.S., Vogel, E.M.: IEEE Trans. Nanotechnol. 6, 256 (2007) 332
48. Popov, I., Gemming, S., Okano, S., Ranjan, N., Seifert, G.: Nano Lett. 8, 4093 (2008) 332
49. Metin, S., Hideki, H.: IEEE/ASME Trans. Mechatron 5, 199 (2000) 332
50. Tong, L., Zhu, T., Liu, Z.: Appl. Phys. Lett. 92, 023109 (2008) 332
51. Shin, C.H., Jeon, I.S., Jeon, S.H., Khim, Z.G.: Appl. Phys. Lett. 94, 163107 (2009) 332
52. Yoshiharu, I., Akihiko, I., Toshio, Y.: Trends in Biotechnology 19, 211 (2001) 332
53. Falvo, M.R., Taylor, R.M.I., Helser, A., Chi, V., Brooks, F.P.J., Washburn, S., Superfine, R.: Nature 397, 236 (1999) 332

54. Tranvouez, E., Boer-Duchemin, E., Comtet, G., Dujardin, G.: Rev. Sci. Instrum. 78, 115103 (2007) 332
55. Strus, M.C., Zalamea, L., Raman, A., Pipes, R.B., Nguyen, C.V., Stach, E.A.: Nano Lett. 8, 544 (2008) 332
56. Ishikawa, M., Harada, R., Sasaki, N., Miura, K.: Appl. Phys. Let. 93, 083122 (2008) 332
57. Xie, H., Haliyo, D.S., Régnier, S.: Appl. Phys. Let. 94, 153106 (2009) 333
58. Xie, H., Rakotondrabe, M., Régnier, S.: Rev. Sci. Instrum. 80, 046102 (2009) 333
59. Xie, H., Régnier, S., Micromech, J.: Microeng. 19, 075009 (2009) 336

Springer Tracts in Advanced Robotics

Edited by B. Siciliano, O. Khatib and F. Groen

Further volumes of this series can be found on our homepage: springer.com

Vol. 71: Xie, H.; Onal, C.; Régnier, S.; Sitti, M.
Atomic Force Microscopy Based Nanorobotics
342 p. 2011 [978-3-642-20328-2]

Vol. 70: Pradalier, C.; Siegwart, R.; Hirzinger, G. (Eds.)
Robotics Research
752 p. 2011 [978-3-642-19456-6]

Vol. 69: Rocon, E.; Pons, J.L.
Exoskeletons in Rehabilitation Robotics
138 p. 2010 [978-3-642-17658-6]

Vol. 68: Hsu, D.; Isler, V.; Latombe, J.-C.; Ming C. Lin (Eds.)
Algorithmic Foundations of Robotics IX
424 p. 2010 [978-3-642-17451-3]

Vol. 67: Schütz, D.; Wahl, F.M. (Eds.)
Robotic Systems for Handling and Assembly
460 p. 2010 [978-3-642-16784-3]

Vol. 66: Kaneko, M.; Nakamura, Y. (Eds.)
Robotics Research
450 p. 2010 [978-3-642-14742-5]

Vol. 65: Ribas, D.; Ridao, P.; Neira, J.
Underwater SLAM for Structured Environments Using an Imaging Sonar
142 p. 2010 [978-3-642-14039-6]

Vol. 64: Vasquez Govea, A.D.
Incremental Learning for Motion Prediction of Pedestrians and Vehicles
153 p. 2010 [978-3-642-13641-2]

Vol. 63: Vanderborght, B.;
Dynamic Stabilisation of the Biped Lucy Powered by Actuators with Controllable Stiffness
281 p. 2010 [978-3-642-13416-6]

Vol. 62: Howard, A.; Iagnemma, K.; Kelly, A. (Eds.):
Field and Service Robotics
511 p. 2010 [978-3-642-13407-4]

Vol. 61: Mozos, Ó.M.
Semantic Labeling of Places with Mobile Robots
134 p. 2010 [978-3-642-11209-6]

Vol. 60: Zhu, W.-H.
Virtual Decomposition Control – Toward Hyper Degrees of Freedom Robots
443 p. 2010 [978-3-642-10723-8]

Vol. 59: Otake, M.
Electroactive Polymer Gel Robots – Modelling and Control of Artificial Muscles
238 p. 2010 [978-3-540-23955-0]

Vol. 58: Kröger, T.
On-Line Trajectory Generation in Robotic Systems – Basic Concepts for Instantaneous Reactions to Unforeseen (Sensor) Events
230 p. 2010 [978-3-642-05174-6]

Vol. 57: Chirikjian, G.S.; Choset, H.; Morales, M., Murphey, T. (Eds.)
Algorithmic Foundations of Robotics VIII – Selected Contributions of the Eighth International Workshop on the Algorithmic Foundations of Robotics
680 p. 2010 [978-3-642-00311-0]

Vol. 56: Buehler, M.; Iagnemma, K.; Singh S. (Eds.)
The DARPA Urban Challenge – Autonomous Vehicles in City Traffic
625 p. 2009 [978-3-642-03990-4]

Vol. 55: Stachniss, C.
Robotic Mapping and Exploration
196 p. 2009 [978-3-642-01096-5]

Vol. 54: Khatib, O.; Kumar, V.; Pappas, G.J. (Eds.)
Experimental Robotics: The Eleventh International Symposium
579 p. 2009 [978-3-642-00195-6]

Vol. 53: Duindam, V.; Stramigioli, S.
Modeling and Control for Efficient Bipedal Walking Robots
211 p. 2009 [978-3-540-89917-4]

Vol. 52: Nüchter, A.
3D Robotic Mapping
201 p. 2009 [978-3-540-89883-2]

Vol. 51: Song, D.
Sharing a Vision
186 p. 2009 [978-3-540-88064-6]

Vol. 50: Alterovitz, R.; Goldberg, K.
Motion Planning in Medicine: Optimization and Simulation Algorithms for Image-Guided Procedures
153 p. 2008 [978-3-540-69257-7]

Vol. 49: Ott, C.
Cartesian Impedance Control of Redundant and Flexible-Joint Robots
190 p. 2008 [978-3-540-69253-9]

Vol. 48: Wolter, D.
Spatial Representation and Reasoning for Robot Mapping
185 p. 2008 [978-3-540-69011-5]

Vol. 47: Akella, S.; Amato, N.; Huang, W.; Mishra, B.; (Eds.)
Algorithmic Foundation of Robotics VII
524 p. 2008 [978-3-540-68404-6]

Vol. 46: Bessière, P.; Laugier, C.; Siegwart R. (Eds.)
Probabilistic Reasoning and Decision Making in Sensory-Motor Systems
375 p. 2008 [978-3-540-79006-8]

Vol. 45: Bicchi, A.; Buss, M.; Ernst, M.O.; Peer A. (Eds.)
The Sense of Touch and Its Rendering
281 p. 2008 [978-3-540-79034-1]

Vol. 44: Bruyninckx, H.; Přeučil, L.; Kulich, M. (Eds.)
European Robotics Symposium 2008
356 p. 2008 [978-3-540-78315-2]

Vol. 43: Lamon, P.
3D-Position Tracking and Control for All-Terrain Robots
105 p. 2008 [978-3-540-78286-5]

Vol. 42: Laugier, C.; Siegwart, R. (Eds.)
Field and Service Robotics
597 p. 2008 [978-3-540-75403-9]

Vol. 41: Milford, M.J.
Robot Navigation from Nature
194 p. 2008 [978-3-540-77519-5]

Vol. 40: Birglen, L.; Laliberté, T.; Gosselin, C.
Underactuated Robotic Hands
241 p. 2008 [978-3-540-77458-7]

Vol. 39: Khatib, O.; Kumar, V.; Rus, D. (Eds.)
Experimental Robotics
563 p. 2008 [978-3-540-77456-3]

Vol. 38: Jefferies, M.E.; Yeap, W.-K. (Eds.)
Robotics and Cognitive Approaches to Spatial Mapping
328 p. 2008 [978-3-540-75386-5]

Vol. 37: Ollero, A.; Maza, I. (Eds.)
Multiple Heterogeneous Unmanned Aerial Vehicles
233 p. 2007 [978-3-540-73957-9]

Vol. 36: Buehler, M.; Iagnemma, K.; Singh, S. (Eds.)
The 2005 DARPA Grand Challenge – The Great Robot Race
520 p. 2007 [978-3-540-73428-4]

Vol. 35: Laugier, C.; Chatila, R. (Eds.)
Autonomous Navigation in Dynamic Environments
169 p. 2007 [978-3-540-73421-5]

Vol. 34: Wisse, M.; van der Linde, R.Q.
Delft Pneumatic Bipeds
136 p. 2007 [978-3-540-72807-8]

Vol. 33: Kong, X.; Gosselin, C.
Type Synthesis of Parallel Mechanisms
272 p. 2007 [978-3-540-71989-2]

Vol. 32: Milutinović, D.; Lima, P.
Cells and Robots: Modeling and Control of Large-Size Agent Populations
130 p. 2007 [978-3-540-71981-6]

Vol. 31: Ferre, M.; Buss, M.; Aracil, R.; Melchiorri, C.; Balaguer C. (Eds.)
Advances in Telerobotics
500 p. 2007 [978-3-540-71363-0]

GPSR Compliance

The European Union's (EU) General Product Safety Regulation (GPSR) is a set of rules that requires consumer products to be safe and our obligations to ensure this.

If you have any concerns about our products, you can contact us on ProductSafety@springernature.com

In case Publisher is established outside the EU, the EU authorized representative is:

Springer Nature Customer Service Center GmbH
Europaplatz 3
69115 Heidelberg, Germany

Batch number: 10755837

Printed by Printforce, the Netherlands